Using PROCOMM PLUS

Walter R. Bruce III

With a Foreword by Stephen Monaco,
Vice President of Marketing
DATASTORM TECHNOLOGIES, INC.

que ®
CORPORATION
LEADING COMPUTER KNOWLEDGE

Using PROCOMM PLUS®

Copyright © 1989 by Que® Corporation

Library of Congress Catalog No.: 89-63568

ISBN 0-88022-530-0

92 91 90   8 7 6 5 4

Interpretation of the printing code: the rightmost double-digit number is the year of the book's printing; the rightmost single-digit number, the number of the book's printing. For example, a printing code of 89-1 shows that the first printing of the book occurred in 1989.

Using PROCOMM PLUS is based on PROCOMM PLUS Version 1.1B, with information also applying to ProComm 2.4.3.

DEDICATION ▼

To Melanie, Lori, and Perry

Publishing Director

Lloyd J. Short

Acquisitions Editor

Karen A. Bluestein

Product Director

Karen A. Bluestein

Production Editor

Jeannine Freudenberger

Editors

Fran Blauw
Kelly Currie
Daniel Schnake
Janet Thrush

Technical Editor

Phil James, Technical Support Manager
DATASTORM TECHNOLOGIES, INC.

Indexer

Sharon Hilgenberg

Book Design and Production

Dan Armstrong
Bill Basham
Brad Chinn
Don Clemons
Sally Copenhaver
Tom Emrick
Dennis Hager
Corinne Harmon
Bill Hurley
Becky Imel
Jodi Jensen

Kathy Keehn
David Kline
Larry Lynch
Lori A. Lyons
Jennifer Matthews
Cindy L. Phipps
Joe Ramon
Dennis Sheehan
Louise Shinault
Bruce Steed
Mary Beth Wakefield

Composed in Garamond and Excellent No. 47
by Que Corporation

ABOUT THE AUTHOR ▼

Walter R. Bruce III

Walter R. Bruce III is a free-lance writer and microcomputer consultant. He is the author of Que's *Using Enable/OA*, *Using Paradox 3*, and *Using DataEase* and has written several instructional texts for use in intermediate and advanced workshops on using popular microcomputer software packages. He also has led workshops for government and private industry clients from coast to coast.

Mr. Bruce is a licensed attorney, who practiced for three years in North Carolina and six years in the United States Air Force. For three years, his duties included acting as an advisor on computer and office automation issues to The Judge Advocate General of the Air Force. Mr. Bruce lives in the Washington, D.C., suburb, Springfield, Virginia, with his wife, two sons, and daughter.

CONTENTS AT A GLANCE

TABLE OF CONTENTS ▼

7   Automating PROCOMM PLUS with Macros and Script Files **167**

III   Becoming a PROCOMM PLUS Expert

8   Tailoring PROCOMM PLUS **193**

It happens to the best of us. As unsuspecting computer users, we justify shelling out a couple hundred dollars for a software program thinking that "If it costs a lot, it must be good." Later, we find ourselves pulling our hair out while we try to get the software to work. The user interface is awkward and the documentation tedious. Unfortunately, too many software titles fall into this frustrating category of "Big Bucks for Mediocre Products."

Our goal with PROCOMM PLUS was to create a robust communications software that combined extreme power with ease of use—at an affordable price. We guessed that by mixing an intuitive feel with powerful features and a reasonable price, PROCOMM PLUS would appeal to data processing professionals, novice computer users, and everyone in between.

We guessed right! Within 18 months after its release, PROCOMM PLUS had established itself as the new standard in communications software and is now the best-selling communications package in the world. DATASTORM has demonstrated that great software doesn't have to be hard to use.

The look and feel of a software package is difficult to put into words. Many PRO-COMM PLUS users have described its interface with words like intuitive, warm, pleasant, friendly, instinctive, and familiar.

Some of those same words came to mind as I read this book. Walter Bruce has done an excellent job writing *Using PROCOMM PLUS*. The book is useful and informative. It's technical, yet easy to read. His writing style is conversational, and he has managed to keep this "technical" book from being dry by making it fun and upbeat. The practical real-world examples keep the text interesting and are very helpful. Walter Bruce effectively shows how to tap the real power that exists in PROCOMM PLUS.

Stephen Monaco

Vice President of Marketing
DATASTORM TECHNOLOGIES, INC.

ACKNOWLEDGMENTS

Thanks to the following individuals for their invaluable help in bringing this book together:

Karen Bluestein, Jeannine Freudenberger, and all the editors at Que Corporation for their outstanding work and tireless patience on this project

Phil James at DATASTORM TECHNOLOGIES, INC., for providing extremely helpful, timely, and greatly appreciated technical advice

Max Arafa of BT Datacom (a member of the British Telecom Group, London, England) of Chantilly, Virginia, for providing the use of a model DM 4245X 2400-bps error-control and data-compression modem

TRADEMARK
ACKNOWLEDGMENTS

Que Corporation has made every effort to supply trademark information about company names, products, and services mentioned in this book. Trademarks indicated below were derived from various sources. Que Corporation cannot attest to the accuracy of this information.

CompuServe Information Service is a registered trademark of CompuServe Incorporated and H&R Block, Inc.

Hayes Smartmodem is a trademark of Hayes Microcomputer Products.

IBM, IBM PC, IBM PC AT, and OS/2 are registered trademarks and IBM PC XT and IBM PC*jr* are trademarks of International Business Machines Corporation.

Lotus and 1-2-3 are registered trademarks of Lotus Development Corporation.

Microsoft is a registered trademark of Microsoft Corporation.

PROCOMM PLUS and DATASTORM are registered trademarks of DATASTORM TECHNOLOGIES, INC.

Telenet is a registered trademark of Telenet Communications Corporation.

Tymnet is a registered trademark of Tymnet, Inc.

CONVENTIONS USED IN THIS BOOK

The conventions used in this book have been established to help you learn to use the program quickly and easily. As much as possible, the conventions correspond with those used in the PROCOMM PLUS documentation.

Commands are written with the letter the user presses to choose the command in boldface: **File**, **A** (Terminal emulation). On-screen messages and prompts appear in a `special typeface`. Words and commands that the user types are written in italic.

Names of files, protocols, and DOS commands are written in all capital letters; names of screens, modes, windows, and menus are written with initial capital letters.

Introduction

Welcome to *Using PROCOMM PLUS*. With PROCOMM PLUS as your magic carpet and with this book as your companion and guide, you are about to explore the exciting world of computer-to-computer communications.

PROCOMM PLUS is a communications program that is both fun to use and full of impressive features. Its exploding screens, vibrant color, and novel sound effects continually remind you that computer software doesn't have to be boring to be useful.

In large measure because of its upbeat personality, PROCOMM PLUS is one of the easiest communications programs to learn to use. It can often transform dull and highly technical procedures into a couple of easy key-strokes. And if you don't remember the keystrokes, PROCOMM PLUS provides an ever-present help facility, never more than a keystroke away.

Because PROCOMM PLUS is so easy to use, you may wonder whether you need a book to help you. Ease of use is a relative concept, and many people find communications in general a difficult area of computing to understand. Although PROCOMM PLUS makes communicating fun, it cannot fully insulate you from the many technical terms and concepts that seem to pop up at every turn. As your companion and guide, the purpose of this book is to teach you how to navigate through these terms and concepts. So with this book at your side, climb on board PROCOMM PLUS. You are about to take an enjoyable and rewarding magic carpet ride into the land of electronic bulletin boards, on-line information services, and electronic mail!

What Is PROCOMM PLUS?

PROCOMM PLUS is a full-featured communications program complete with terminal emulation, automated dialing directory, error-checking protocols, and a powerful script language. Published by DATASTORM TECHNOLOGIES, INC., Columbia, Missouri (previously known as PIL Software Systems), PROCOMM PLUS is a commercially marketed descendent of the popular user-supported communications program ProComm, first released in 1985.

The program is designed to enable you to connect your computer with other computers through standard telephone lines. Using PROCOMM PLUS, you can transfer and receive computer files, access an on-line information service, sign on to a powerful mainframe computer, or simply chat with another computer operator by typing messages on your computer's keyboard. PROCOMM PLUS provides these capabilities in a way that seems natural and familiar and is easy to comprehend. Indeed, one of the advertising slogans and trademarks used by PROCOMM PLUS is the phrase *Intuitive Communications*. After you have used PROCOMM PLUS for a short while, you will agree that this slogan accurately describes the program and its features.

Even though the program is designed to be easy and intuitive to use, PRO-COMM PLUS doesn't scrimp on powerful features. It has 13 built-in file-transfer protocols, can emulate 16 different types of computer terminals, contains a dialing directory facility that can store and speed dial an unlimited number of phone numbers, and includes a bulletin-board-like Host mode complete with electronic mail. PROCOMM PLUS even enables you to record your keystrokes so that you can play them back later—to sign on to an information service, for example. For the power user, PROCOMM PLUS is also completely programmable through the script command language, ASPECT. In short, PROCOMM PLUS is all the communications program you will ever need.

What's New in PROCOMM PLUS?

Since the introduction of the user-supported program ProComm in 1985, DATASTORM has progressively enhanced the program while remaining true to its original design concepts. The latest version of PROCOMM PLUS (as of this writing, Version 1.1B) carries the program's capabilities to new heights. The most recent ProComm (the user-supported program) release is Version 2.4.3. DATASTORM continues to support both PROCOMM

PLUS and ProComm, but only PROCOMM PLUS is available through commercial retail channels. Table I.1 lists many of the features of PROCOMM PLUS that are not available in ProComm 2.4.3.

Table I.1
New Features in PROCOMM PLUS

Category	Enhancement
User interface	Lotus 1-2-3 style menus in addition to Alt-key commands and pop-up menus
Dialing directory	Up to 200 telephone numbers per directory
	Unlimited number of dialing directories
	"Point-and-shoot" dialing
	Call history
	Default protocol and default terminal emulation for each directory entry
	Call activity audit
Terminal emulation	DEC VT102
	TeleVideo 910, 925, 955
	Wyse 50
	IBM 3270/950 (asynchronous)
Error-checking protocols	SEALINK
	IMODEM
	3 user-defined protocols
Host mode	Electronic mail
	Individualized password protection
	Open or closed system

Table I.1—*continued*

Category	Enhancement
ASPECT script language	49 new commands
	Full file input/output
	String handling
	Math commands
	Display handling, including windows
	New user input commands
	Get system date or time
	Temporary exit to DOS
	Suspend execution until a predetermined time
	Make a sound of specified frequency and duration
	Display a file on-screen
	Terminate script, exit PROCOMM PLUS, but leave connection open
	Dial a phone number using dialing codes from dialing directory
Miscellaneous	Two user-definable "hot keys" to run other DOS programs from within PROCOMM PLUS
	350-page user manual
	Context-sensitive help facility
	Support for COM1 through COM8
	Script Record mode
	Custom keyboard mapping for terminal emulation
	90-day free voice technical support

What Your System Should Have

To run PROCOMM PLUS, you must be using an IBM PC, XT, AT, or PS/2 computer or compatible. Your system must have at least 192K of available memory (RAM) and DOS 2.0 or higher. PROCOMM PLUS can be run from a floppy disk drive or from a hard disk drive. You also usually must have a modem installed in your computer (referred to as an internal modem) or connected to one of your system's serial ports (usually called an external modem). If you have a choice, the modem should recognize the Hayes AT command set (usually referred to as a Hayes-compatible modem), but virtually any modem can be used with PROCOMM PLUS. (Note: A modem is not an absolute necessity. Many people use PROCOMM PLUS to connect directly to a mainframe or another PC. In this case, you will need a special cable called a null modem cable. Refer to Appendix B for more about null modem cables.)

Who Should Read This Book?

Using PROCOMM PLUS is for you if you are a new PROCOMM PLUS user who wants to get the most from this fantastic product. Experienced Pro-Comm users will also find this book helpful in gaining a clear understanding of the PROCOMM PLUS enhancements.

This book assumes that you are using PROCOMM PLUS Version 1.1B. However, because many of the features of ProComm are either identical to or similar to features of PROCOMM PLUS, ProComm users will be able to use this book as well. Discussions involving features that are new in PROCOMM PLUS are marked with an icon ◇ in the margin. ProComm owners who have not yet chosen to upgrade to PROCOMM PLUS can skip these sections.

The approach of this book reflects the software it describes. You don't have to be a computer whiz to learn to use PROCOMM PLUS. This book likewise makes no assumptions about your background in computers or communications. If you are a new user of communications software, start at the beginning of the book and move through it at a comfortable pace. More experienced users can skim Part I, "Introducing PROCOMM PLUS," and study the text more closely beginning in Part II, "Getting Acquainted with PROCOMM PLUS." Power ProComm users may want to breeze through Parts I and II and concentrate on Part III, "Becoming a PROCOMM PLUS Expert."

What Is Covered in This Book?

This book is divided into three major parts: Part I, "Introducing PROCOMM PLUS"; Part II, "Getting Acquainted with PROCOMM PLUS"; and Part III, "Becoming a PROCOMM PLUS Expert."

Introducing PROCOMM PLUS

Part I is intended to help you quickly develop a basic understanding of communications concepts and terminology and the fundamentals of PROCOMM PLUS.

Chapter 1, "A Communications Primer," is a basic introduction to communications, with an emphasis on how it relates to PROCOMM PLUS. You will learn what it means to use PROCOMM PLUS to "communicate" with another computer and learn how to recognize the various hardware components necessary to use PROCOMM PLUS. This chapter also presents a few simple rules for selecting the right modem for your needs.

Chapter 2, "Getting Around in PROCOMM PLUS," gives you a quick tour of PROCOMM PLUS's most fundamental features. You learn how to start the program, understand the basic PROCOMM PLUS screen, use PROCOMM PLUS menus, and access context-sensitive help screens.

Getting Acquainted with PROCOMM PLUS

The largest amount of information is presented in Part II. Here you learn how to use the major features of the program, those that you will use daily as you become a proficient PROCOMM PLUS user.

Chapter 3, "Building Your Dialing Directory," explains how to use the many features of the PROCOMM PLUS dialing directory, including the new capability to have multiple directories.

Chapter 4, "A Session with PROCOMM PLUS," introduces you to the most commonly used features of the PROCOMM PLUS Terminal mode—the commands you need to know when you are on-line.

Chapter 5, "Transferring Files," explains how to use PROCOMM PLUS to send and receive electronic files to and from another computer. The chapter includes explanations of how to use the file-transfer protocols available in PROCOMM PLUS.

Chapter 6, "Using the PROCOMM PLUS Editor," describes how to use PCEDIT, the quick but simple editor supplied with PROCOMM PLUS. This chapter also explains how you can attach another text editor so that it can be activated from within PROCOMM PLUS.

The last chapter in Part II, Chapter 7, "Automating PROCOMM PLUS with Macros and Script Files," explains how you can "teach" PROCOMM PLUS to perform many tasks automatically by use of the Record mode. This chapter also introduces you to some of the basic concepts relating to the creation and use of PROCOMM PLUS scripts.

Becoming a PROCOMM PLUS Expert

Part III, "Becoming a PROCOMM PLUS Expert," covers the most technical features of the program. Once you are comfortable with the topics presented in Parts I and II of the book, Part III shows you how to fine-tune the program and take advantage of some of its most powerful capabilities.

Chapter 8, "Tailoring PROCOMM PLUS," discusses how to change COM port and line settings, as well as the many options in the PROCOMM PLUS Setup Utility. The Setup Utility enables you to make adjustments to more than 100 different characteristics of PROCOMM PLUS.

Chapter 9, "Using Host Mode," explains how to use the bulletin-board-like Host mode. The chapter discusses how to prepare and manage a simple company electronic bulletin board, including how to assign and manage user privileges and how to administer the electronic mail facility.

Chapter 10, "Terminal Emulation," offers a description of the terminal emulation capabilities of PROCOMM PLUS. The chapter includes an explanation of how you can customize the translation table and keyboard mapping of your terminal.

Chapter 11, "An Overview of the ASPECT Script Language," is the last chapter in Part III—and in *Using PROCOMM PLUS*. This chapter provides an overview of the programming capabilities of the powerful script command language, ASPECT. An in-depth discussion of all the commands and capabilities of the ASPECT script language is beyond the scope of this book.

Using the Appendixes

For those of you who have not yet installed PROCOMM PLUS, Appendix A, "Installing and Starting PROCOMM PLUS," describes the necessary steps to install and start PROCOMM PLUS. Specific instructions are included for installing PROCOMM PLUS on a floppy disk system, as well as on a hard disk system. Steps for upgrading to PROCOMM PLUS from ProComm are also provided. This appendix also explains how to use several special start-up parameters.

Appendix B, "Installing a Modem," explains the basics of installing a modem for use with PROCOMM PLUS. This appendix includes a discussion of connecting internal modems and external modems, as well as connecting two computers with a null modem cable. A special section covers using PROCOMM PLUS with the new error-control and data-compression modems.

Appendix C, "Using External Protocols," describes how to install an external error-checking protocol program, such as the user-supported program ZMODEM, so that it can be used from within PROCOMM PLUS.

Appendix D, "Patching PROCOMM PLUS," explains how to use the PROCOMM PLUS patch utility program to update your copy of PROCOMM PLUS with patch files, distributed periodically by DATASTORM.

Now that you know what to expect, you're ready to get started with PROCOMM PLUS. With this book as your companion and guide, climb on board the PROCOMM PLUS magic carpet and let it carry you to new and exciting computing heights.

Part I

Introducing PROCOMM PLUS

Includes

A Communications Primer

Getting Around in PROCOMM PLUS

A Communications
Primer

I f you jump right into PROCOMM PLUS or any similar communications program without a little preparation, you may feel as if you have been transported to a foreign country where everyone speaks a language you don't understand. This chapter is intended as a brief travel brochure on "communications land," and a mini-tutorial to help you learn the native dialect. The more you understand about your destination and its lingo, the more likely you are to enjoy your stay.

This chapter first describes the most common uses of communications programs. It then introduces you to many of the terms and concepts fundamental to understanding communications. If you are not sure whether you know the kinds of things you can do with PROCOMM PLUS or you are not yet comfortable with terms like *modem* and *bits per second*, take a few moments to go through this chapter. Then you will be ready to move on to Chapter 2, "Getting Around in PROCOMM PLUS."

Using Your Computer
To Communicate

This book focuses on using your PC and the PROCOMM PLUS program to communicate with other computers over telephone lines. To keep the concepts relatively simple, *Using PROCOMM PLUS* divides your potential uses for communications programs like PROCOMM PLUS into three major categories: PC to PC, electronic bulletin board systems, and on-line services. Communications programs do have other uses, and some overlap may exist in these categories, but for the purposes of this book and to help you think about ways you can use PROCOMM PLUS, this division is useful. The following paragraphs give you a brief synopsis of each usage category.

PC-to-PC Communicating

PROCOMM PLUS enables you to communicate from one PC to another over telephone lines. The purpose of this communication is usually to transfer computer files, such as word processing, spreadsheet, and database files, between the computers. Transferring files between PCs is covered in Chapter 5. A special mode of PROCOMM PLUS called Host mode allows you to authorize other PC operators to connect by telephone line to your PC for the purpose of transferring files, even when no one is at your PC's keyboard. Host mode is described in Chapter 9.

For example, picture yourself as the plaintiff's attorney in a multimillion-dollar product liability lawsuit. You have just flown into Chicago to interview the chief executive officer of the defendant company, Deep Pocket, Inc. To your utter dismay, however, you forgot to pack a copy of the 30-page list of questions you worked on all week. The list is still tucked away on your desktop PC's hard disk back in your office, 1,500 miles away. But you're in luck. You had the incredible foresight to leave your office PC turned on with PROCOMM PLUS running in Host mode. Using the laptop PC that you brought on the trip and PROCOMM PLUS, you call and connect to the PC in your office. In just a few minutes, you retrieve the list of questions and are back in business.

Many companies that routinely have to send computer files across town or across the country dedicate a computer to file transfer. Using PROCOMM PLUS, for example, each district sales manager in a large sales force can send the monthly sales report spreadsheet to a PC at the regional headquarters. Each regional manager consolidates the district figures into a regional report and transfers it to the PC at corporate headquarters. Not only faster and cheaper than express mail or even facsimile (fax), this method provides live data in a working spreadsheet at every level, without the necessity of entering the same numbers several times.

Reading Electronic Bulletin Boards

Perhaps the most popular use of communicating with a PC is to connect to electronic bulletin board systems (BBSs). As the name implies, *bulletin board systems* provide a medium through which computer users can obtain and even share information.

Nearly all BBSs provide traditional style bulletins—announcements or information posted by the operator of the bulletin board to be read by the various BBS users. BBSs also provide an important medium for computer users who want to share information. BBSs usually have areas, often called *conferences*, where users can leave public messages—analogous to tacking three-by-five cards on a cork bulletin board. Typically, the message areas are divided by topic or subject matter. When you find an area that interests you, you can choose merely to read the messages others have left, or you can participate in the discussion by leaving messages of your own. BBS conferences can be found on nearly any topic you can think of—astronomy, ham radio, finance, gun-control legislation, computer programming, engineering, science, mathematics, travel, and many others. Sometimes entire bulletin boards are devoted to the exchange of ideas in a particular area of common interest.

Electronic bulletin boards are most often run on PCs or other microcomputers, usually by a computer hobbyist entirely at his or her own expense. Some bulletin boards, however, charge subscription fees or ask for donations to defray the costs of buying and maintaining the sophisticated computer systems often needed to run an active bulletin board.

Many BBSs are run on PCs with huge storage capacity, often over 100M. These boards enable you to send and receive complete files, including program files. In fact, bulletin boards often are your best source for useful tips and practical software, especially when you are just learning to use the PC. BBSs frequently have hundreds and sometimes thousands of application programs, utility programs, games, and text files that can make your PC both easier and more fun to use. Figure 1.1 shows the menu of a typical bulletin board file system.

Some of the software on bulletin boards has been released to the public domain by the software authors, and anyone is free to use the software at no charge. Most of the more powerful programs, however, are referred to as *shareware*, *freeware*, or *user-supported software*. You are permitted to use these user-supported programs, including ProComm 2.4.3, without charge, but only on a trial basis. If you decide to continue using a program, you are obligated to pay a registration fee to the program's author in order to license your copy.

The PROCOMM PLUS Host mode has several features found in typical bulletin board software, including a file-transfer system and a message system. The Host mode may be all you need to set up a small bulletin board for your personal or office use. Host mode is not, however, adequate for use as a full-fledged public BBS. This facet of PROCOMM PLUS is covered extensively in Chapter 9, "Using Host Mode."

Fig. 1.1.

A typical PC bulletin board menu, from the Capital PC Users Group Member Exchange.

```
Use MEMBER option under Doors to increase your access if you just joined CPCUG.
All messages are in conferences (use J option to get to messages).
Leave comments to SYSOP in MIX conference (NOT general messages).

MAIN command <?,A,B,D,E,F,H,I,J,K,P,Q,R,S,T,U,V,W,X>? f

   185 min left

RBBS-PC FILE SYSTEM
TRANSFER ------ INFORMATION --- UTILITIES ----- ELSEWHERE
Download file   List files     Help (or ?)      Goodbye
Upload file     New files      eXpert on/off    Quit
                Search files
                View archive

Uploaded files are not available for general downloading until they have been
reviewed by the SYSOP. Please include identification of the author of any
information file that you upload, either in the file itself or in the file
description. Uploaded software is forwarded to the CPCUG Software Library.

Monitor articles must be uploaded with an extension of .MON. Also indicate
in the file description that it is a Monitor article and give the subject.

FILE command <?,D,G,H,L,M,P,Q,S,U,V,X>?
 Alt-Z FOR HELP│ VT102 │    FDX │ 2400 N81 │ LOG CLOSED │ PRINT OFF │ ON-LINE
```

Using On-line Services

On-line services are generally run on large computer systems that can handle simultaneous use by multiple computers connected over multiple phone lines. Often you connect to an on-line service through a public data network (PDN), a special type of long-distance network for communication by computers. Figure 1.2 shows a menu from the popular on-line service CompuServe.

The primary business of most on-line services is to sell information —news, stock quotes, airline schedules, and so on. Some on-line services have specialized information, such as medical journals or law reporters, but many services make available information on a wide variety of subjects. The most popular services offer numerous other features, including electronic shopping malls, which let you order merchandise right from your PC's keyboard, and CB-simulators, which enable you to converse through your computer with many people at once.

Some on-line services specialize in *electronic mail*—an electronic substitute for paper or voice communication through which you send private correspondence to another subscriber. Rather than sort through a stack of mail in your "in" box, you log on to the electronic mail system to check your electronic mailbox. The system greets you with the message You

```
CompuServe                                                          TOP


   1 Member Assistance (FREE)
   2 Find a Topic (FREE)
   3 Communications/Bulletin Bds.
   4 News/Weather/Sports
   5 Travel
   6 The Electronic MALL/Shopping
   7 Money Matters/Markets
   8 Entertainment/Games
   9 Hobbies/Lifestyles/Education
  10 Reference
  11 Computers/Technology
  12 Business/Other Interests




Enter choice number !
 Alt-Z FOR HELP| VT102  |  FDX  |  2400 E71  | LOG CLOSED | PRINT OFF | ON-LINE
```

Fig. 1.2.

A menu from CompuServe, an on-line service.

have mail. You proceed to read your mail on-screen, saving to disk the important correspondence, tossing everything else into the "bit-bucket" (that is, deleting it).

Similar to bulletin boards, but on a larger scale, the most popular on-line services provide dozens of electronic forums for the exchange of ideas among individuals and groups. Many computer hardware and software manufacturers, including DATASTORM, the publisher of PROCOMM PLUS, host one or more of these on-line forums in order to be more accessible to users and to distribute information about patches and upgrades. Vendor-hosted forums provide valuable feedback to the manufacturer and allow users to share tips and techniques on using the vendor's products. Many vendors consider user input gathered from on-line forums an important factor in future product development decisions.

Learning the Lingo

Countless other reasons exist for communicating with your PC over telephone lines, from electronic banking to playing interactive games. But before you can begin to learn how to use PROCOMM PLUS to these ends, you need to become familiar with a few new words and phrases.

Communications terminology has developed over the short history of computing in a way not unlike that of a spoken language. Some terms have been coined by communications experts when necessary to describe a new apparatus or a new communications concept, but other terms seem to have sprung up on their own and are kept alive by continual usage. Like the words of any language, the definitions of many communications terms are universally accepted, but the meanings of some other terms seem to have changed over time or location. The remainder of this chapter introduces you to a number of the communications-related terms and concepts you need to understand in order to get the most from PRO-COMM PLUS.

Understanding Modems

Because you probably already know how to use your PC for word processing, spreadsheets, and information management, and you are certainly familiar with telephones, learning to use a PC to communicate over telephone lines should be an easy next step. The problem is, however, that telephones are made to transmit voice, but computers need to send a different type of signal. Most of the complexity of communications boils down to this rather subtle distinction.

Analog and Digital Signals

Understanding a little bit about how telephones transmit voice signals can help you more easily comprehend how computer data is sent. When you make a sound with your voice, your vocal chords cause the air to vibrate in a wavelike motion. The vibrating air, called a sound wave, causes a listener's eardrum to vibrate, enabling the listener to hear the sound you made. The frequency of the sound wave (the number of wave cycles in a given period of time) is heard as pitch. The amplitude (size) of the wave is heard as volume.

When you speak into your telephone, the telephone electronically converts the sound wave of your voice into an electromagnetic wave that can be transmitted over telephone lines. The frequency and amplitude of this electromagnetic wave correlate directly to the frequency and amplitude of the sound wave. As the sound wave's frequency varies up or down, the electromagnetic wave's frequency varies up or down in the same proportion; as the sound wave's amplitude varies, so does the electromagnetic wave's amplitude. In other words, the electromagnetic wave is an *analog* of the sound wave. Indeed, the signal your telephone sends over the telephone lines is often referred to as an *analog signal* (see fig. 1.3). When this signal reaches the other end of the line, the phone at that end converts the signal back into a sound wave for the benefit of the person to whom you are talking.

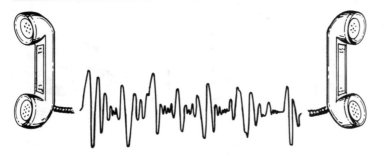

Fig. 1.3.

The analog signal transmitted by telephones.

Computers do not, however, communicate by sound waves. They understand only discrete electrical pulses that represent numbers. Computers can communicate directly with one another by sending data over a cable. All data is encoded as a stream of 1s and 0s called *bits*. A *bit* (short for *binary digit*) is the most basic form in which a PC stores information. To transmit a bit with a value of 1, for instance, the computer may set the line voltage to − 12 volts (direct current) for a set length of time. This set length of time determines the *transmission speed*. The computer may transmit a bit with a value of 0 by setting the voltage to + 12 volts for a set length of time. A typical transmission speed is 1,200 bits per second. At that speed, each voltage pulse is 1/1200th of a second in duration. Some PCs and compatibles are capable of sending as many as 115,200 bits per second, with each pulse lasting only 1/115200th of a second.

When one of the communicating computers needs to send the code 01001011, for example, the computer sets the voltage to + 12 volts for one unit of time, − 12 volts for one unit of time, + 12 volts for two units of time, and so on. (The unit of time can be any "set length of time" agreed on by the two computers. For example, 2,400 units of time per second is typical.) The square-shaped waves shown in figure 1.4 illustrate the voltage pulses that the computer uses to transmit this code. Because a signal of this sort is transmitting bits—binary digits—it is usually called a *digital signal*.

In the not-so-distant future (within the next five years), Integrated Services Digital Networks (ISDNs) will be able to carry simultaneous voice, data, and image transmissions. This capability is already available in some areas of the country. Until ISDNs are more widely available, however, your PC's digital signal has to be converted (*modulated*) to an analog signal in order to be sent over the phone lines. The analog signal must then be converted (*demodulated*) back to a digital signal in order to communi-

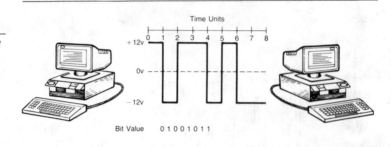

Fig. 1.4.

The digital signal 01001011 transmitted directly between two computers, without using modems or a phone line.

cate with another computer at the other end. The piece of hardware that accomplishes both *mod*ulation and *dem*odulation is the aptly named *modem*, short for modulator-demodulator (see fig. 1.5). Without a modem at each end of the phone line, the two computers cannot communicate.

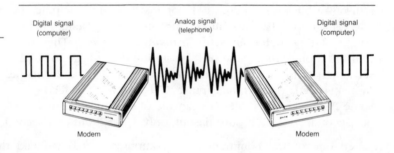

Fig. 1.5.

Two modems transmitting computer data over telephone lines.

Modem Speed

Although all modems modulate and demodulate transmitted signals, all modems are not created equal. The most important difference among the various types of modems available for use with your PC is the maximum speed at which they can transmit data.

Only a few years ago, a common practice was to use modems that sent data at a maximum rate of 300 bits per second (bps). To put this transmission rate into perspective, consider the number of bits required to represent meaningful information. All information stored and used in your PC is represented by combining eight bits at a time, referred to as a *byte*. When you are transmitting text, or typing at your keyboard on-line (that is, when connected to another computer), each character sent to the other computer is represented in ASCII (American Standard Code for Information Interchange). The IBM version of ASCII used by the PC is a code that uses one byte, or eight bits, to represent each character.

At 8 bits per character and 300 bits per second, you would think that the modem could transfer 37.5 characters per second, or 375 five-letter words per minute. PCs, however, normally add to each byte two extra bits, called *start* and *stop* bits (see "Understanding Serial Ports," later in this chapter). Consequently, a speed of 300 bps results in a data transmission speed of approximately 30 characters per second, or about 300 five-letter words per minute (even a fast typist usually averages less than 80 words per minute).

This rate is fast enough when you simply want to type an instruction to a remote computer or "chat" (type messages back and forth) with another user on-line, but this rate is not exactly speedy when you are sending a long document you have already typed. At about 1,000 words a page, sending a 10-page document could take nearly half an hour. Because you may be sending the document over long-distance lines, use of a 300-bps modem may translate into high telephone bills.

Fortunately, modem speeds have significantly increased in recent years, while prices have decreased. Soon after IBM introduced the PC in August 1981, the Smartmodem 300 and Smartmodem 1200 by Hayes Microcomputer Products became the standard modems for business use. Since 1987 the entry-level standard for PC modems seems to have become the 2400-bps modem. With its Smartmodem 2400, Hayes has again set the standard against which other modems are measured. Technological advances in the manufacture of integrated circuits and fierce competition have brought the prices of this category of modems down so far that you have little if any reason to buy a less capable modem.

Smartmodems have certainly not been the only modems available, but they possess a number of built-in features that, if properly driven by communications software, can substantially automate many communications functions. For example, with the Hayes Smartmodems, you can store a list of telephone numbers in your computer and then have the modem dial a selected number when you want to call a particular computer. This feature is usually referred to as *auto-dial* capability. With older, less "smart" modems, you had to dial the number manually each time, using a telephone hand set, and then push a button to activate the modem. Hayes Smartmodems also can be programmed to answer the phone and automatically connect to an incoming call from another modem—an *auto-answer*

TIP You may sometimes hear the term *baud* used synonymously with *bits per second*, but they do not mean the same thing. Although 300-bps modems are also correctly referred to as 300-baud modems, 1200-bps modems do *not* operate at 1200 baud (they usually operate at 600 baud).

Baud is a technical term that means the number of symbols per second sent over a communication line. (The term is named after J. M. E. Baudot, a French telegraphy expert.) Each symbol may be represented by a certain voltage, a certain combination of frequencies (tones), or a certain wave phase (angle). Each symbol may, however, be able to represent more than one bit. In 2400-bps modems, for example, each symbol represents 4 bits. So even though the modem transmits only 600 symbols per second (600 baud), data is transmitted at 2400 bits per second (4 × 600).

The term *baud* has been misused so often, however, that its technical meaning is largely ignored. You can therefore assume that the salesman who wants to sell you a 1200-"baud" modem means a 1200-bps modem. Similarly, when PROCOMM PLUS uses the term *baud*, read it to mean bits per second.

capability. Many earlier modems required you to answer the phone yourself and then press a button as soon as you heard the carrier tone (a high-pitched tone generated by the calling modem).

What Is "Hayes-Compatible"?

The series of software commands that activate the Hayes Smartmodem's smart features nearly all begin with the letters *AT* (short for *attention*). This command set has therefore come to be known as the *AT command set*. For example, to dial a telephone number by using touch-tone, PROCOMM PLUS sends to the modem (through a serial port) the command *ATDT* followed by the telephone number.

Because of the early lead that Hayes built in sales of modems to PC users, nearly all communications software written for use on a PC assumes that your modem understands the AT software command set. Consequently, the AT command set has been adopted by almost every other modem manufacturer as the de facto standard command language for modems intended for use with PCs. Modems that recognize this command set are often called *Hayes-compatible* modems. Not all so-called Hayes-compatible modems, however, implement the entire command set, so in that sense some modems are more compatible than others.

When Hayes introduced the Smartmodem 2400, the company also introduced new commands to the AT command set. Other manufacturers have implemented this so-called *extended AT command set* to varying degrees. You can use PROCOMM PLUS effectively whether or not your modem supports the extended AT command set. Using a modem that supports the AT command set is most convenient, however.

New Standards

The majority of 2400-bps modems available today are so similar in basic function that they practically constitute a commodity, and nearly all 2400-bps modems can communicate with each other. Technology, however, marches on. Just as 2400-bps modems have gradually replaced the 1200-bps versions in general business usage, 2400-bps modems with error-control and data-compression capabilities are beginning to take hold as the next low-end standard.

Hayes and other modem manufacturers have also introduced even faster modems, some capable of speeds up to 19,200 bps. The CCITT (Consultative Committee on International Telephone and Telegraph), an international communications standards organization and an agency of the United Nations, has adopted a standard for 9600-bps modems, referred to as V.32. Prices for these modems are substantially higher than prices of 2400-bps modems, but the use of 9600-bps modems is increasing among PC users who do a substantial amount of data transfer over telephone lines. To take advantage of these speed demons, you are usually forced to buy them in pairs and install one on each end of the phone line.

Error Control

Transmission of computer data must be error free to be effective. The detection and elimination of errors in computer data transmissions is usually called *error checking*, or *error control*. Although a telephone line doesn't have to be perfectly clear of static or interference for voice transmission to be effective, a slight variation or interruption of a computer signal can completely change the meaning of the data the signal carries.

All communications programs such as PROCOMM PLUS are capable of running checks on incoming data to determine whether any errors were introduced into the data during transmission. The techniques used to perform this error checking are called *error-checking protocols*. (A *protocol* is a set of agreed-on rules.) Chapter 5, "Transferring Files," discusses at length the various protocols available in PROCOMM PLUS and how best to use them.

Several manufacturers now produce modems that perform error control, as opposed to requiring that the communications software perform this chore. When error control is accomplished by the modem, both error detection and the overall speed of transmission are improved, compared to the results of software error checking. For this feature to work, however, the modems on both ends of the connection have to be using the same error-control protocol.

Two different types of error-control protocols have developed a significant following. Fortunately, an industry standard has emerged that incorporates the two competing error-control schemes.

Hayes Microcomputer Products produces a line of modems called the Hayes V-series modems. At the low end of this line is a 2400-bps modem that performs error control by using a method called Link-Access Procedure for Modems (LAPM). Nearly all other PC modem manufacturers, however, produce modems that support a different set of error-control protocols called the Microcom Networking Protocol (MNP) Classes 1 through 4, developed by Microcom, Inc. The MNP protocols are progressive. A modem can support Class 1, Classes 1 and 2, Classes 1 through 3, or Classes 1 through 4. The higher the class, the faster the transmission. These protocols are not, however, compatible with LAPM. Two 2400-bps modems using the two different standards can communicate, but only as 2400-bps modems without modem-based error control. (Your software must take care of error detection to ensure that line impairment during transmission doesn't corrupt the transmitted data.)

In 1989, the CCITT established an error-control standard referred to as the *V.42*. This standard includes the protocol used by Hayes V-series (LAPM) as well as MNP Classes 1 through 4 error-control protocols, with a bias toward LAPM. When connected to another modem, a V.42-compliant modem attempts to use the LAPM protocol. If this approach fails, the V.42 modem then attempts to use the MNP protocols. If the other modem supports none of these error-control protocols, the V.42 modem acts like a standard 2400-bps modem without error control.

Data Compression

Many of the modems that provide built-in error control also compress the data as it is sent. A compatible modem on the other end decompresses the data. Compressing data is analogous to sitting on a loaf of bread, squeezing out the "air" but leaving the nourishment. In fact, a modem compresses data by matching long strings (sequential patterns) of characters in the data with entries in a "dictionary" of known strings. Each entry in the dictionary has an index value or code. The sending modem finds a code for each string of characters in the data and transmits only the codes. The receiving modem in turn converts these codes back into the original data. This process can often more than double the effective transmission speed by reducing the number of characters the modem has to send. In other words, a 2400-bps modem using data compression can sometimes send as much information in the same length of time as a 4800-bps modem that is not using data compression.

As is the case with error-control methods, two different data-compression standards have developed in the PC modem market; but this time a CCITT standard will replace them both. Hayes Microcomputer Products V-series modems perform data compression by using a proprietary algorithm. Many other PC-modem manufacturers follow Microcom's lead and use the MNP Class 5 data-compression algorithm developed by Microcom, Inc. Again, these two competing data-compression methods are not compatible.

In September 1989, the CCITT ratified the *V.42bis* standard, which adds data compression to the existing V.42 error-control standard. The data-compression scheme included in this new standard is neither the Hayes algorithm nor the MNP algorithm. The CCITT V.42bis proposes, instead, the use of an algorithm known as the British Telecom Lempel-Ziv (BTLZ) compression algorithm.

The V.42bis standard does not include MNP data compression (referred to as MNP Class 5), but many manufacturers plan to produce modems that support V.42bis data compression as well as MNP data compression. Hayes has announced a plan to upgrade V-series modems to the V.42bis standard for a nominal fee.

Appendix B includes an explanation of how to use PROCOMM PLUS to take full advantage of the advanced capabilities of Hayes V-series, V.42-compliant, and MNP modems.

External versus Internal Modems

In addition to modems capable of transmitting data at different speeds, modem manufacturers often produce modems in both external and internal versions. Each type offers a number of advantages.

An *external modem* is typically a metal or plastic box about 10 inches by 6 inches by 2 inches, with a panel of LED (light emitting diodes) on the front; the modem is connected to your PC by a serial cable and powered by an AC adapter. The external modem has at least one telephone jack for connecting the modem to the telephone line and often a second telephone jack for connecting a telephone. Figure 1.6 shows a drawing of a Hayes V-series Smartmodem 9600.

Fig. 1.6.

An external Hayes V-series Smartmodem 9600.

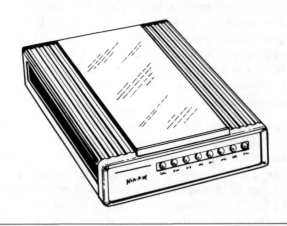

A relatively new subset of external modems are small enough to fit in your pocket and can run on batteries. These mini-modems typically plug directly into a serial port on your computer (see fig. 1.7) and are particularly handy for laptop computer users.

Fig. 1.7.

A pocket-sized modem.

The most obvious advantage of an external modem is that it can be easily moved from one computer to another. Indeed, an external modem can be used with nearly any type or brand of computer that has an asynchronous serial port (see "Understanding Serial Ports," later in this chapter). An external modem can even be shared by several computers through a serial switch box (with only one computer using the modem at a time). And external modems usually sport a panel of LEDs (lights) that enable you to monitor continuously the state of certain modem parameters. Table 1.1 lists the meanings of the LEDs on the front panel of external Hayes modems and most external Hayes-compatible modems.

Table 1.1
Hayes Smartmodem LEDs

LED	Meaning
HS	High Speed
AA	Auto-Answer
CD	Carrier Detect
OH	On Hook
RD	Receive Data
SD	Send Data
TR	Terminal Ready
MR	Modem Ready

On the negative side, an external modem is usually a little more expensive than an internal modem with otherwise identical features from the same manufacturer. An external modem also requires your computer to have an available serial port and a serial cable. Although an external modem takes up some space on your desk, often you can stash the modem under your desk telephone or on top of your PC. And pocket-sized modems usually can plug directly into a serial port, taking up no room on your desk.

Each *internal modem* is built on a circuit board that plugs into an empty expansion slot inside your PC, thus taking up no room on your desk. Some internal modems are long enough to fill up a long expansion slot, and others need only a half-size slot.

Because an internal modem is designed to work only with a PC (or PS/2—the same internal modem does not work in both), internal modems are usually bundled and sold with PC communications software. This setup is an advantage unless you don't like the software.

The major disadvantage of an internal modem is that it is inconvenient to use with several different computers. If you need to move a modem among several computers or want to share a modem, an external modem is the better choice. Also, because most IBM PS/2s do not accept expansion boards designed for PCs (PCs, XTs, ATs, or compatibles), you cannot move an internal modem from a PC into a PS/2 if you ever decide to take that upgrade path. For some users, the lack of status LEDs is also an annoyance.

Understanding Serial Ports

Before data can be transmitted by your modem, your PC must send the data from your keyboard or disk to the modem. The computer sends data to the modem through a *serial port* in your PC. A serial port, often called a *COM port*, is an outlet through which your computer can send data as a stream of bits, one bit at a time (that is, in *serial*). Data normally moves around the computer, however, eight bits at a time (referred to as in *parallel*).

TIP | When you use an internal modem rather than an external modem, you don't connect the modem to a serial port. A special chip on the modem, called the UART (universal asynchronous receiver/transmitter), performs the same function a serial port does—converting the computer's parallel signal into a serial signal. When you install the modem, you must configure it as one of your computer's COM ports (see Appendix B, "Installing a Modem"). So for all intents and purposes, you can think of the internal modem as having its serial port built in.

On IBM PC, PC/XT, and PS/2 computers, as well as most compatibles, each serial port is a D-shaped connector that has 25 protruding metal pins and is located on the back of the computer. This type of connector is called a *DB-25 M* (male) connector. The connector on the serial cable that attaches the modem to this serial port has 25 holes to match the male connector's 25 pins. This connector is called a *DB-25 F* (female) connector.

The serial ports on IBM PC AT computers and most compatibles use D-shaped connectors with 9 protruding pins. These connectors are referred to as *DB-9 M* connectors. To connect a serial cable to such a port, the cable must end in a *DB-9 F* connector.

Depending on the brand of computer, each port is marked, if at all, with the label *COM*, *Serial*, or *RS-232*. (RS-232 is a published communications hardware standard with which PC serial ports comply.)

DOS 3.3 and higher can support as many as four serial ports. Earlier versions of DOS support only two serial ports. Serial ports are referred to by DOS as COM1, COM2, COM3, and so on. Even though your version of DOS may not be able to support more than two or perhaps four serial ports, PROCOMM PLUS still enables you to use up to COM8. In other words, PROCOMM PLUS goes around the DOS limitation. Appendix A explains how to inform PROCOMM PLUS, in the initial setup, which COM port is connected to the modem. Chapter 8, "Tailoring PROCOMM PLUS," also describes how to change this setting later.

Asynchronous versus Synchronous Transmission

Data sent through a serial port comes out as a stream of bits in single file, but each bit means nothing by itself. Because a PC stores information in bytes, each consisting of eight bits, the receiving computer must be able to reconstruct the bytes of data from the stream of bits in order to make sense of the transmission. Computers use two ways to identify clearly each byte of data sent through a serial port: *asynchronous* transmission and *synchronous* transmission.

PCs use the asynchronous method to send data through a serial port. The sending PC marks with a *start bit* the beginning of each byte that is to be transmitted. This start bit informs the receiving computer that a byte of data follows. To mark the end of the byte, the PC sends either one or two *stop bits* (the number of stop bits is a user option). The stop bits inform the receiving computer that it has just received the entire byte. Using this procedure, timing is not critical.

When you set up PC communications software such as PROCOMM PLUS to communicate with another computer, you have to specify whether to use one or two stop bits. The number of stop bits used on your end must match the number used by the computer on the other end. In PROCOMM PLUS, you select the number of stop bits when you establish a list of phone numbers you want to call with your modem. This list is called the dialing directory and is described in Chapter 3.

TIP	When in doubt, use one stop bit. This choice is the default in PRO-COMM PLUS (the setting PROCOMM PLUS uses if you don't specify another one). You will seldom if ever see two stop bits used by another computer.

When computers use the synchronous method of transmitting data, bytes of data are sent at precise intervals. Both the sending and receiving modems must be perfectly synchronized for this method to work properly.

Because a PC uses the asynchronous method to send data through its serial port, a *protocol converter* is required before a PC can communicate with a computer using the synchronous method. A protocol converter converts the signal from asynchronous to synchronous. Many mainframe computers communicate only in a synchronous mode, but they often have an asynchronous dial-in port with a built-in protocol converter. This design allows PCs and other computers to connect, using an asynchronous signal.

TIP | Synchronous transmission can achieve higher transmission rates but requires higher-quality telephone lines. Telephone lines used for synchronous transmission are often specially prepared and used exclusively for that purpose. These lines usually must be leased from the telephone company, making synchronous transmission prohibitively expensive for use in everyday PC communications.

Data Bits

As mentioned previously, the PC uses eight bits at a time to represent a byte of data. Indeed, because exactly 256 ways are available to arrange eight 1s and 0s (2 to the 8th power), eight bits are needed to represent each of the 256 characters in IBM's extended ASCII character set. Many other types of computers, however, including mainframes used by on-line services, use only seven data bits per byte. The ASCII character set used by these computers includes only the first 128 characters of the IBM extended ASCII character set (2 to the 7th power equals 128). These 128 characters include all the numbers and letters (upper- and lowercase), punctuation marks, and some extra control characters. Special characters such as foreign letters and box-drawing characters require all eight bits.

When you create the PROCOMM PLUS dialing directory, for each computer you plan to call, you specify whether data occupies the first seven bits of each byte or all eight bits. PROCOMM PLUS calls this specification the number of *data bits*. Some programs use the term *word length* or *character length* for the same specification. You may sometimes use your PC to communicate with another PC that needs eight data bits per byte. At other times, you may communicate with a type of computer that can use only seven data bits per byte. The only way to know for sure which setting you need is to ask the operator of the other computer.

TIP | A simple rule of thumb is to use eight data bits for PC-to-PC or PC-to-bulletin-board connections. All PCs need eight-bit bytes to represent the full IBM character set as well as to transmit program files. When calling an on-line service, such as CompuServe, or another mainframe-based system, use seven data bits. Most such systems are run on computers that can handle only seven data bits per byte.

Parity Checking

As mentioned earlier, an error in transmission of even a single bit can completely change the meaning of the byte that includes that bit. To help detect these errors as they occur, some computers use the eighth bit of each byte as a *parity bit*. The *parity* of an integer (whole number) is whether it is odd or even. You can use a parity bit in two ways. Both methods add up the other seven digits in each byte and check whether the sum is an even number or an odd number.

When using the *even parity* method, the computer sets the value of the parity bit (either 1 or 0) so that the total of all eight bits is even. If the sum of the first seven bits is odd, such as 0000001, the parity bit becomes a 1. The eight-bit byte that is transmitted is therefore 00000011. If the sum of the first seven bits is even, such as 0010001, the parity bit becomes 0, and the eight-bit transmitted byte is 00100010. In both cases, adding all eight bits results in an even number (2 in both of the examples). The receiving computer then adds the digits of each byte as it is received. If the sum is odd, an error must have been introduced during transmission, so the computer asks that the byte be sent again.

Similarly, the *odd parity* method assigns the value of the parity bit so that the sum of all digits in each byte always is an odd number.

One parity bit method always sets the parity bit to the value 1. This method is called *mark parity*. Another method, known as *space parity*, always sets the parity bit to the value 0.

Parity checking provides rather minimal error checking when compared to the many other more sophisticated error-checking methods now available in both software (see Chapter 5, "Transferring Files") and hardware (see "New Standards," earlier in this chapter). The parity-check method is not available if the eighth bit in each byte is needed to represent data, such as when transmitting the full IBM extended ASCII character set, or when transmitting non-ASCII files, such as all programs that run on your PC and most data files used with word processing, spreadsheet, and database programs. As with the number of stop bits and the number of data bits, you specify parity in PROCOMM PLUS on the Dialing Directory screen.

TIP For PC-to-PC communication, including to bulletin boards, use NONE (no parity). When connecting to an on-line service, you usually have to use EVEN parity. If you are connected to an on-line service and are receiving nothing but strange-looking characters, you probably still have parity set to NONE. Try changing the parity to EVEN. (Changing the parity setting after you are connected is explained in Chapter 8.)

Understanding Telephone Line Requirements

Both external and internal modems must be connected to a working telephone line in order to be effective, but modems do not require a special type of telephone line. A voice-quality line is sufficient. Either touch-tone or pulse dial service can be used because nearly all auto-dial modems are capable of dialing either type of signal. Refer to Appendix B for information on installing your modem and connecting it to your telephone line.

TIP One relatively new telephone service can cause problems for a modem transmission. The feature, usually referred to as call waiting, is available from an ever-increasing number of telephone companies. It uses an audible click or beep to alert you of an incoming call while you are already talking on the line. If your modem is on the line, this call-waiting signal may disrupt and even disconnect your transmission. If you have this type of service and plan to use your modem during periods when you may receive incoming calls, you should consider having the service removed. In some areas, you can temporarily disable the feature by typing a special code on your telephone keypad or even by transmitting the code in a PRO-COMM PLUS dial command. (For more information, see Chapter 8 on setup.) Check with your local telephone company to determine whether such a code is available and, if so, how to use it.

Chapter Summary

This chapter has briefly described the most common uses of communications programs. It also has introduced you to many of the terms and concepts fundamental to understanding communications. As you use this book to help you learn to use PROCOMM PLUS, you will certainly run across a number of other new communications concepts and terms, but you now have a solid foundation on which to build. Turn now to Chapter 2 to begin learning how to get around in PROCOMM PLUS.

Getting Around in
PROCOMM PLUS

N ow that you have read Chapter 1, the travel brochure on "communications land," you are ready to start finding your way around PROCOMM PLUS. This chapter helps you get started by giving you an idea of how PROCOMM PLUS looks on-screen and by explaining how the program makes requests and responds to your actions. This chapter introduces you to the Terminal mode screen, the keyboard, the Command Menu, and the menu line. You also learn how to display the PROCOMM PLUS context-sensitive help system and how to exit from PROCOMM PLUS.

Before you can use PROCOMM PLUS, of course, you must have installed it properly on your computer, and you must know how to activate the program. Both installation and start-up are discussed in Appendix A. If you need help installing PROCOMM PLUS on to a working disk or getting the program up and running, turn to that appendix before you go any farther in this portion of the book.

Once you start the program, PROCOMM PLUS displays its logo screen, shown in figure 2.1. Notice that this screen tells you to PRESS ANY KEY TO ENTER TERMINAL MODE. When you press a key, the program, as promised, displays the Terminal mode screen, shown in figure 2.2. PROCOMM PLUS is under way.

Fig. 2.1.

The PROCOMM PLUS logo screen.

Intuitive Communications (tm)

* PROCOMM PLUS - Version 1.1B *
Copyright (C) 1987, 1988 DATASTORM TECHNOLOGIES, INC.
All Rights Reserved
UNAUTHORIZED DUPLICATION PROHIBITED

PRESS ANY KEY TO ENTER TERMINAL MODE

Fig. 2.2.

The Terminal mode screen.

PROCOMM PLUS Ready!

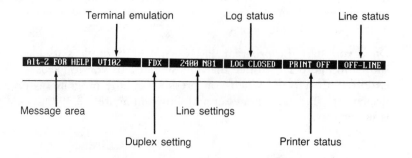

Terminal emulation   Log status   Line status

Alt-Z FOR HELP | VT102 | FDX | 2400 N81 | LOG CLOSED | PRINT OFF | OFF-LINE

Message area   Line settings

Duplex setting   Printer status

Understanding the Terminal Mode Screen

You need to become familiar with the Terminal mode screen because it is the program's main screen. PROCOMM PLUS displays on this screen all text transmitted by the computer with which you are communicating.

PROCOMM PLUS greets you at the upper left of the Terminal mode screen with the message PROCOMM PLUS Ready! and places a blinking under-score, referred to as the *cursor*, just beneath the P in PROCOMM. With the exception of the *status line* at the bottom of the screen, the remainder of the screen is blank.

The Terminal mode screen's status line is divided into seven sections:

- The *message area* is the first section on the left end of the status line. This area usually contains instructions for displaying the PROCOMM PLUS help system, as shown in figure 2.2, but occasionally is used to display other status messages. Two messages that can be shown in the message area are RECORDING..., when you have instructed PROCOMM PLUS to record keystrokes, or SNAPSHOT, when you tell PROCOMM PLUS to make a copy of the current screen contents to a disk file. This area also shows the name of any ASPECT script file you may be running.

- The second section on the status line displays the current *terminal emulation*. *Terminals* are special "slave" computers whose sole purpose is to connect to a larger computer for data entry and retrieval. When PROCOMM PLUS *emulates* a particular type of terminal, the program "impersonates" the terminal type in order to communicate with a large computer that expects only terminals to be connected.

  When you install PROCOMM PLUS using the PCINSTAL Installation Utility (see Appendix A), you choose which type of terminal you want the program to emulate. You can change this *default* setting (a setting that is used unless you specify otherwise) by using the Setup Utility (covered in Chapter 8). The default is overridden when you choose a different terminal emulation in a dialing directory entry. (See Chapter 3 for a complete discussion of the dialing directory, and Chapter 10 for details about terminal emulation.) Figure 2.2 indicates that the current terminal emulation is VT102.

- The third section of the status line indicates whether PROCOMM PLUS is set to *full duplex (FDX)* or *half duplex (HDX)*. The meaning of this setting is described in "Toggling Duplex," in Chapter 4. As with the terminal emulation setting, the Installation Utility establishes a default setting, which you can change by using the Setup Utility and override by changing the setting in a dialing directory entry or by pressing Alt-E.

- The fourth status line section shows the current *line settings*—the transmission speed setting as well as the current parity, data bits, and stop bits settings. (These settings are discussed in Chapter 1.) For example, 2400 N81 means that PROCOMM PLUS is set to communicate at a transmission rate of 2400 bps, with no parity bit, 8 data bits, and 1 stop bit. As with the other default settings, you establish them initially with the Installation Utility, can change them with the Setup Utility, and can override them with a dialing directory entry.

- The next section on the status line, the fifth from the left, indicates whether you have instructed PROCOMM PLUS to save the current session to a *log file* on disk. By default, PROCOMM PLUS does not save to a disk file (*capture*) all the information that scrolls across your screen. The system therefore usually displays LOG CLOSED in the status line, as shown in figure 2.2. If you turn on the Log File feature, PROCOMM PLUS changes the message to LOG OPEN. You can also temporarily suspend the capture of the session to disk. The message in the status line then reads LOG ON HOLD. Refer to the section "Capturing the Session to Disk," in Chapter 4, for a complete discussion of this feature.

- The sixth section in the status line indicates the *printer status*—whether PROCOMM PLUS is sending data to your printer. The message PRINT OFF, as in figure 2.2, means that nothing is being sent to the printer. PRINT ON means that the characters you type on the screen and any ASCII characters transmitted to your PC by a remote computer are printed on your printer. This feature is described in Chapter 4.

- The last section on the PROCOMM PLUS status line reminds you of the *line status*—whether you are currently connected to another computer. When you first start PROCOMM PLUS, the message indicates OFF-LINE, meaning that your PC is not communicating with another computer. Later, when you connect to another computer, this status line message changes to ON-LINE.

TIP | If the last message on the status line always incorrectly displays ON-LINE, even when you first start PROCOMM PLUS and before you have called another computer, refer to the discussion in Appendix B on installing a modem. The Carrier Detect (CD) setting on your modem is probably inadvertently set so that PROCOMM PLUS "thinks" that a carrier signal (the signal generated by another modem) is always present. Change this modem setting so that the modem monitors the existence of a carrier and raises the CD signal high (tells PROCOMM PLUS that a carrier exists) when a carrier from another modem is detected.

Although the largest portion of the Terminal mode screen is blank, this screen comes alive once you connect to another computer. The first 24 lines are where all the action takes place.

Using the Keyboard

PROCOMM PLUS works with any PC, AT, or PS/2 style keyboard or equivalent (see fig. 2.3). The program uses practically every key on the keyboard in one way or another, so you need to be familiar with your keyboard. This book uses the labels found on an IBM Enhanced keyboard to describe keys you should press.

PROCOMM PLUS translates some keys in special ways when the program is emulating a particular type of terminal. Refer to Chapter 10 for a description of PROCOMM PLUS terminal emulation. That chapter includes tables that show how PROCOMM PLUS translates the various *programmable keys*, such as cursor-movement keys and function keys, for each type of terminal.

Using the Command Menu as a Reminder

The PROCOMM PLUS Command Menu is similar to the Help menu found in ProComm 2.4.3 and earlier versions of ProComm. The new menu includes several new commands, however, and the entire screen is rearranged. To display the Command Menu from the Terminal mode screen, press Alt-Z (Help). PROCOMM PLUS displays the screen shown in figure 2.4.

Fig. 2.3.

The IBM PC keyboard (top), the IBM Personal Computer AT keyboard (center), and the IBM Enhanced keyboard (bottom).

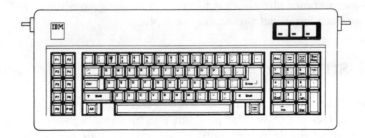

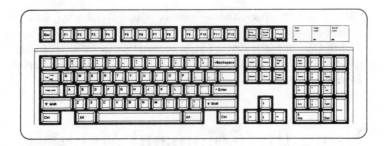

PROCOMM PLUS Ready!

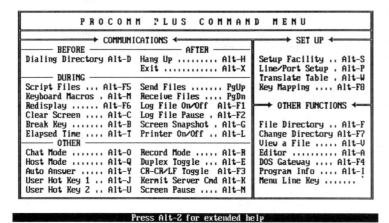

Fig. 2.4.

The PROCOMM PLUS Command Menu.

To execute any command listed on this menu, you press the key or key-stroke combination indicated. For instance, the first command in the menu is the Dialing Directory command, and Alt-D is listed to its right. This entry tells you that you must press the keystroke combination Alt-D to display the PROCOMM PLUS dialing directory.

Even though the screen shown in figure 2.4 is referred to as a "menu," you don't have to display it before using the commands listed there. If you can remember the PROCOMM PLUS command, you can use it directly from the Terminal mode screen, thus saving yourself a keystroke. But the Command Menu is an excellent learning tool and memory aid and is one of the reasons that people find PROCOMM PLUS so easy to learn. When you are new to PROCOMM PLUS, you can routinely display the Command Menu before executing a command. And any time you forget a command, you still have the option of displaying this screen to remind you.

The Command Menu is divided into three main sections: Communications, Set Up, and Other Functions. These divisions are intended to help you find the commands you need. Commands listed in the Communications section of the Command Menu are used in connection with a communications session. Those listed in the Set Up section, on the other hand, relate to configuring PROCOMM PLUS. The third group of commands is a catchall category for commands that don't fit neatly into either of the other two groups.

The Communications group of commands is further divided into four subgroups: Before, During, After, and Other. These groupings merely suggest the stage of communications in which each command is most likely to be used. But you can use any of the commands at any time during any stage of communications, from the Terminal mode screen or from the Command Menu.

Because all but three of the commands listed in the Command Menu are Alt-key combination keystrokes, this book and the PROCOMM PLUS documentation refer to this group of commands collectively as the *Alt-key commands*. Only PgUp (Send Files), PgDn (Receive Files), and the menu-line key (`) do not use the Alt key.

When this book instructs you to press one of these Alt-key commands, the command name is included in parentheses, as in Alt-C (Clear Screen). This method can help you learn to identify the keystroke with the command.

Using the Menu Line

In addition to the commands listed in the Command Menu, you can invoke most of PROCOMM PLUS's features by selecting options from a Lotus-style top-line menu. This feature is entirely new in PROCOMM PLUS and is not found in ProComm.

As you have already learned, PROCOMM PLUS doesn't automatically display a menu at the top of the Terminal mode screen. To access the top-line menu system, you must be at the Terminal mode screen. Then when you press the menu-line key (`), PROCOMM PLUS displays the *menu line* (consisting actually of *two* lines) at the top of the screen (see fig. 2.5).

Note: When you first install PROCOMM PLUS, the menu-line key is the left apostrophe key (`), on the same key as the tilde (~). You can reassign the menu-line key to a different key by using the Setup Utility. For more information, refer to Chapter 8, "Tailoring PROCOMM PLUS." If you forget which key is the menu-line key, you can refer to the Command Menu, which lists the menu-line key as the last command. You do not have to display the Command Menu, however, before you can display the menu line.

While the menu line is displayed, you execute a PROCOMM PLUS command by choosing the appropriate option from the menu. This menu operates in a manner familiar to most PC users. You can use one of two equally effective methods to select an option: either type the first letter of

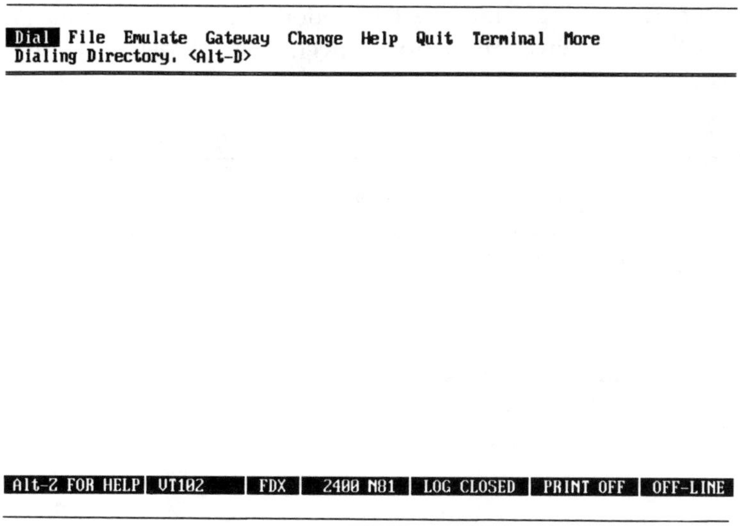

```
Dial File Emulate Gateway Change Help Quit Terminal More
Dialing Directory. <Alt-D>
```

Fig. 2.5.

The menu line.

```
Alt-Z FOR HELP | VT102    | FDX |  2400 N81 | LOG CLOSED | PRINT OFF | OFF-LINE
```

the option or use the right arrow or left arrow to move the highlighted menu-selection bar to the appropriate option and then press Enter. The first-letter technique is usually the quicker of the two methods because in most cases it requires fewer keystrokes.

If you use the second method, sometimes referred to as the "point-and-shoot" method, PROCOMM PLUS offers some additional information to help you decide on the proper menu selection. For example, when you first display the menu, **D**ial, the leftmost option on the top line, is highlighted. The second line of the menu contains a description of this menu option:

```
Dialing Directory. <Alt-D>
```

TIP As a cross-reference, PROCOMM PLUS lists the equivalent Alt-key command (that is, the command listed in the Command Menu for performing the same function) in each description of a menu line option. This message does not mean, however, that you can use this Alt-key command *while* the menu line is displayed. The command for accessing the help system, Alt-Z (Help), is the only Alt-key command that works while the menu line is displayed.

If you then press the right arrow, PROCOMM PLUS moves the highlighted bar to the second menu choice, File, and displays a different message in the second line of the menu:

```
Send, Receive, Directory, Aspect, View, Toggle Log, Hold Log
```

This message means that selecting the File option displays another menu that includes Send, **R**eceive, **D**irectory, **A**spect, **V**iew, **T**oggle Log, and Hold Log as choices (see fig. 2.6).

Fig. 2.6.

The File menu.

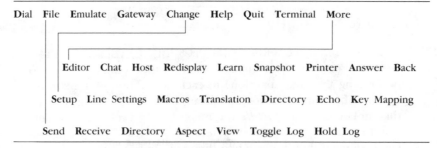

Table 2.1 depicts the PROCOMM PLUS menu structure as a tree, showing all the commands available through this top-line menu system.

Table 2.1
The PROCOMM PLUS Menu Tree

Dial File Emulate Gateway Change Help Quit Terminal More

    Editor Chat Host Redisplay Learn Snapshot Printer Answer Back

    Setup Line Settings Macros Translation Directory Echo Key Mapping

    Send Receive Directory Aspect View Toggle Log Hold Log

When you select a menu option, PROCOMM PLUS executes the command and removes the menu line from the screen—unless, of course, the purpose of the menu option is to display another menu. To remove the menu from the screen without executing any option, press the Esc key. If you are at a second-level menu when you press Esc, PROCOMM PLUS returns to the first menu. Press Esc a second time, and PROCOMM PLUS removes the menu line altogether.

As a general rule, the Alt-key commands are the fastest way to execute PROCOMM PLUS features. Once you commit the commonly used Alt-key commands to memory, you may decide to forgo use of the menu line. To encourage you to learn the Alt-key commands quickly, the instructions in this book do not include the menu line method unless it presents a particular advantage over the equivalent Alt-key method. But if you find using top-line menus more familiar and user friendly than memorizing keyboard commands, by all means don't hesitate to use the PROCOMM PLUS menu line.

Getting Help

One of the most significant enhancements of PROCOMM PLUS, when compared to its sibling ProComm 2.4.3, is the context-sensitive help system. *Context sensitive* means that when you ask PROCOMM PLUS for help, the program provides information that is pertinent to the task you are currently attempting to perform. To access the PROCOMM PLUS help system from any screen, press Alt-Z (Help).

You may recall that this command is the same one that displays the Command Menu from the Terminal mode screen. If you need help about the Terminal mode, you need to press Alt-Z (Help) again when you are at the Command Menu. PROCOMM PLUS then displays the first of two screens of information about the Terminal mode (see fig. 2.7).

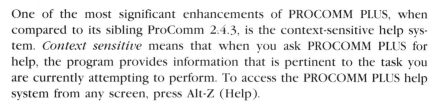

```
 General Help - Terminal Mode (1 of 2)          PROCOMM PLUS On-Line Help

                        - TERMINAL MODE -

 You are now in Terminal Mode.  All of the PROCOMM PLUS functions are just a
 few keystrokes away.  The screen you just saw is a list of those keystrokes.

                        - ONLINE HELP -

 Pressing [Alt-Z] will usually get you detailed help for a particular function.
 When you are not sure what to do, try pressing [Alt-Z].

                        - STATUS LINE -

 In Terminal Mode, the 25th line of the screen is a dynamic display of current
 PROCOMM PLUS settings:

                        Log File Status──┐   Printer
              Line Settings──┐    │    Status──┐
        Duplex──┐       │    │    │       │              Line Status
 Emulation──┐   │       │    │    │       │                   │
        ↓   ↓   ↓       ↓    ↓    ↓       ↓                   ↓
 | MESSAGE AREA | VT102 | FDX | 2400 N81 | LOG CLOSED | PRINT OFF | OFFLINE

                   PgDn: Next Page    Esc: Exit Help
```

Fig. 2.7.

The first of two help screens on the Terminal mode.

When you press Alt-Z (Help) at screens other than the Terminal mode screen, PROCOMM PLUS does not display the Command Menu. Instead, the program displays the appropriate context-sensitive help screen.

Often a topic has multiple help screens, the presence of which is indicated by a message in the top line of the help screen. For example, the screen shown in figure 2.7 displays the message General Help - Terminal Mode (1 of 2), indicating that this help screen is the first of two General Help screens on the Terminal mode. When multiple help screens are available, you can move from page to page by using the PgUp and PgDn keys.

For example, you may be working in the dialing directory and cannot remember how to set up a *dialing queue*, a list of several phone numbers for the modem to dial in sequence. You press Alt-Z (Help), and PRO-COMM PLUS displays the first of five help screens on the dialing directory.

You cannot find the information you need on the first screen, so you press PgDn to display help screen 2 of 5. This screen is still not the one you need, so you press PgDn once more to display the help screen shown in figure 2.8. The second paragraph on this screen describes how to dial a "circular list" of numbers—in other words, a dialing queue.

Fig. 2.8.

The third of five help screens on the dialing directory.

```
Dialing Directory (3 of 5)                    PROCOMM PLUS On-Line Help

To dial a number not in your Dialing Directory press [M], enter the number and
press [Enter].

To dial a series of numbers in a "circular list" use [↑] and [↓] to highlight
an entry to be dialed then press [Space] to "mark" it.  The entry will be
marked by a ► at the left.  You may mark up to 15 entries.  To dial the marked
entries, press [Enter].

You may also dial one or more numbers with or without dialing codes by using
the [D] command.  Press [D] and at the prompt enter one or more entry numbers
separated by spaces.  You may include dialing codes with each entry specified.

Other commands are available in the Dialing Directory as well.

    Press ...       In order to ...

    [R]             Revise the highlighted entry.  More help is available on
                    revising by pressing [Alt-Z] after selecting the [R]
                    command.


        PgUp: Previous Page    PgDn: Next Page    Esc: Exit Help
```

Once you have finished reading the help screens, press Esc to return to the screen in which you were working when you pressed Alt-Z (Help).

Keep in mind that the help system is not intended to be a substitute for the PROCOMM PLUS documentation. Instead, the context-sensitive help screens provide a convenient quick reference useful to jog your memory or clear up uncertainty. The feature is particularly handy when you are away from your home or office, using PROCOMM PLUS on a portable or laptop computer, and don't have your PROCOMM PLUS manual or this book with you.

Exiting PROCOMM PLUS

One of the first things you should learn about any powerful tool is how to turn it off properly. Before you attempt to turn off PROCOMM PLUS, you should first sign off or otherwise disconnect from any active communications session (refer to "Ending a Call," in Chapter 4). Once the status line indicates that PROCOMM PLUS is OFF-LINE, you are ready to quit the program. From the Terminal mode screen, press Alt-X (Exit). Near the top of the screen, PROCOMM PLUS opens a small window that contains the question EXIT TO DOS? (Y/N) (see fig. 2.9). A default answer of Yes is already entered as a response to this prompt. You can affirm this response by pressing Enter or the letter Y. In either case, PROCOMM PLUS exits and returns you to the operating system. Press N and then press Enter, instead, when you want to continue with your PROCOMM PLUS session.

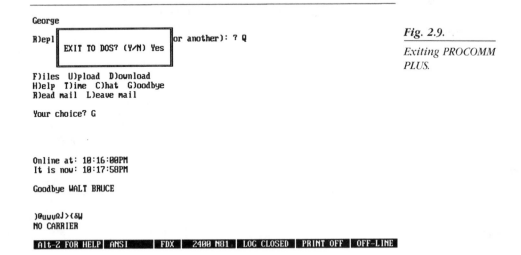

Fig. 2.9.

Exiting PROCOMM PLUS.

TIP | If you forget to disconnect from an active communications session before attempting to exit PROCOMM PLUS, the program reminds you with the prompt HANG-UP LINE? (Y/N). Normally, you respond in the affirmative to this query. If you press N, however, PROCOMM PLUS quits, returns to DOS, but does not drop the line. This feature enables you to exit completely from PROCOMM PLUS without disconnecting your communications session. You may need to take advantage of this feature if you need to run another DOS program quickly (the program will not run in a DOS shell, described in the "Accessing the Operating System" section of Chapter 5), but you don't want to lose the line. When you restart PROCOMM PLUS later, your connection is still intact.

CAUTION: Don't forget that you are still on-line, especially if you or your company is paying for the connect time by the minute.

Chapter Summary

This chapter has continued your introduction to PROCOMM PLUS, giving you a look at the Terminal mode screen, the keyboard, the Command Menu, and the menu line, as well as the PROCOMM PLUS context-sensitive help system. You have also learned how to exit PROCOMM PLUS. Now that you have completed Part I of the book, you should move on to Part II, "Getting Acquainted with PROCOMM PLUS."

Part II

Getting Acquainted with PROCOMM PLUS

Includes

Building Your Dialing Directory

A Session with PROCOMM PLUS

Transferring Files

Using the PROCOMM PLUS Editor

Automating PROCOMM PLUS
with Macros and Script Files

3

Building Your
Dialing Directory

As explained in Chapter 1, most PC communication is accomplished over telephone lines. Consequently, the first step toward communicating with another computer—whether next door, across town, or across the country—is usually to dial the telephone number. The most obvious purpose of the PROCOMM PLUS Dialing Directory screen is to help you take this first step.

Phone dialing, however, is not the only purpose of the Dialing Directory screen. This screen helps you take care of several other chores necessary for successful PC communications. Through the Dialing Directory screen, you establish your computer's communications settings, such as transmission speed, parity, data bits, and so on.

This chapter describes how to use the Dialing Directory screen to control the process of dialing and connecting to remote computers. You learn how to build dialing directories and how to add, edit, and delete directory entries and dialing codes. The chapter then shows you how to dial another computer by using a single dialing directory entry, a dialing queue, and the Manual Dial feature. The last portion of this chapter describes features that make managing one or more dialing directory files easier. You learn how to print a dialing directory, how to switch between multiple dialing directories, and how to sort dialing directory entries.

Refer also to Appendix A for instructions on how to convert a ProComm 2.4.3 dialing directory into a directory that you can use with PROCOMM PLUS.

Examining the Dialing Directory Screen

To display the Dialing Directory screen, press Alt-D (Dialing Directory) from the Terminal mode screen. As shown in figure 3.1, when you first install PROCOMM PLUS, the dialing directory contains no entries.

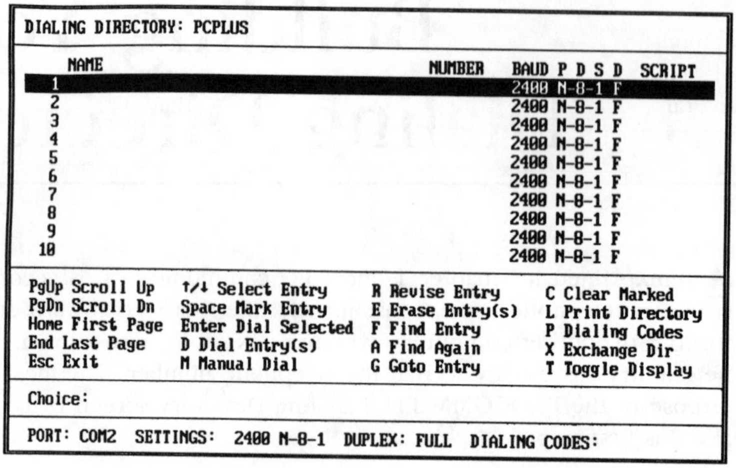

Fig. 3.1.

The empty Dialing Directory screen.

```
┌─────────────────────────────────────────────────────────────────┐
│ DIALING DIRECTORY: PCPLUS                                         │
│       NAME                              NUMBER   BAUD P D S D  SCRIPT │
│    1                                             2400 N-8-1 F    │
│    2                                             2400 N-8-1 F    │
│    3                                             2400 N-8-1 F    │
│    4                                             2400 N-8-1 F    │
│    5                                             2400 N-8-1 F    │
│    6                                             2400 N-8-1 F    │
│    7                                             2400 N-8-1 F    │
│    8                                             2400 N-8-1 F    │
│    9                                             2400 N-8-1 F    │
│   10                                             2400 N-8-1 F    │
│  PgUp Scroll Up    ↑/↓ Select Entry    R Revise Entry    C Clear Marked │
│  PgDn Scroll Dn    Space Mark Entry    E Erase Entry(s)  L Print Directory │
│  Home First Page   Enter Dial Selected F Find Entry      P Dialing Codes │
│  End Last Page     D Dial Entry(s)     A Find Again      X Exchange Dir │
│  Esc Exit          M Manual Dial       G Goto Entry      T Toggle Display │
│  Choice:                                                          │
│  PORT: COM2  SETTINGS:  2400 N-8-1  DUPLEX: FULL  DIALING CODES:   │
└─────────────────────────────────────────────────────────────────┘
```

Note that the Dialing Directory screen is divided horizontally into five sections. For convenience and clarity, this chapter names these sections, from top to bottom, the *name line*, the *entry section*, the *command section*, the *choice line*, and the *status line*.

The Name Line

The first section, the *name line*, displays the name of the current dialing directory. As shown in figure 3.1, the default directory is named PCPLUS. This directory is automatically created when you install PROCOMM PLUS and is contained in a disk file named PCPLUS.DIR.

PROCOMM PLUS enables you to create and use multiple directories in the Dialing Directory screen. Refer to "Building and Using Other Dialing Directories," in this chapter, for information on how to use multiple dialing directories.

The Entry Section

The main portion of the Dialing Directory screen is the *entry section*. This part of the screen contains the list of phone numbers you want PROCOMM PLUS to dial. PROCOMM PLUS displays only 10 entries at a time, but you can scroll this portion of the screen to display any of the other entries in the directory. Each directory can contain up to 200 different entries.

When you first display the Dialing Directory screen, PROCOMM PLUS highlights (displays in inverse video) the first entry, as shown in figure 3.1. You move the highlight to other entries in order to *select* an entry. Several dialing directory commands require that you first highlight (select) the entry to which the command should apply. This chapter's section "Selecting an Entry" describes several methods of moving the highlight to a particular directory entry.

In addition to the telephone number, each entry in the directory stores the transmission speed; line settings (parity, data bits, and stop bits); error-checking protocol; and terminal-emulation setting for one of the up to 200 different computers you plan to call. These settings have to match those of the computer at the other end of the telephone line in order for communication to be successful. You can also assign to each dialing directory entry a script that can be used to automate some or all of the session with the other computer.

Initially, the entry section contains the column headings NAME, NUMBER, BAUD, P, D, S, D, and SCRIPT. These headings label the communications settings (name, telephone number, transmission speed, parity setting, number of data bits, number of stop bits, duplex setting, and script file name if a script has been specified) stored in the directory entries. Several other settings stored with each directory entry, however, such as the error-checking protocol and the terminal-emulation setting, don't show up as headings in figure 3.1. Refer to "Toggling the Display," in this chapter, to find out how to display these other settings on the Dialing Directory screen.

The Command Section

Below the entry section of the screen is the *command section*. It lists the keyboard commands available while the Dialing Directory screen is displayed. These commands can be divided into four groups according to their function: *editing* commands, *entry-selection* commands, *dialing* commands, and *global* commands (see table 3.1).

Table 3.1
PROCOMM PLUS Dialing Directory Commands

Key	Name	Function
Editing Commands		
R	Revise Entry	Revise highlighted entry
E	Erase Entry(s)	Erase highlighted entry
P	Dialing Codes	Display Dialing Codes window
Entry-Selection Commands		
PgUp	Scroll Up	Scroll entry section up 10 rows
PgDn	Scroll Dn	Scroll entry section down 10 rows
Home	First Page	Move highlight to row 1
End	Last Page	Move highlight to row 191
↑	Select Entry	Move highlight up one row
↓	Select Entry	Move highlight down one row
F or /	Find Entry	Search for entry by name or phone number
A	Find Again	Search for next matching entry
G	Goto Entry	Select an entry by entry number
Dialing Commands		
Enter	Dial Selected	Dial highlighted entry or dialing queue
D	Dial Entry(s)	Dial one or more entries by entry number
M	Manual Dial	Dial number you type manually
Space bar	Mark Entry	Add entry to dialing queue
C	Clear Marked	Clear all dialing queue marks
Global Commands		
L	Print Directory	Print current directory
X	Exchange Dir	Load a different directory
T	Toggle Display	Display other communications settings
Esc	Exit	Return to Terminal mode screen

All these commands are explained in this chapter. The editing commands are discussed in "Adding an Entry" and "Modifying an Entry." Entry-selection commands are covered in "Selecting an Entry." The "Dialing a Number" portion of the chapter describes how to use the dialing commands, and the global commands are covered in "Managing Your Dialing Directory."

The Choice Line

Beneath the command section is the *choice line*. It contains only the word Choice followed by a blinking cursor. This line is a prompt asking you to enter a command. PROCOMM PLUS never displays anything else in this section.

The Status Line

The last section of the Dialing Directory screen is the *status line*. It shows you the current communications settings: serial port, transmission speed, parity, data bits, stop bits, and duplex. At the right end of this line, PROCOMM PLUS also lists the dialing codes defined in the current directory. (See "Working with Dialing Codes" for more information on dialing codes.)

Working with Dialing Directory Entries

Although the primary purpose of a dialing directory is to dial the telephone, this screen is more than just a speed dialer. The dialing directory is more like having your own private secretary. When you use a directory entry to dial another computer, PROCOMM PLUS automatically changes your computer's communications settings to those specified in the entry (if necessary), instructs the modem to dial the phone, and runs any script file specified in the entry.

The next several sections discuss how to add entries to the Dialing Directory screen, including how to specify communications settings and a script file; how to use the entry-selection commands to move to particular entries; and how to modify and delete directory entries.

Adding an Entry

To add an entry to the directory currently displayed on the Dialing Directory screen, use the up-arrow or down-arrow key to highlight a blank entry line, and press **R** (Revise Entry). PROCOMM PLUS displays a pop-up Revise Entry window similar to the one shown in figure 3.2.

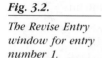

Fig. 3.2.

The Revise Entry window for entry number 1.

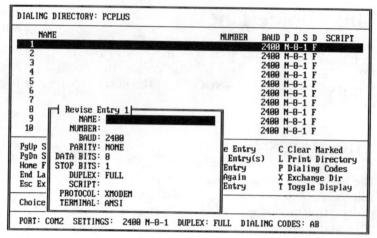

When PROCOMM PLUS first displays the Revise Entry window, the first line, NAME, is highlighted. Use this line to enter the name of the computer system you want to call. Type up to 24 characters (letters, numbers, symbols, and spaces), and press Enter. The cursor-movement and editing keys listed in table 3.2 are available for you to use while entering or editing a name in this line. In fact, you can use these keys in any line in the Revise Entry window.

Table 3.2
Dialing Directory Cursor-Movement and Editing Keys

Key	Function
←	Move cursor left one character
→	Move cursor right one character
Home	Move cursor to left end of line
End	Move cursor to right end of line
Ins	Toggle Insert/Overtype mode
Del	Delete character at cursor
Backspace	Delete character left of cursor
Tab	Delete entire line
Ctrl-End	Delete characters from cursor to right end

When you press Enter at the NAME line, PROCOMM PLUS moves the highlight to the second line, labeled NUMBER. Type the telephone number of the system you want to call, and press Enter. Be sure to include all the digits you would dial if you picked up the phone to call the other system manually. You can include parentheses and hyphens, but they are not required.

Letters and symbols are also acceptable, but keep in mind that some letters and symbols have special meanings when used in the NUMBER line. PROCOMM PLUS uses each of the letters *A* through *J* as a *dialing code*, a shorthand way to represent a string of numbers or other characters. (See "Working with Dialing Codes," in this chapter.) In addition, Hayes-compatible modems interpret certain letters and symbols as commands when they are included in the phone number. For example, the letter *W* causes the modem to wait for a second dial tone before continuing to dial the number, and a comma in a phone number tells the modem to pause the dialing for two seconds. Most letters and symbols, however, have no meaning to your modem.

TIP | The Revise Entry window is a one-way street. Pressing Enter moves you from line to line, but only downward. For example, you can move from the NAME line to the NUMBER line, but not from the NUMBER line to the NAME line. If you notice that you made a mistake in the Revise Entry window but have already moved to another line, continue entering the other settings. Then, when you get to the question ACCEPT THIS ENTRY? (Y/N), press N. PROCOMM PLUS moves the highlight back to the NAME line but does not erase any entries you have made. Refer to this chapter's section "Modifying an Entry" for help in correcting the mistake.

After you type the telephone number and press Enter, PROCOMM PLUS moves the highlight to the line labeled BAUD. This line already contains the default transmission speed selected when you installed PROCOMM PLUS. When this line is highlighted, PROCOMM PLUS pops up another window that lists the available transmission speeds. If you don't need to change the setting, just press Enter. Otherwise, use the up arrow or down arrow to highlight the appropriate speed, and press Enter. PROCOMM PLUS removes the list of transmission speeds from the screen and moves the highlight to the next line in the Revise Entry window, PARITY.

You complete the remaining lines in the Revise Entry window in a similar manner. For each line, the default setting is already filled in, and a pop-up window appears from which you can choose another setting. You press Enter to accept the default, or highlight another choice in the pop-up window and then press Enter.

The pop-up window for the PARITY line contains the options NONE, ODD, EVEN, MARK, and SPACE. The meanings of these options are given in Chapter 1, "A Communications Primer." To be certain which setting you need, ask the operator of the other computer how that system's parity is set. When in doubt, use NONE if connecting to another PC (including PC bulletin boards), and use EVEN if connecting to any other type of computer.

The only two choices in the DATA BITS line are the numbers 7 and 8. As a rule of thumb, when you are connecting to another PC (including PC bulletin boards), use 8. Use 7 when connecting to an on-line service or another mainframe system.

The next line is the STOP BITS line. The only two choices available are the numbers 1 and 2. The most commonly used setting is 1. In fact, systems that require 2 stop bits are rare.

The pop-up window for the DUPLEX line contains the choices FULL and HALF. The duplex setting refers to a communications feature known as *echoplex*. Normally, you should choose FULL. If this choice turns out to be the wrong setting, you can easily change it after you are connected to the other computer. The significance of each of the settings and the procedure for making an on-line adjustment is explained in Chapter 4, "A Session with PROCOMM PLUS."

PROCOMM PLUS next moves the highlight to the SCRIPT line. In its most basic form, a *script* is a series of keystrokes recorded in a file to be "played back" later. More complex scripts are programs that can control the entire PC communications process. For example, you can create a script to sign on to an electronic mail system, check your mail, download and save to disk any mail that is in your electronic mailbox, and then sign off. Chapter 7 introduces you to the automation of PROCOMM PLUS with scripts, and Chapter 11 gives you an overview of the powerful script programming language, ASPECT.

When you have created a script and want PROCOMM PLUS to run it as soon as the program connects to a particular computer, type the name of the script in the SCRIPT line for that computer's dialing directory entry, and press Enter. If you haven't created such a script, just press Enter at the blank line.

After you press Enter at the SCRIPT line, PROCOMM PLUS moves the highlight to the line labeled PROTOCOL. This line already contains the default file-transfer protocol you indicated when you installed the program. This setting is of interest only if you intend to send or receive files to or from the other computer. When the highlight is in the PROTOCOL line, PROCOMM PLUS displays a list of 17 options, including 13 built-in file-transfer protocols, 3 external protocols, and NONE (no protocol). The most commonly used protocol for transferring files between PCs is XMODEM. Refer to Chapter 5, "Transferring Files," for more information.

The last line in the Revise Entry window is labeled TERMINAL. The pop-up window for this line includes a list of 16 terminal-emulation options. Chapter 10 explains the differences among these options. If you are connecting to a mainframe computer through either an on-line service or some other gateway or protocol converter, you need to know what type of terminal your PC should emulate. The VT102 emulation is probably the most popular. If you are connecting to another PC, particularly when you want to access a PC bulletin board, try the ANSI emulation first.

Once you select the terminal emulation, PROCOMM PLUS pops up a second, much smaller, Revise Entry window containing the prompt CLEAR LAST DATE AND TOTAL? (Y/N). This question has no significance when you are adding a new directory entry, so just press Enter to accept the default value of No.

PROCOMM PLUS next displays the question ACCEPT THIS ENTRY? (Y/N). Press Y to accept all the settings you have entered. This action adds the entry to the directory currently loaded in memory but does not add the entry to the directory file on the disk. If you answer N, PROCOMM PLUS moves the highlight back to the first line in the Revise Entry window, without erasing any entries you have made.

Once you accept the entry, PROCOMM PLUS displays one final question: SAVE ENTRY TO DISK? (Y/N) (see fig. 3.3). Press Y, and PROCOMM PLUS saves the entry to the current directory file on disk (the default is PCPLUS.DIR) and returns to the Dialing Directory screen.

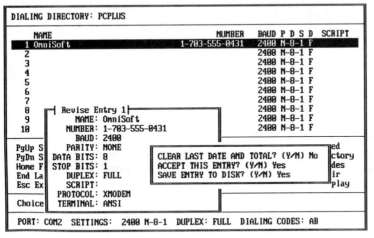

Fig. 3.3.

The completed Revise Entry window.

You can, of course, respond No to the last question in the Revise Entry window. PROCOMM PLUS then returns to the Dialing Directory screen without saving the new entry to disk. The entry is listed on-screen but is available for the current session only. Once you quit PROCOMM PLUS, this temporary entry is erased from memory. Using this feature, you can create a temporary directory entry for a computer system with which you expect to connect only once.

Selecting an Entry

When PROCOMM PLUS displays the Dialing Directory screen for the first time in a particular session, directory entry number 1 is highlighted, as shown in figure 3.4[1]. This highlighted entry is the currently *selected* entry. While an entry is selected, you can use it in several ways: to dial the phone number, to modify the entry, to erase the entry, and so on.

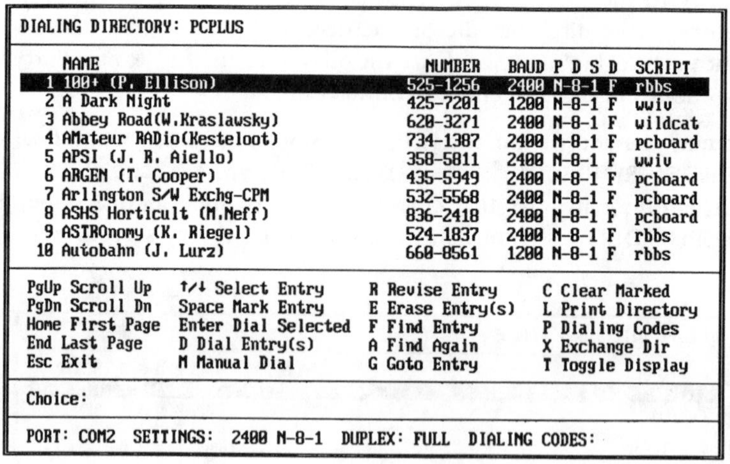

Fig. 3.4.

The Dialing Directory screen with entry number 1 selected.

```
DIALING DIRECTORY: PCPLUS

        NAME                            NUMBER    BAUD P D S D  SCRIPT
      1 100+ (P. Ellison)              525-1256   2400 N-8-1 F  rbbs
      2 A Dark Night                   425-7201   1200 N-8-1 F  wwiv
      3 Abbey Road(W.Kraslawsky)       620-3271   2400 N-8-1 F  wildcat
      4 AMateur RADio(Kesteloot)       734-1387   2400 N-8-1 F  pcboard
      5 APSI (J. R. Aiello)            350-5811   2400 N-8-1 F  wwiv
      6 ARGEN (T. Cooper)              435-5949   2400 N-8-1 F  pcboard
      7 Arlington S/W Exchg-CPM        532-5568   2400 N-8-1 F  pcboard
      8 ASHS Horticult (M.Neff)        836-2418   2400 N-8-1 F  pcboard
      9 ASTROnomy (K. Riegel)          524-1837   2400 N-8-1 F  rbbs
     10 Autobahn (J. Lurz)             660-8561   1200 N-8-1 F  rbbs

   PgUp Scroll Up    ↑/↓ Select Entry     R Revise Entry    C Clear Marked
   PgDn Scroll Dn    Space Mark Entry     E Erase Entry(s)  L Print Directory
   Home First Page   Enter Dial Selected  F Find Entry      P Dialing Codes
   End  Last Page    D Dial Entry(s)      A Find Again      X Exchange Dir
   Esc  Exit         M Manual Dial        G Goto Entry      T Toggle Display

   Choice:

   PORT: COM2  SETTINGS:  2400 N-8-1  DUPLEX: FULL  DIALING CODES:
```

The simplest way to select a directory entry is with the up-arrow and down-arrow keys. The up arrow moves the highlight up one entry, and the down arrow moves the cursor down one entry. Because a PROCOMM PLUS dialing directory can contain as many as 200 entries, however, finding a particular entry not currently displayed would be tedious if you

---

[1] The bulletin board list that makes up the directory displayed in the figures of this chapter was compiled by Mike Focke of Oakton, Virginia. This list is available on many bulletin boards in the Washington, D.C., area. This particular list was current in July 1989 and was converted to a PROCOMM PLUS directory with CvtFon, a program written by Steve Linhart of New Brunswick, New Jersey. CvtFon is capable of converting dialing directories from many different formats and is also available on many PC bulletin boards. The phone numbers may no longer be accurate when you are reading this book.

could move up or down only one entry at a time. PROCOMM PLUS therefore provides several other ways to select a directory entry.

Using the PgUp and PgDn keys, you can scroll quickly through the directory, 10 entries at a time. The first time you press PgDn (Scroll Dn), PROCOMM PLUS replaces entries 1 through 10 with entries 11 through 20, highlighting entry 11. Each subsequent time you press this key, PROCOMM PLUS displays the next 10 entries, with the first entry selected. The last 10 entries in the directory are numbered 191 through 200. If you press PgDn at this last set of entries, PROCOMM PLUS returns to entries 1 through 10, with entry 1 highlighted.

The effect of the PgUp key is the reverse of PgDn. Each time you press PgUp (Scroll Up), PROCOMM PLUS scrolls the screen up 10 entries. If you start at the first screen (entries 1 through 10) and press PgUp, PROCOMM PLUS goes to the end of the directory and displays entries 191 through 200, with entry 191 selected.

When you want to move to the end of the directory, press End (Last Page). Regardless of which entry is current when you press this key, PROCOMM PLUS displays entries 191 through 200, with entry 191 selected. To move directly to the top of the directory, press Home (First Page). PROCOMM PLUS displays entries 1 through 10 and selects entry 1.

PROCOMM PLUS provides two methods that enable you to go directly to an entry without scrolling. When you know the entry number of the entry you want to select, press **G** (Goto Entry) at the Dialing Directory screen. PROCOMM PLUS displays a narrow horizontal window containing the prompt ENTRY TO GOTO (see fig. 3.5). At this prompt, type the number of the entry you want to select, and press Enter. PROCOMM PLUS moves the highlight directly to the specified entry.

```
DIALING DIRECTORY: PCPLUS

        NAME                      NUMBER    BAUD P D S D  SCRIPT
   1 100+ (P. Ellison)            525-1256  2400 N-8-1 F  rbbs
   2 A Dark Night                 425-7281  1200 N-8-1 F  wwiv
   3 Abbey Road(W.Kraslawsky)     620-3271  2400 N-8-1 F  wildcat
   4 AMateur RADio(Kesteloot)     734-1387  2400 N-8-1 F  pcboard
   5 APSI (J. R. Aiello)          358-5811  2400 N-8-1 F  wwiv
   6 ARGEN (T. Cooper)            435-5949  2400 N-8-1 F  pcboard
   7 Arlington S/W Exchg-CPM      532-5568  2400 N-8-1 F  pcboard
   8 ASHS Horticult (M.Neff)      836-2418  2400 N-8-1 F  pcboard
   9 ASTROnomy (K. Riegel)        524-1837  2400 N-8-1 F  rbbs

 ENTRY TO GOTO:

 PgDn Scroll Dn     Space Mark Entry      E Erase Entry(s)    L Print Directory
 Home First Page    Enter Dial Selected   F Find Entry        P Dialing Codes
 End Last Page      D Dial Entry(s)       A Find Again        X Exchange Dir
 Esc Exit           M Manual Dial         G Goto Entry        T Toggle Display

 Choice:

 PORT: COM2  SETTINGS:  2400 N-8-1  DUPLEX: FULL  DIALING CODES: A
```

Fig. 3.5.

Selecting an entry by entry number.

More often than not, however, you don't know the number of the entry. PROCOMM PLUS provides a quick and easy way to search for the entry by name or telephone number. At the Dialing Directory screen, press **F** (Find Entry). At the SEARCH FOR prompt, type the *search criterion*—any portion of either the entry's name or the entry's telephone number. You can use either upper- or lowercase letters. Press Enter, and PROCOMM PLUS searches the directory for the first *match*, an entry in which the name or phone number contains the characters specified in the search criterion.

When searching, the program starts at the currently selected entry and searches entries numerically toward the end of the directory. If PROCOMM PLUS does not find a match by the time it reaches the end of the directory, the program starts at entry number 1 and searches down toward the current entry. If no match is found, PROCOMM PLUS displays the message TARGET NOT FOUND and then removes the search window from the screen.

For example, suppose that you want to call the bulletin board for the Capital PC User Group, which you remember is listed as CPCUG in your dialing directory. At the Dialing Directory screen, press **F** (Find Entry). When PROCOMM PLUS displays the SEARCH FOR prompt, type *cpcug*, and press Enter (see fig. 3.6). PROCOMM PLUS finds the first entry that contains the letters you typed, as shown in figure 3.7. Notice that PROCOMM PLUS finds the first entry that contains CPCUG (uppercase) even though you typed *cpcug* (lowercase).

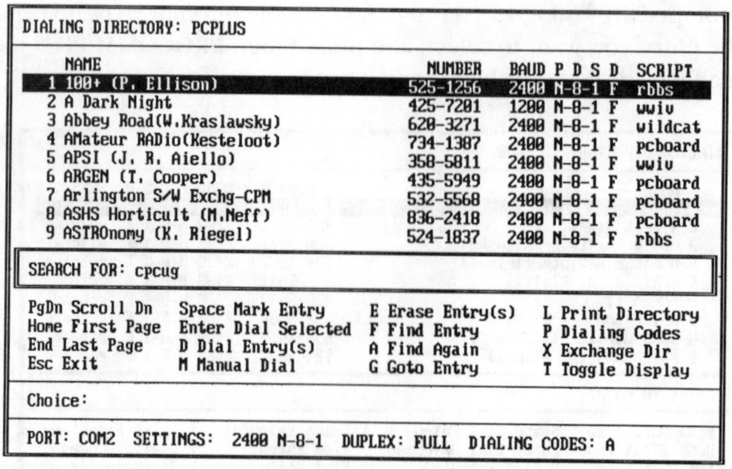

Fig. 3.6.

Entering a search criterion.

```
DIALING DIRECTORY: PCPLUS

         NAME                        NUMBER    BAUD P D S D  SCRIPT
   1 100+ (P. Ellison)               525-1256  2400 N-8-1 F  rbbs
   2 A Dark Night                    425-7281  1200 N-8-1 F  wwiv
   3 Abbey Road(W.Kraslawsky)        628-3271  2400 N-8-1 F  wildcat
   4 AMateur RADio(Kesteloot)        734-1387  2400 N-8-1 F  pcboard
   5 APSI (J. R. Aiello)             358-5811  2400 N-8-1 F  wwiv
   6 ARGEN (T. Cooper)               435-5949  2400 N-8-1 F  pcboard
   7 Arlington S/W Exchg-CPM         532-5568  2400 N-8-1 F  pcboard
   8 ASHS Horticult (M.Neff)         836-2418  2400 N-8-1 F  pcboard
   9 ASTROnomy (K. Riegel)           524-1837  2400 N-8-1 F  rbbs

 SEARCH FOR: cpcug

 PgDn Scroll Dn    Space Mark Entry     E Erase Entry(s)   L Print Directory
 Home First Page   Enter Dial Selected  F Find Entry       P Dialing Codes
 End Last Page     D Dial Entry(s)      A Find Again        X Exchange Dir
 Esc Exit          M Manual Dial        G Goto Entry        T Toggle Display

 Choice:

 PORT: COM2  SETTINGS:  2400 N-8-1  DUPLEX: FULL  DIALING CODES: A
```

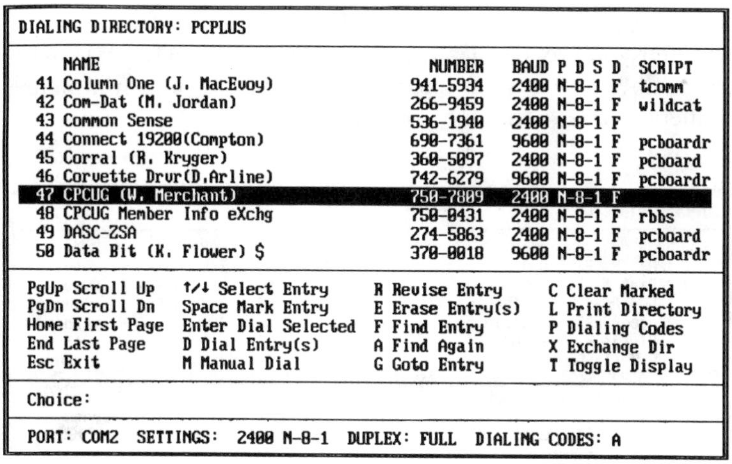

```
┌─────────────────────────────────────────────────────────────────┐
│ DIALING DIRECTORY: PCPLUS                                          │
│                                                                    │
│     NAME                              NUMBER   BAUD P D S D  SCRIPT │
│  41 Column One (J. MacEvoy)           941-5934 2400 N-8-1 F  tcomm  │
│  42 Com-Dat (M. Jordan)               266-9459 2400 N-8-1 F  wildcat│
│  43 Common Sense                      536-1940 2400 N-8-1 F         │
│  44 Connect 19200(Compton)            690-7361 9600 N-8-1 F  pcboardr│
│  45 Corral (R. Kryger)                360-5897 2400 N-8-1 F  pcboard │
│  46 Corvette Drvr(D.Arline)           742-6279 9600 N-8-1 F  pcboardr│
│ ▓47 CPCUG (W. Merchant)               750-7009 2400 N-8-1 F        ▓│
│  48 CPCUG Member Info eXchg           750-0431 2400 N-8-1 F  rbbs   │
│  49 DASC-ZSA                          274-5863 2400 N-8-1 F  pcboard │
│  50 Data Bit (K. Flower) $            370-0018 9600 N-8-1 F  pcboardr│
│                                                                    │
│  PgUp Scroll Up    ↑/↓ Select Entry   R Revise Entry   C Clear Marked│
│  PgDn Scroll Dn    Space Mark Entry    E Erase Entry(s) L Print Directory│
│  Home First Page   Enter Dial Selected F Find Entry    P Dialing Codes│
│  End Last Page     D Dial Entry(s)     A Find Again    X Exchange Dir│
│  Esc Exit          M Manual Dial       G Goto Entry    T Toggle Display│
│                                                                    │
│  Choice:                                                           │
│                                                                    │
│  PORT: COM2  SETTINGS:  2400 N-8-1  DUPLEX: FULL  DIALING CODES: A  │
└─────────────────────────────────────────────────────────────────┘
```

Fig. 3.7.

The first directory entry found by the search criterion.

> **TIP** Although not listed in the command section of the Dialing Directory screen, the slash key (/) has the same effect as **F** (Find Entry).

Of course, several entries in your directory may contain a given string of characters. Figure 3.7, for instance, shows two entries that contain the string *CPCUG*. The first found entry may not be the one you want. PROCOMM PLUS, however, provides another command that enables you to continue the search, using the same criterion, until you locate the appropriate entry. This command is **A** (Find Again).

Before you can use the **A** (Find Again) command, you must have used the **F** (Find Entry) command to establish the search criterion. When the **F** (Find Entry) command does not select the entry you want, press **A** (Find Again). PROCOMM PLUS then continues the search, using the same criterion and searching from the current entry to the bottom of the directory and then from the top of the directory to the current entry. PROCOMM PLUS moves the highlight to the next matching entry, if one is found. If PROCOMM PLUS finds no more matches, it beeps but displays no message and doesn't move the highlight.

For example, suppose that you want to dial a bulletin board but are not sure of its name. All you can remember is that the name includes the word *Exchange*. To find this entry, first press **F** (Find Entry), type *exch* in the search window, and press Enter. The first entry found is entry 7, Arlington S/W Exchg-CPM, as shown in figure 3.8. Because you are sure that this entry is not the one you want, press **A** (Find Again) to continue the search. This time PROCOMM PLUS selects entry 48, CPCUG Member

Info eXchg (see fig. 3.9), which is still not the bulletin board you have in mind. Finally, after you press **A** (Find Again) once more, PROCOMM PLUS selects entry number 156, System Exchange (Andrus), as shown in figure 3.10. This bulletin board is the one you want to call.

Fig. 3.8.

*Using **F** (Find Entry) to select the first entry that matches the search criterion exch.*

```
DIALING DIRECTORY: PCPLUS

         NAME                         NUMBER    BAUD P D S D   SCRIPT
      1 100+ (P. Ellison)             525-1256  2400 N-8-1 F   rbbs
      2 A Dark Night                  425-7201  1200 N-8-1 F   wwiv
      3 Abbey Road(W.Kraslawsky)      620-3271  2400 N-8-1 F   wildcat
      4 AMateur RADio(Kesteloot)      734-1387  2400 N-8-1 F   pcboard
      5 APSI (J. R. Aiello)           358-5811  2400 N-8-1 F   wwiv
      6 ARGEN (T. Cooper)             435-5949  2400 N-8-1 F   pcboard
      7 Arlington S/W Exchg-CPM       532-5568  2400 N-8-1 F   pcboard
      8 ASHS Horticult (M.Neff)       836-2418  2400 N-8-1 F   pcboard
      9 ASTROnomy (K. Riegel)         524-1837  2400 N-8-1 F   rbbs
     10 Autobahn (J. Lurz)            660-8561  1200 N-8-1 F   rbbs

    PgUp Scroll Up    ↑/↓ Select Entry    R Revise Entry      C Clear Marked
    PgDn Scroll Dn    Space Mark Entry    E Erase Entry(s)    L Print Directory
    Home First Page   Enter Dial Selected F Find Entry        P Dialing Codes
    End Last Page     D Dial Entry(s)     A Find Again        X Exchange Dir
    Esc Exit          M Manual Dial       G Goto Entry        T Toggle Display

    Choice:

    PORT: COM2  SETTINGS:  2400 N-8-1  DUPLEX: FULL  DIALING CODES: A
```

Fig. 3.9.

*Using **A** (Find Again) to select the next entry that matches the search criterion exch.*

```
DIALING DIRECTORY: PCPLUS

         NAME                         NUMBER    BAUD P D S D   SCRIPT
     41 Column One (J. MacEvoy)       941-5934  2400 N-8-1 F   tcomm
     42 Com-Dat (M. Jordan)           266-9459  2400 N-8-1 F   wildcat
     43 Common Sense                  536-1940  2400 N-8-1 F
     44 Connect 19200(Compton)        698-7361  9600 N-8-1 F   pcboardr
     45 Corral (R. Kryger)            368-5897  2400 N-8-1 F   pcboard
     46 Corvette Drvr(D.Arline)       742-6279  9600 N-8-1 F   pcboardr
     47 CPCUG (W. Merchant)           758-7009  2400 N-8-1 F
     48 CPCUG Member Info eXchg       758-0431  2400 N-8-1 F   rbbs
     49 DASC-ZSA                      274-5863  2400 N-8-1 F   pcboard
     50 Data Bit (K. Flower) $        370-0018  9600 N-8-1 F   pcboardr

    PgUp Scroll Up    ↑/↓ Select Entry    R Revise Entry      C Clear Marked
    PgDn Scroll Dn    Space Mark Entry    E Erase Entry(s)    L Print Directory
    Home First Page   Enter Dial Selected F Find Entry        P Dialing Codes
    End Last Page     D Dial Entry(s)     A Find Again        X Exchange Dir
    Esc Exit          M Manual Dial       G Goto Entry        T Toggle Display

    Choice:

    PORT: COM2  SETTINGS:  2400 N-8-1  DUPLEX: FULL  DIALING CODES: A
```

```
┌─────────────────────────────────────────────────────────────┐
│ DIALING DIRECTORY: PCPLUS                                     │
│                                                               │
│      NAME                      NUMBER   BAUD P D S D  SCRIPT   │
│  151 Split Infinity            841-1859 2400 N-8-1 F          │
│  152 Springfld Bypass $(Hill)  941-5815 9600 N-8-1 F pcboardr │
│  153 Star Pirates II           644-2347 2400 N-8-1 F          │
│  154 StarShip                  522-9568 2400 N-8-1 F wwiv     │
│  155 Swap Shop (R. Siddiqui)   385-3114 2400 N-8-1 F rbbs     │
│ ▐156 System Exchange (Andrus)  323-7654 2400 N-8-1 F tpboard▌ │
│  157 Tackless $                764-9735 2400 N-8-1 F tcomm    │
│  158 Tax Assist (R.Stanley)    237-8430 2400 N-8-1 F pcboard  │
│  159 Tech Connect (T.Reardon)  556-0266 2400 N-8-1 F pcboard  │
│  160 TechMail (B.Hardin)       430-2535 2400 N-8-1 F rbbs     │
│                                                               │
│ PgUp Scroll Up    ↑/↓ Select Entry   R Revise Entry   C Clear Marked    │
│ PgDn Scroll Dn    Space Mark Entry    E Erase Entry(s) L Print Directory │
│ Home First Page   Enter Dial Selected F Find Entry     P Dialing Codes   │
│ End Last Page     D Dial Entry(s)     A Find Again     X Exchange Dir    │
│ Esc Exit          M Manual Dial       G Goto Entry     T Toggle Display  │
│                                                               │
│ Choice:                                                       │
│                                                               │
│ PORT: COM2  SETTINGS: 2400 N-8-1  DUPLEX: FULL  DIALING CODES: A │
└─────────────────────────────────────────────────────────────┘
```

Fig. 3.10.

Using A (Find Again) to find another entry that matches the search criterion exch.

Modifying an Entry

Modifying an entry in the Dialing Directory screen is nearly identical to adding a new entry. First, you must select the entry you want to modify. Then press **R** (Revise Entry). PROCOMM PLUS displays the Revise Entry window with the values from the selected entry already filled in.

To change the entry's name, use the cursor-movement and editing keys listed in table 3.2, or just type a new name over the old one. To leave a setting as it is and move to the next setting, press Enter. To select a new setting, highlight your choice in the accompanying pop-up window, and press Enter.

For example, you may decide to change the error-checking protocol of entry 156 in your directory from XMODEM to YMODEM (the latter protocol results in faster data transmissions). Select entry 156, and press **R** (Revise Entry). PROCOMM PLUS displays the Revise Entry window with the values from entry 156 filled in. Press Enter at each of the first eight settings until the highlight is at the PROTOCOL line. PROCOMM PLUS opens a window that lists the available protocols, as shown in figure 3.11. Press the up-arrow key twice until the highlight is on YMODEM, and press Enter. PROCOMM PLUS changes the value in the PROTOCOL line to YMODEM and moves the highlight to the TERMINAL line. Press Enter four times to accept the TERMINAL setting and the default responses to PROCOMM PLUS's three questions, including saving the new version of the entry to disk. PROCOMM PLUS then returns to the Dialing Directory screen, with entry 156 still selected.

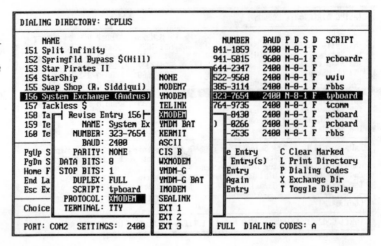

Fig. 3.11.

Modifying the PROTOCOL line in a directory entry.

The only item that requires additional thought when you are modifying an entry is the first question listed on the second Revise Entry window (again see fig. 3.3), CLEAR LAST DATE AND TOTAL? (Y/N). As you use a particular entry to connect to another computer, PROCOMM PLUS maintains a simple call history. The program records the date of the last successful connection made with the entry and keeps a tally of the number of connections. This Revise Entry window prompt is asking whether you want to set the call-history date back to the default value of 00/00/00 and set the tally to 0. The default answer is No. If you press Enter, PROCOMM PLUS does not clear the date or tally. When you want to start running a new tally, press Y in response to the question. For example, if you are modifying the entry so that it dials a different remote computer, you probably will want to abandon the old LAST DATE value and start the tally again at 0.

TIP In addition to the call-history data maintained in the dialing directory, PROCOMM PLUS also keeps a detailed log of all the calls you make that result in a connection. This log is kept in a disk file named PCPLUS.FON. Refer to "Viewing a File," in Chapter 4, for more information on how to take a look at the contents of this file.

Deleting an Entry

Occasionally you may want to purge your directory of entries you no longer use. To delete a single directory entry, select it and press **E** (Erase Entry(s)). PROCOMM PLUS displays the prompt ERASE ENTRY NUMBER n?

(Y/N), where *n* is the selected entry number. Press Y, and PROCOMM PLUS deletes the entry from the directory. PROCOMM PLUS does not, however, move other entries up to fill the empty space.

If you want to remove two or more entries from the directory, you do not have to delete each one separately. Select the first entry you want to delete, and press the space bar. PROCOMM PLUS *marks* the entry by placing a triangular-shaped pointer on the Dialing Directory screen just to the left of the entry number. Repeat this procedure for each entry you want to delete. Finally, press **E** (Erase Entry(s)). PROCOMM PLUS displays the prompt ERASE n MARKED ENTRIES? (Y/N), where *n* is the number of entries you have marked. Press Y, and PROCOMM PLUS deletes the marked entries. As is the case when you delete only one entry, PROCOMM PLUS does not move other entries into the lines vacated by the deleted entries.

Working with Dialing Codes

As you build your PROCOMM PLUS dialing directory, you may find that you are entering a certain string of numbers over and over again. For example, if your long-distance carrier requires that you dial a special sequence of numbers before dialing the regular telephone number, you must include that access sequence for every directory entry's telephone number. With PROCOMM PLUS's special shorthand *dialing codes*, you can incorporate a string of numbers by reference without having to type the entire string. You can establish up to 10 dialing codes in each dialing directory.

Adding or Modifying a Dialing Code

Press **P** (Dialing Codes) from the Dialing Directory screen to display the Dialing Codes window, as shown in figure 3.12. This window contains 10 lines, each labeled with a letter from *A* through *J*, and each line can hold one dialing code. The cursor blinks near the lower left corner of the window, to the right of a small triangular pointer. To enter a new dialing code, use the up or down arrow to move the cursor to one of the 10 lines, and press **R** (Revise).

In the selected dialing code line, type the number you want to establish as a dialing code. (Keep in mind that while the cursor is in a dialing code line, the cursor-movement and editing keys listed in table 3.2 are available and have the same effect they do when you are editing the Revise Entry window.) Each code can contain up to 34 characters, which can be any

Fig. 3.12.

The Dialing Codes window.

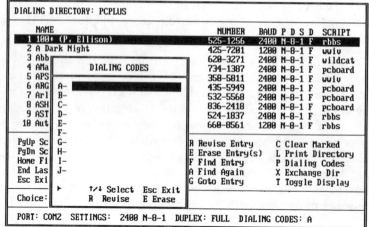

displayable characters but must be recognizable by your modem to be of any value (consult your modem's manual for available dialing codes). Most typically, you type numbers, hyphens, and parentheses in dialing codes. The numbers instruct your modem to dial the telephone, and the hyphens and parentheses make the phone number easier for you to read but have no effect on the dialing.

Sometimes you may want the modem to pause during the dialing procedure. For example, if the dialing code is intended to dial a long-distance telephone carrier, you may need the modem to pause after dialing the long-distance access code. If you have a Hayes-compatible modem, you can add commas to the dialing code to tell the modem to pause. (If you don't have a Hayes-compatible modem, check your modem manual for a pause code.) Each comma pauses the modem for two seconds. So if you want your modem to pause the dialing for six seconds—between the dialing of a phone number and five-digit extension, for example—you add three commas. Figure 3.13 shows a number assigned to dialing code *A*. When used in a dialing directory telephone number, this code causes the modem to dial the telephone number *1112222*, pause for six seconds, and then dial the extension *33333*.

Once you have finished adding dialing codes, press Esc (Exit) to return to the Dialing Directory screen.

If you later decide that you need to modify a dialing code, just press **P** (Dialing Codes) to redisplay the Dialing Codes window. Use the up or down arrow to select the dialing code that you want to modify, and use

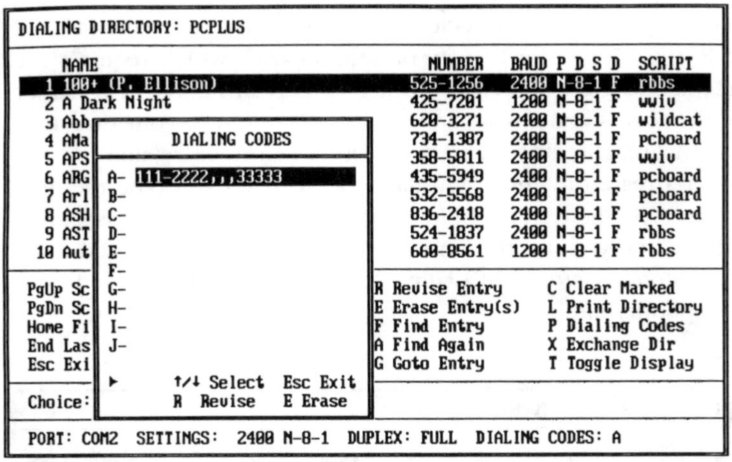

```
DIALING DIRECTORY: PCPLUS

       NAME                              NUMBER   BAUD P D S D   SCRIPT
 1 100+ (P. Ellison)                    525-1256  2400 N-8-1 F   rbbs
 2 A Dark Night                         425-7281  1200 N-8-1 F   wwiv
 3 Abb                                  620-3271  2400 N-8-1 F   wildcat
 4 AMa       DIALING CODES              734-1387  2400 N-8-1 F   pcboard
 5 APS                                  358-5811  2400 N-8-1 F   wwiv
 6 ARG  A- 111-2222,,,33333             435-5949  2400 N-8-1 F   pcboard
 7 Arl  B-                              532-5568  2400 N-8-1 F   pcboard
 8 ASH  C-                              836-2418  2400 N-8-1 F   pcboard
 9 AST  D-                              524-1837  2400 N-8-1 F   rbbs
10 Aut  E-                              660-8561  1200 N-8-1 F   rbbs
       F-
PgUp Sc G-                             R Revise Entry      C Clear Marked
PgDn Sc H-                             E Erase Entry(s)    L Print Directory
Home Fi I-                             F Find Entry        P Dialing Codes
End Las J-                             A Find Again        X Exchange Dir
Esc Exi                               G Goto Entry        T Toggle Display
        ►      ↑/↓ Select  Esc Exit
Choice:        R  Revise   E Erase

PORT: COM2  SETTINGS:  2400 N-8-1  DUPLEX: FULL  DIALING CODES: A
```

Fig. 3.13.

A sample dialing code.

the cursor-movement and editing keys to make your changes. Press Enter to accept the changes, and Esc (Exit) to return to the Dialing Directory screen.

TIP Many office telephone systems require that you dial *9* or some other digit to get an outside line. You then have to wait for a second dial tone before you continue to dial. For this reason, most Hayes-compatible modems recognize the letter *W* as a command to pause and wait for a second dial tone before continuing to dial the telephone number. (Check your modem documentation to make sure that your modem recognizes this command.) For example, you can create a dialing code that contains the following:

9W111-2222,,,33333

When used in a directory entry telephone number, this code causes the modem to dial *9* and then wait for a second dial tone. As soon as the modem detects the dial tone, it continues dialing the rest of the number, pauses for six seconds, and dials a five-digit extension.

Using Dialing Codes

You can use a dialing code in any entry in the dialing directory. Typically, you will use dialing codes in many entries, or you probably would not have gone to the trouble of creating the dialing code in the first place.

For example, suppose that you decide to create a directory entry to call BCSNET (a Boston Computer Society bulletin board), which requires the long-distance access code shown in dialing code *A* in figure 3.13. Rather than retype the access code, you can use dialing code *A* in your directory entry. In the Dialing Directory screen, select an empty entry, and press **R** (Revise Entry). Create the entry as usual, except when you type the telephone number, type the letter *A* before you type the area code, as shown in figure 3.14. When you use this entry to dial BCSNET, PROCOMM PLUS first sends the number stored in dialing code *A* to the modem and then sends the telephone number.

Fig. 3.14.

Using a dialing code.

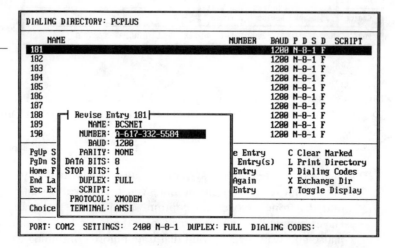

```
DIALING DIRECTORY: PCPLUS

         NAME                              NUMBER     BAUD P D S D   SCRIPT
 101                                                  1200 N-8-1 F
 102                                                  1200 N-8-1 F
 103                                                  1200 N-8-1 F
 104                                                  1200 N-8-1 F
 105                                                  1200 N-8-1 F
 106                                                  1200 N-8-1 F
 107                                                  1200 N-8-1 F
 108      ┌─┤ Revise Entry 101 ├─────────────┐       1200 N-8-1 F
 109      │    NAME: BCSNET                   │       1200 N-8-1 F
 110      │  NUMBER: A-617-332-5584           │       1200 N-8-1 F
          │    BAUD: 1200                     │
 PgUp S   │  PARITY: NONE            e Entry  │    C Clear Marked
 PgDn S   │ DATA BITS: 8            Entry(s)  │    L Print Directory
 Home F   │ STOP BITS: 1           Entry     │    P Dialing Codes
 End La   │  DUPLEX: FULL          Again      │    X Exchange Dir
 Esc Ex   │  SCRIPT:               Entry      │    T Toggle Display
          │ PROTOCOL: XMODEM                  │
 Choice   │ TERMINAL: ANSI                    │
          └──────────────────────────────────┘

 PORT: COM2  SETTINGS:  2400 N-8-1  DUPLEX: FULL  DIALING CODES:
```

Erasing a Dialing Code

Erasing a dialing code is similar to erasing a dialing directory entry. First, from the Dialing Directory screen, press **P** (Dialing Codes) to display the Dialing Codes window. Use the up or down arrow to select the code you want to erase. Finally, press **E** (Erase). PROCOMM PLUS erases the dialing code without asking for any confirmation. Press Esc (Exit) to return to the Dialing Directory screen.

Dialing a Number

The ultimate purpose of each entry in a PROCOMM PLUS dialing directory is to set up your computer for communication with another computer and then to instruct your modem to dial that computer's telephone number. This section of the chapter describes several ways to use the Dialing Directory screen to that end, including two ways to dial a number by using a single directory entry, and two ways to create a *dialing queue* —a series of entries for PROCOMM PLUS to use in sequence. In this section, you also review the Manual Dial feature.

When you use one of the methods described in this chapter to initiate a call, PROCOMM PLUS first sets the transmission speed, line settings (parity, data bits, and stop bits), duplex setting, error-checking protocol, and terminal emulation according to the values stored in the entry's fields. PROCOMM PLUS then instructs the modem to dial the telephone number stored in the entry's NUMBER field.

Using a Directory Entry

Once you have added an entry to your PROCOMM PLUS dialing directory, you can use that entry when you want to connect to a remote computer. Two different methods are available.

When you can remember the number of the entry, display the Dialing Directory screen and press **D** (Dial Entry(s)). PROCOMM PLUS displays a narrow window containing the prompt Entry(s) to dial, with the cursor blinking to the right of this prompt (see fig. 3.15).

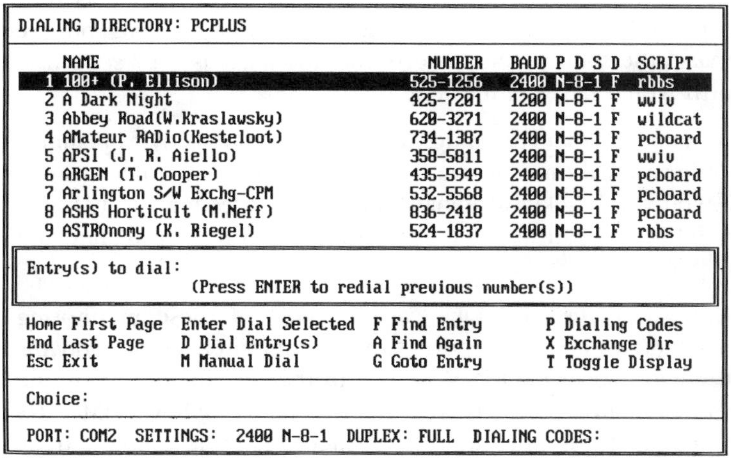

```
DIALING DIRECTORY: PCPLUS

        NAME                          NUMBER    BAUD P D S D  SCRIPT
    1 100+ (P. Ellison)               525-1256  2400 N-8-1 F  rbbs
    2 A Dark Night                    425-7201  1200 N-8-1 F  wwiv
    3 Abbey Road(W.Kraslawsky)        620-3271  2400 N-8-1 F  wildcat
    4 AMateur RADio(Kesteloot)        734-1387  2400 N-8-1 F  pcboard
    5 APSI (J. R. Aiello)             358-5811  2400 N-8-1 F  wwiv
    6 ARGEN (T. Cooper)               435-5949  2400 N-8-1 F  pcboard
    7 Arlington S/W Exchg-CPM         532-5568  2400 N-8-1 F  pcboard
    8 ASHS Horticult (M.Neff)         836-2418  2400 N-8-1 F  pcboard
    9 ASTROnomy (K. Riegel)           524-1837  2400 N-8-1 F  rbbs

 Entry(s) to dial:
            (Press ENTER to redial previous number(s))

 Home First Page  Enter Dial Selected  F Find Entry    P Dialing Codes
 End  Last Page   D Dial Entry(s)      A Find Again    X Exchange Dir
 Esc  Exit        M Manual Dial        G Goto Entry    T Toggle Display

 Choice:

 PORT: COM2  SETTINGS:  2400 N-8-1  DUPLEX: FULL  DIALING CODES:
```

Fig. 3.15.

Specifying a dialing directory entry to dial.

Type the entry number, and press Enter. PROCOMM PLUS moves the highlight to the entry; adjusts the transmission speed and line settings, if necessary; and instructs the modem to begin dialing the number. When the program successfully connects to the other computer, the Terminal mode screen appears. The next time you display the Dialing Directory screen during the same PROCOMM PLUS session, the entry that was selected when the connection was made is still selected.

The second method of using a dialing directory entry is to select it, and press Enter. PROCOMM PLUS adjusts the transmission speed and line settings, if necessary, and instructs the modem to begin dialing the number. Even if PROCOMM PLUS is unsuccessful in connecting to the other computer, if you return to the Terminal mode screen and then later redisplay the Dialing Directory screen, this entry is still selected. You can use it again by pressing Enter.

Monitoring Call Progress

Most modems have a built-in speaker so that you can monitor each call audibly. If your modem has a speaker, and if everything goes as planned, you hear the dial tone, the number being dialed, and then the ringing. When the modem on the other end answers the line, you hear a high-pitched tone called the *carrier tone*, then a hissing sound, and finally nothing. The hissing sound is the sound made by the modems as they "shake hands" and begin communicating. Your modem turns off the speaker because the sounds generated by the modems as they send data back and forth would have no meaning to you.

As you might expect, however, things don't always go exactly as planned. If you do not hear a dial tone, or you hear a busy signal or a voice answering the call, you can quickly terminate the call by pressing Esc (Abort). PROCOMM PLUS sends an instruction to the modem to stop dialing and hang up the phone.

Many modems, however, also take care of most of these monitoring chores for you and send status reports to your computer. These modems can recognize and inform PROCOMM PLUS of the absence of a dial tone and the presence of a busy signal. They can also inform PROCOMM PLUS when a connection to another modem is successfully accomplished, as well as the transmission speed at which the modems are communicating. PROCOMM PLUS monitors these status reports and informs you of the call's progress in the PROCOMM PLUS Dialing window, shown in figure 3.16.

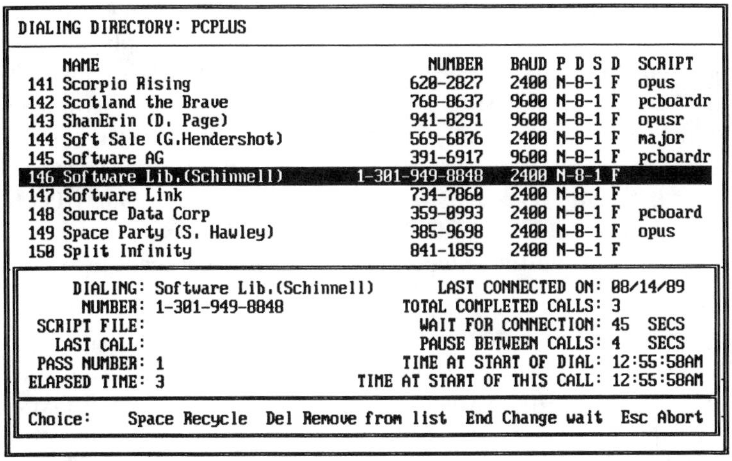

```
DIALING DIRECTORY: PCPLUS

         NAME                        NUMBER    BAUD P D S D  SCRIPT
141 Scorpio Rising                   628-2827  2400 N-8-1 F  opus
142 Scotland the Brave               768-8637  9600 N-8-1 F  pcboardr
143 ShanErin (D. Page)               941-8291  9600 N-8-1 F  opusr
144 Soft Sale (G.Hendershot)         569-6876  2400 N-8-1 F  major
145 Software AG                      391-6917  9600 N-8-1 F  pcboardr
146 Software Lib.(Schinnell)  1-301-949-8848   2400 N-8-1 F
147 Software Link                    734-7860  2400 N-8-1 F
148 Source Data Corp                 359-8993  2400 N-8-1 F  pcboard
149 Space Party (S. Hawley)          385-9698  2400 N-8-1 F  opus
150 Split Infinity                   841-1059  2400 N-8-1 F

       DIALING: Software Lib.(Schinnell)    LAST CONNECTED ON: 08/14/89
        NUMBER: 1-301-949-8848       TOTAL COMPLETED CALLS: 3
   SCRIPT FILE:                       WAIT FOR CONNECTION: 45   SECS
     LAST CALL:                       PAUSE BETWEEN CALLS: 4    SECS
   PASS NUMBER: 1                   TIME AT START OF DIAL: 12:55:58AM
  ELAPSED TIME: 3            TIME AT START OF THIS CALL: 12:55:58AM

 Choice:   Space Recycle  Del Remove from list  End Change wait  Esc Abort
```

Fig. 3.16.

The Dialing window.

The Dialing window is intended to help you visually monitor the status of the call. In this window PROCOMM PLUS lists the name and telephone number of the system your modem is dialing, and keeps you abreast of several other parameters.

For example, if your modem detects no dial tone when beginning to dial the telephone, the modem stops dialing and sends the code NO DIAL TONE to PROCOMM PLUS. PROCOMM PLUS in turn displays this message in the LAST CALL line of the Dialing window. Similarly, the call may proceed to the point at which the modem detects a busy signal. The modem sends the message BUSY, which PROCOMM PLUS then displays in the LAST CALL line.

When the modem sends either of these messages, it also stops dialing the number and hangs up the phone line (often referred to in modem documentation as *going on book*). PROCOMM PLUS then displays the message PAUSING... in the DIALING line.

TIP In some special circumstances, you don't want your modem to wait for a dial tone before trying to complete the call. This dialing method is often referred to as *blind dialing*. Perhaps your phone system generates a dial tone that your modem cannot detect, so you want the modem to dial "blindly." In such a case, refer to "Setting Modem Options," in Chapter 8, and to your modem manual for instructions on how to modify the initialization command to enable blind dialing.

While pausing the dialing procedure, PROCOMM PLUS counts the number of seconds that have elapsed since the last attempted call, displaying the count in the ELAPSED TIME line. PROCOMM PLUS waits the number of seconds indicated in the PAUSE BETWEEN CALLS line, increments the PASS NUMBER line by 1, and sends an instruction to the modem to dial the number again. The PAUSE BETWEEN CALLS setting is necessary to let the modem get ready to redial. The length of time PROCOMM PLUS pauses between calls is set by default to four seconds. You can increase or decrease this time by using the Setup Utility (refer to Chapter 8 for more details).

For example, figure 3.17 shows PROCOMM PLUS pausing because the modem detected a busy signal on the first two attempted calls. Two seconds have elapsed. When the ELAPSED TIME value reaches 4, PROCOMM PLUS will increment the PASS NUMBER line to 3 and dial the number once more. PROCOMM PLUS continues this procedure until successfully connecting to the other modem, or until you stop the dialing by pressing Esc (Abort).

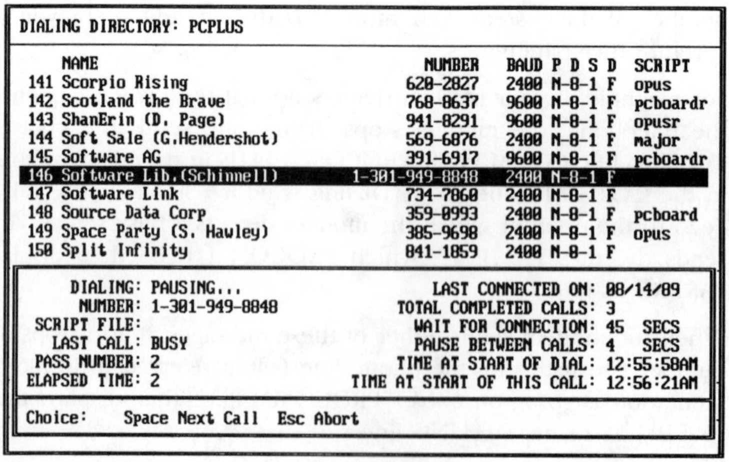

Fig. 3.17.

Monitoring call progress in the Dialing window.

```
DIALING DIRECTORY: PCPLUS

      NAME                       NUMBER      BAUD P D S D   SCRIPT
141 Scorpio Rising              620-2827    2400 N-8-1 F   opus
142 Scotland the Brave          768-8637    9600 N-8-1 F   pcboardr
143 ShanErin (D. Page)          941-8291    9600 N-8-1 F   opusr
144 Soft Sale (G.Hendershot)    569-6876    2400 N-8-1 F   major
145 Software AG                 391-6917    9600 N-8-1 F   pcboardr
146 Software Lib.(Schinnell)  1-301-949-8848 2400 N-8-1 F
147 Software Link               734-7860    2400 N-8-1 F
148 Source Data Corp            359-0993    2400 N-8-1 F   pcboard
149 Space Party (S. Hawley)     385-9698    2400 N-8-1 F   opus
150 Split Infinity              841-1859    2400 N-8-1 F

      DIALING: PAUSING...              LAST CONNECTED ON: 08/14/89
       NUMBER: 1-301-949-8848    TOTAL COMPLETED CALLS: 3
  SCRIPT FILE:                      WAIT FOR CONNECTION: 45   SECS
    LAST CALL: BUSY                 PAUSE BETWEEN CALLS: 4    SECS
  PASS NUMBER: 2                   TIME AT START OF DIAL: 12:55:58AM
 ELAPSED TIME: 2             TIME AT START OF THIS CALL: 12:56:21AM

Choice:    Space Next Call  Esc Abort
```

The Dialing window shown in both figures 3.16 and 3.17 also indicates in the LAST CONNECTED ON line that you last communicated with Software Lib.(Schinnell) on 08/14/89. The TOTAL COMPLETED CALLS line indicates that three calls have been completed during the current PROCOMM PLUS session. The last two lines of the Dialing window keep track of the time of day when you initially started the dialing procedure, as well as the time that dialing for the most recent call started.

Even if your modem does not recognize a busy signal, PROCOMM PLUS has another feature that can help you. After sending the instruction to the modem to dial the telephone number, PROCOMM PLUS waits the length of time listed in the WAIT FOR CONNECTION line of the Dialing window. If the modem does not complete a connection in that length of time, either because the line is busy or because the other end does not answer, PROCOMM PLUS automatically puts the phone line back on hook and pauses the dialing. After the time span indicated in the PAUSE BETWEEN CALLS line, PROCOMM PLUS tries the call again. The program repeatedly attempts to connect to the modem on the other end of the line until the connection is successful or until you press Esc (Abort), whichever comes first.

TIP At times you may decide that you want to increase or decrease the WAIT FOR CONNECTION value in the Dialing window. You can change this setting permanently by using the PROCOMM PLUS Setup Utility, explained in "Setting Modem Options," in Chapter 8. While PROCOMM PLUS is displaying the Dialing window, however, you can easily change this setting temporarily. Just press End. PROCOMM PLUS pauses dialing and places a cursor in the WAIT FOR CONNECTION line, where you can edit the number and press Enter. PROCOMM PLUS uses this new wait time until you change it or until the end of the current PROCOMM PLUS session.

PROCOMM PLUS takes advantage of another handy call-progress reporting feature available on many modems. Your modem may be able to detect when someone or something other than a modem answers your call. If so, your modem refrains from sending a carrier signal to the ear of the individual who answers the line. Instead, the modem puts the phone line back on hook and sends a signal to PROCOMM PLUS that a VOICE was encountered. PROCOMM PLUS then displays this message in the LAST CALL line of the Dialing window.

When your modem is finally successful in connecting to the modem of the computer with which you want to communicate, the modem sends a message to PROCOMM PLUS. Older Hayes-compatible modems (most 1200-bps and some 2400-bps modems) send only the message CONNECT, regardless of the speed at which they connect. Most newer Hayes-compatible modems, however, are more specific. When these modems connect at 300 bps, they send the message CONNECT, but when they connect at 1200 bps, the message is CONNECT 1200. Connecting at 2400 bps results in the message CONNECT 2400.

Whatever the message, PROCOMM PLUS displays it in flashing characters in the LAST CALL line of the Dialing window. In case you don't notice this visual connect signal, PROCOMM PLUS also sounds several beeps on your computer to alert you. This feature allows you to leave the room and get a snack while you wait for your modem to get a clear line and connect to the remote computer. Just don't forget to be listening for the "alarm."

TIP For any of these modem-generated call-progress monitoring features to work properly, your modem must send progress reports (usually referred to as *result codes*) that PROCOMM PLUS can understand. Most Hayes-compatible modems can send messages back to PROCOMM PLUS as words, such as CONNECT, NO DIAL TONE, or BUSY. These messages are what PROCOMM PLUS expects. These same modems, however, are also capable of sending numeric result codes instead, such as 1, 6, or 7. If the call-progress features don't seem to work on your system, check to make sure that the initialization command for the modem is set correctly in the Modem Options screen of the PROCOMM PLUS Setup Utility, and that the Modem Result Messages screen in the PROCOMM PLUS Setup Utility contains the correct result codes. Refer to "Setting Modem Options," in Chapter 8, and to your modem manual for instructions on how to modify the initialization command to match the capabilities of your modem and how to complete the Modem Result Messages screen.

Once your modem successfully makes a connection, PROCOMM PLUS removes the Dialing window and the Dialing Directory screen, returns to the Terminal mode screen, and displays the following message:

    PROCOMM PLUS on-line to entry_name at speed baud

where *entry_name* is the entry's name, and *speed* is the transmission speed at which the modems are communicating. Figure 3.18 shows an example. Typically, you are then greeted by a welcome message from the remote computer (see fig. 3.19).

Using a Dialing Queue

As you become familiar with PROCOMM PLUS and start using it to connect with PC bulletin boards in your area, you will soon discover that PC communications is a popular hobby. You will also find that many bulletin boards are next to impossible to connect with because of their popularity.

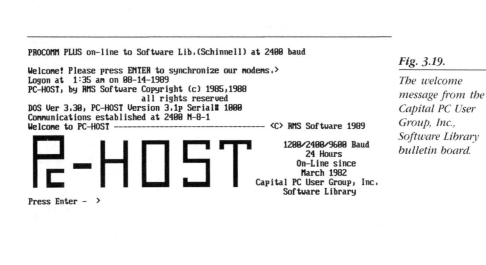

```
PROCOMM PLUS on-line to Software Lib.(Schinnell) at 2400 baud
```

Fig. 3.18.

Connecting to a remote computer at a transmission speed of 2400 bps.

```
Alt-Z FOR HELP  VT102    FDX   2400 N81  LOG CLOSED  PRINT OFF  ON-LINE
```

```
PROCOMM PLUS on-line to Software Lib.(Schinnell) at 2400 baud

Welcome! Please press ENTER to synchronize our modems.>
Logon at  1:35 am on 08-14-1989
PC-HOST, by RMS Software Copyright (c) 1985,1988
                all rights reserved
DOS Ver 3.30, PC-HOST Version 3.1p Serial# 1000
Communications established at 2400 N-8-1
Welcome to PC-HOST ------------------------------ <C> RMS Software 1989

PC-HOST                       1200/2400/9600 Baud
                                   24 Hours
                                 On-Line since
                                   March 1982
                          Capital PC User Group, Inc.
                              Software Library

Press Enter -  >
```

Fig. 3.19.

The welcome message from the Capital PC User Group, Inc., Software Library bulletin board.

```
Alt-Z FOR HELP  VT102    FDX   2400 N81  LOG CLOSED  PRINT OFF  ON-LINE
```

Using either of the two methods described in "Using a Directory Entry," in this chapter, to try to connect to one of these busy bulletin boards may soon result in your frustration. You may wait for hours trying to connect to a single bulletin board by using PROCOMM PLUS's automatic redial capability. Or if you try to call several different boards in succession, you will be tied to the keyboard, searching for a number that doesn't ring busy.

Fortunately, PROCOMM PLUS frees you from this tedium by enabling you to use one of two methods to create a list of entries you want PROCOMM PLUS to dial in succession. If the first entry is busy or doesn't answer, PROCOMM PLUS tries the second. If the second is busy or doesn't answer, the program tries the third, and so on. Such a list of entries is usually called a *dialing queue*.

The first way to create a dialing queue uses the **D** (Dial Entry(s)) command. You have to know the entry numbers of the entries you want PROCOMM PLUS to use. From the Dialing Directory screen, press **D** (Dial Entry(s)), and PROCOMM PLUS displays a window similar to the one shown in figure 3.15. To the right of the Entry(s) to dial prompt, type a list of from 1 to 15 entry numbers. Separate the entry numbers with spaces, commas (,), or semicolons (;). Press Enter after you type the last entry number.

PROCOMM PLUS moves the highlight to the first entry in your list and begins to dial that entry's telephone number. If the number dialed is busy or doesn't answer within the time specified in the WAIT FOR CONNECTION line of the Dialing window, PROCOMM PLUS puts the line back on hook. The program then displays the message PAUSING... in the DIALING line and begins the ELAPSED TIME counter, as shown in figure 3.17. When the counter reaches the number of seconds listed in the PAUSE BETWEEN CALLS line, PROCOMM PLUS moves the highlight to the next entry listed in the queue and begins to dial that entry's telephone number.

If a connection is successful, PROCOMM PLUS removes that entry from the queue. Once you finish communicating with this computer, you can easily restart the dialing queue. Display the Dialing Directory screen, and press **D** (Dial Entry(s)). Then, without typing anything in the displayed window, press Enter. PROCOMM PLUS starts again with the first number in the dialing queue but skips any entry for which a connection was successful. This process continues until a connection is made or until you press Esc (Abort).

The second method of specifying a dialing queue uses the same *marking* feature you use to delete entries. Highlight an entry you want included in the queue, and press the space bar. PROCOMM PLUS places a small triangular pointer to the left of the entry number, *marking* the entry. If you make a mistake and want to remove the mark, highlight the marked entry and press the space bar again. When you have marked all entries to be included in the dialing list, press Enter. PROCOMM PLUS begins dialing the telephone number of the marked entry nearest the top of the dialing directory (that is, the entry with the smallest entry number).

A busy signal or no answer causes PROCOMM PLUS to pause the dialing for the specified period of time and then move to the next entry in the queue. When a connection is made, PROCOMM PLUS removes the mark from the entry that connected, and returns to the Terminal mode screen. Once your connection with the remote computer is finished, restarting the dialing queue is even easier than when using the **D** (Dial Entry(s)) method. Simply display the Dialing Directory screen, and press Enter. PROCOMM PLUS starts with the marked entry nearest the top of the directory.

Regardless of which dialing queue method you use, PROCOMM PLUS enables you to skip an entry in the dialing queue. Press the space bar at any time during the dialing process for that entry, but before a connection. PROCOMM PLUS stops the dialing, displays PAUSING... in the DIALING line of the Dialing window, and begins to count off elapsed time in the ELAPSED TIME line. When the elapsed time reaches the value listed in the PAUSE BETWEEN CALLS line, PROCOMM PLUS resumes dialing but uses the next entry in the queue. The program skips the entry that was being dialed when you pressed the space bar.

You can also delete an entry from the dialing queue. While PROCOMM PLUS is dialing the entry, press the Del key. PROCOMM PLUS displays the message REMOVED in the LAST CALL line and begins the elapsed time counter. When the counter reaches the current PAUSE BETWEEN CALLS setting, PROCOMM PLUS continues with the dialing queue, starting at the next entry.

Using the Manual Dial Feature

On occasions you may decide to call a computer but don't expect ever to need to call the same computer again. Rather than go to the effort of creating a dialing directory entry that you will just erase later, you can use PROCOMM PLUS's Manual Dial feature.

To dial another computer without first creating an entry, press **M** (Manual Dial) from the Dialing Directory screen. PROCOMM PLUS displays a narrow window across the middle of the screen, prompting you for the Number to dial (see fig. 3.20).

At the prompt, type the phone number you want the modem to dial. Include any digits necessary to get an outside line. You can also use commas (to pause dialing) and dialing codes. When you finish typing the number, double-check it to make sure that you typed it correctly. Then press Enter.

Fig. 3.20.

*Using the Manual
Dial feature.*

```
DIALING DIRECTORY: PCPLUS

     NAME                          NUMBER   BAUD P D S D   SCRIPT
  1 100+ (P. Ellison)             525-1256  2400 N-8-1 F   rbbs
  2 A Dark Night                  425-7201  1200 N-8-1 F   wwiv
  3 Abbey Road(W.Kraslausky)      628-3271  2400 N-8-1 F   wildcat
  4 AMateur RADio(Kesteloot)      734-1387  2400 N-8-1 F   pcboard
  5 APSI (J. R. Aiello)           358-5811  2400 N-8-1 F   wwiv
  6 ARGEN (T. Cooper)             435-5949  2400 N-8-1 F   pcboard
  7 Arlington S/W Exchg-CPM       532-5568  2400 N-8-1 F   pcboard
  8 ASHS Horticult (M.Neff)       836-2418  2400 N-8-1 F   pcboard
  9 ASTROnomy (K. Riegel)         524-1837  2400 N-8-1 F   rbbs

Number to dial:

PgDn Scroll Dn      Space Mark Entry     E Erase Entry(s)    L Print Directory
Home First Page     Enter Dial Selected  F Find Entry        P Dialing Codes
End Last Page       D Dial Entry(s)      A Find Again        X Exchange Dir
Esc Exit            M Manual Dial        G Goto Entry        T Toggle Display

Choice:

PORT: COM2  SETTINGS:  2400 N-8-1  DUPLEX: FULL  DIALING CODES:
```

PROCOMM PLUS replaces the Number To Dial window with the Dialing
window. Instead of displaying an entry name in the DIALING line,
PROCOMM PLUS displays the message MANUAL DIAL and begins dialing the
number. Because PROCOMM PLUS has no entry from which it can take
transmission speed, line settings, and other communications settings, the
program uses whatever settings are current, as indicated in the Dialing
Directory screen status line. The current settings are determined by the
last dialing directory entry used. If no entry has yet been used during
the current PROCOMM PLUS session, the program uses the default communications settings. Refer to Chapter 8, "Tailoring PROCOMM PLUS," for
instructions on how to modify the default settings.

Except for the message in the DIALING line, the Dialing window operates
in the same fashion as when you are using a directory entry to dial a
remote computer.

If you use the **M** (Manual Dial) feature again during the same PROCOMM
PLUS session, the telephone number you typed still displays in the
Number To Dial window. Press Enter to use the same number, or edit
the number if you want to call a different computer.

Managing Your Dialing Directory

A PROCOMM PLUS dialing directory is a small database—a collection
of information arranged in *fields* (columns) and *records* (rows). In
PROCOMM PLUS, each dialing directory entry is a record in the database;
and each column (NAME, NUMBER, BAUD, and so on) is a field. Up to this

point, this chapter has concentrated on how to create, edit, delete, and use directory entries, either one at a time or in groups (as a dialing queue). But as is true with any database program, PROCOMM PLUS enables you to perform certain functions on the entire database—the entire dialing directory. The remainder of the chapter describes how to use these directory-wide functions to manage your dialing directory database. You first learn how to change the display in the Dialing Directory screen so that it shows a set of fields different from what you normally see. The chapter next shows you how to print an entire directory, use other dialing directories, and sort directory entries.

Toggling the Display

You have probably noticed by now that you enter more information in the Revise Entry window about a particular entry (see fig. 3.2) than is displayed in the Dialing Directory screen itself. Each entry's name, telephone number, transmission rate, parity, data bits, stop bits, duplex setting, and script are normally shown in the Dialing Directory screen. But the Revise Entry window includes a setting for a default file-transfer protocol and for a terminal-emulation setting. Neither of these settings shows up in the Dialing Directory screen. You may also recall that PROCOMM PLUS keeps a call history on each directory entry—the date of the last connection and the total number of connections made with the entry. This call-history information does not normally display in the Dialing Directory screen. PROCOMM PLUS does, however, provide a way to see all this usually hidden information. This procedure is called *toggling* the display.

To display in the Dialing Directory screen the call-history data, file-transfer protocol, and terminal-emulation setting, press **T** (Toggle Display). PROCOMM PLUS removes the NUMBER, BAUD, P (parity), D (data bits), S (stop bits), and D (duplex) columns from the Dialing Directory screen, replacing them with the TOTAL, LAST, PROTOCOL, and TERMINAL columns, as shown in figure 3.21. Compare this figure with figure 3.4.

The TOTAL column in this toggled version of the Dialing Directory screen lists the total number of times you have used each directory entry to make a connection. The LAST column contains the date on which each entry made its last connection. The column PROTOCOL shows the default file-transfer protocol for each entry. The TERMINAL column indicates which terminal is to be emulated when the entry is used to connect to a remote computer. The other two columns displayed in the directory screen, NAME and SCRIPT, contain the same data they showed before you toggled the display.

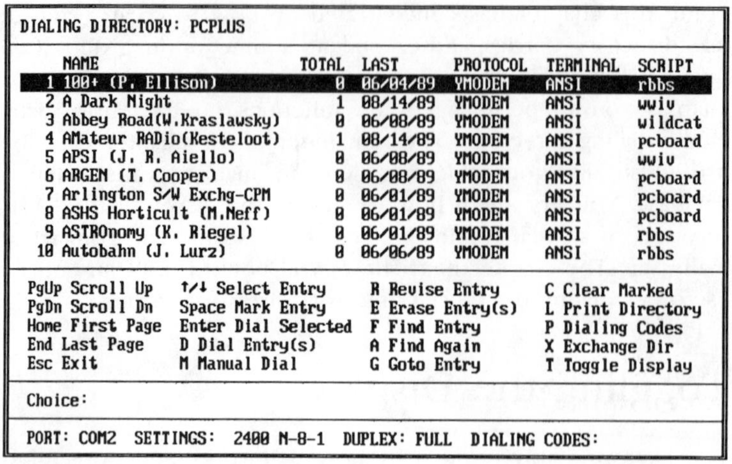

Fig. 3.21.

Toggling the
Dialing Directory
screen.

```
┌─────────────────────────────────────────────────────────────────────┐
│ DIALING DIRECTORY: PCPLUS                                             │
├─────────────────────────────────────────────────────────────────────┤
│      NAME                  TOTAL  LAST     PROTOCOL  TERMINAL  SCRIPT │
│  1 100+ (P. Ellison)         0   06/04/89  YMODEM    ANSI     rbbs    │
│  2 A Dark Night              1   08/14/89  YMODEM    ANSI     wwiv    │
│  3 Abbey Road(W.Kraslawsky)  0   06/08/89  YMODEM    ANSI     wildcat │
│  4 AMateur RADio(Kesteloot)  1   08/14/89  YMODEM    ANSI     pcboard │
│  5 APSI (J. R. Aiello)       0   06/08/89  YMODEM    ANSI     wwiv    │
│  6 ARGEN (T. Cooper)         0   06/08/89  YMODEM    ANSI     pcboard │
│  7 Arlington S/W Exchg-CPM   0   06/01/89  YMODEM    ANSI     pcboard │
│  8 ASHS Horticult (M.Neff)   0   06/01/89  YMODEM    ANSI     pcboard │
│  9 ASTROnomy (K. Riegel)     0   06/01/89  YMODEM    ANSI     rbbs    │
│ 10 Autobahn (J. Lurz)        0   06/06/89  YMODEM    ANSI     rbbs    │
├─────────────────────────────────────────────────────────────────────┤
│ PgUp Scroll Up    ↑/↓ Select Entry    R Revise Entry    C Clear Marked│
│ PgDn Scroll Dn    Space Mark Entry     E Erase Entry(s)  L Print Directory│
│ Home First Page   Enter Dial Selected  F Find Entry      P Dialing Codes│
│ End Last Page     D Dial Entry(s)      A Find Again      X Exchange Dir│
│ Esc Exit          M Manual Dial        G Goto Entry      T Toggle Display│
├─────────────────────────────────────────────────────────────────────┤
│ Choice:                                                               │
├─────────────────────────────────────────────────────────────────────┤
│ PORT: COM2  SETTINGS:  2400 N-8-1  DUPLEX: FULL  DIALING CODES:       │
└─────────────────────────────────────────────────────────────────────┘
```

To return to the standard Dialing Directory screen, press **T** (Toggle Display) again. PROCOMM PLUS "hides" the four settings. Keep in mind, however, that these settings are there and can be modified, even if you don't normally see them in the Dialing Directory screen.

Printing the Directory

Even though finding and displaying an entry in the dialing directory is easy, sometimes having a hard copy to look at is more convenient. You can print a copy of the directory on paper, using your printer, and also can send a copy of the directory to a disk file so that you can print it later.

To print your dialing directory, press **L** (Print Directory) from the Dialing Directory screen. PROCOMM PLUS displays a narrow window containing the prompt PRINT TO and the default response PRN. PRN is the DOS device name that represents the printer connected to the first parallel printer port on your computer. If you have a printer connected to your computer in this fashion, press Enter. If your printer is connected to the second parallel printer port on your PC, type the value *LPT2* in place of *PRN*.

PROCOMM PLUS displays the message, PREPARE YOUR PRINTER, AND PRESS ANY KEY TO BEGIN. Make sure that your printer is turned on, properly connected, and loaded with paper; and then press any key. PROCOMM PLUS prints the directory and displays the message Printing dirname directory – Press ESC to quit, where *dirname* is the name of your directory. When PROCOMM PLUS finishes printing the dialing directory, the Dialing Directory screen reappears.

Note: PROCOMM PLUS prints the entire directory, including all 200 entries—even the empty ones.

Rather than print a copy of the dialing directory, you may want to create a copy of the directory as a disk file. To do so, simply type a valid DOS file name to the right of the PRINT TO prompt that appears when you press **L** (Print Directory); then press Enter. The next two prompts are the same as if PROCOMM PLUS were printing the directory, but the output goes to a file rather than the printer. PROCOMM PLUS creates a disk image of the report, using ASCII characters.

To view this file, first press Esc (Exit) to return to the Terminal mode screen. Then press Alt-V (View a File). PROCOMM PLUS displays the View a File window containing the prompt Please enter filename. Type the name of the file that contains the dialing directory print image, and press Enter. PROCOMM PLUS displays the file on-screen, as shown in figure 3.22.

```
                ProComm Plus (tm) Dialing Directory
                -------- ---- ---- ------- ---------


        NAME                         NUMBER    BAUD P D S D  SCRIPT
    1- 100+ (P. Ellison)             525-1256  2400 N-8-1 F  rbbs
    2- A Dark Night                  425-7201  1200 N-8-1 F  wwiv
    3- Abbey Road(W.Kraslawsky)      620-3271  2400 N-8-1 F  wildcat
    4- AMateur RADio(Kesteloot)      734-1387  2400 N-8-1 F  pcboard
    5- APSI (J. R. Aiello)           358-5811  2400 N-8-1 F  wwiv
    6- ARGEN (T. Cooper)             435-5949  2400 N-8-1 F  pcboard
    7- Arlington S/W Exchg-CPM       532-5568  2400 N-8-1 F  pcboard
    8- ASHS Horticult (M.Neff)       836-2418  2400 N-8-1 F  pcboard
    9- ASTROnomy (K. Riegel)         524-1837  2400 N-8-1 F  rbbs
   10- Autobahn (J. Lurz)            660-8561  1200 N-8-1 F  rbbs
   11- B. O. S.- MAC Hyper RPG       329-8024  2400 N-8-1 F  wildcat
   12- Barneysville                  528-0124  2400 N-8-1 F  pconnect
   13- BCSNET                  A-617-332-5504  2400 N-8-1 F
   14- Bear's Den                    671-8598  2400 N-8-1 F  fido
   15- BearFacts Lk Braddock HS      323-9434  9600 N-8-1 F  pcboardr
   16- Beda Board (P. Beda)          893-8262  2400 N-8-1 F  pcboard
   17- Beehive - 2600 Magazine       823-6591  9600 N-8-1 F  srchlite
   18- Bob's #1 (B. Allen)           271-9036  1200 N-8-1 F  wildcat
 Home: Top of file      PgUp: Previous page    PgDn: Next page      ESC: Exit
```

Fig. 3.22.

A PROCOMM PLUS dialing directory printed to a disk file.

You can easily print a hard copy of this file later. From the DOS prompt, type

    COPY *filename* PRN

where *filename* is the file name of the dialing directory file. If your printer is connected to the second parallel printer port, substitute *LPT2* for PRN.

Alternatively, you can use your word processor to print the file. Most word processors can import ASCII files. You can therefore load the file into your favorite word processor and use the word processor's printing capabilities.

Building and Using Other Dialing Directories

Although PROCOMM PLUS limits you to 200 entries per dialing directory, the program does not limit the number of directories you can build. In effect, then, you can manage as many directory entries as you want. PROCOMM PLUS can, however, work with only one directory file at a time.

Creating a new directory may be necessary to hold all your entries. For example, suppose that you live in the Washington, D.C., metropolitan area, which encompasses parts of two states and the District of Columbia. You could easily compile a list of area bulletin boards that is longer than 200 entries. One convenient way to divide the entries into groups is by state or district. You can create one directory for Virginia bulletin boards named VA.DIR, another for Maryland named MD.DIR, and a third for Washington, D.C., named WASHDC.DIR. Or because you live in Virginia, you may decide to give the directory for bulletin boards in that state the name PCPLUS.DIR so that PROCOMM PLUS uses it by default every time you start the program.

When you want to create another dialing directory, in addition to the default PCPLUS directory, display the Dialing Directory screen and press **X** (Exchange Directory). PROCOMM PLUS displays a window containing the prompt DIRECTORY TO LOAD. Type a valid DOS file name but do not type a period or an extension. Press Enter. PROCOMM PLUS creates a file on disk with the specified name and the extension .DIR, removes the current directory from the screen, and displays the new directory with blank entries. You then can add, modify, delete, and use entries in this new directory just as you did in the PCPLUS dialing directory.

Note: valid DOS file name can contain one to eight characters, including the letters A through Z, the numbers 0 through 9, and the following symbols:

! @ # $ % ^ & ( ) - _ { } '

Spaces are not permitted.

To switch among your existing directories, display the Dialing Directory screen and press **X** (Exchange Dir). In the window that displays, type the file name of the directory you want to access, omitting the period and extension, and press Enter. PROCOMM PLUS loads the specified dialing directory. You can use an unlimited number of dialing directories in this manner, each with the capacity to hold up to 200 entries.

DATASTORM TECHNOLOGIES, INC., distributes four supplemental directories with PROCOMM PLUS 1.1B. These directories are distributed on the Supplemental Diskette in files named ATLANTA.DIR, BOSTON.DIR, SANFRAN.DIR, and WASHDC.DIR. As is clearly denoted by their names, these directories contain entries for popular bulletin boards in the Atlanta; Boston; San Francisco; and Washington, D.C., areas, respectively. If you live in one of these areas, you should copy the appropriate supplemental dialing directory file into the DOS directory that contains the PROCOMM PLUS program files. (Refer to Appendix A for instructions on how to copy files.) You can then switch to the dialing directory by pressing **X** (Exchange Dir) from the Dialing Directory screen and typing the directory's file name.

Sorting the Directory

Typically, you build your dialing directory one entry at a time. In doing so, you are unlikely to add dialing directory entries in any particular order. Often, however, a long directory is easier to use if the entries are listed alphabetically or perhaps sorted by phone number. To assist you in putting your dialing directories in order, PROCOMM PLUS includes a special utility program that enables you to sort a dialing directory by name or phone number. This program, SORTDIR, is contained in the file named SORTDIR.EXE.

Before you can use SORTDIR, you must copy SORTDIR.EXE from the Supplemental Diskette to the DOS directory that contains the dialing directory you want to sort (see Appendix A).

CAUTION: Do not use SORTDIR from the PROCOMM PLUS DOS Gateway (described in "Accessing the Operating System," in Chapter 5). If you sort the current dialing directory from the DOS Gateway, you can render the dialing directory file unusable and possibly corrupt the DOS file allocation table. Instead, quit PROCOMM PLUS, and the run SORTDIR from the DOS prompt.

When you want to use SORTDIR, exit PROCOMM PLUS to DOS. Make current the DOS directory that contains both SORTDIR.EXE and the dialing directory file you want to sort. For example, if these files are contained in the DOS directory C:\PCPLUS, type the following command at the DOS C› prompt:

    CD \PCPLUS

and press Enter. Then start SORTDIR by typing the following command at the DOS prompt:

    SORTDIR *dirname*

where *dirname* is the file name of the dialing directory you want to sort. You can but don't have to type the file extension .DIR. If you don't type a dialing directory name at all, SORTDIR uses PCPLUS.DIR as a default. Press Enter, and SORTDIR displays a screen similar to figure 3.23.

```
SORTDIR v 1.1  Copyright (C) 1988 DATASTORM TECHNOLOGIES, INC.
               All Rights Reserved.

Usage: SORTDIR [directory name]     (Defaults to PCPLUS.DIR)

Sorts PROCOMM PLUS directory by name or phone number.
Current directory will be renamed PCPDIR.OLD.

WARNING: You should not use this routine when in PROCOMM PLUS'
DOS Gateway.

File to be sorted: PCPLUS.DIR

1- Sort by name, ascending (A-Z)
2- Sort by name, decending (Z-A)
3- Sort by phone number, ascending
4- Sort by phone number, decending
Q- Return to DOS

Option:
```

Fig. 3.23.

Using SORTDIR to sort a dialing directory.

This screen is the only SORTDIR screen. It presents four sorting options:

- 1- Sort by name, ascending (A-Z). Press 1 to cause SORTDIR to sort the dialing directory in alphabetical order by entry name.

- 2- Sort by name, decending [*sic*] (Z-A). Press 2 when you want SORTDIR to put the dialing directory in reverse alphabetical order by entry name.

- 3- Sort by phone number, ascending. Press 3 to arrange the dialing directory in order by telephone number, with entries that have phone numbers starting with lower numbers listed before entries that have phone numbers starting with higher numbers. Entries with phone numbers that include area codes are listed after those having phone numbers without area codes.

- 4- Sort by phone number, decending [*sic*]. Press 4 to cause SORTDIR to sort the dialing directory in reverse order by telephone number. Entries having phone numbers with area codes are listed first. Entries having phone numbers that start with high numbers are listed before entries with phone numbers that start with low numbers.

As SORTDIR performs the sort, it issues status reports by building the following message one word at a time:

    Reading...Sorting...Writing...Done!

When the program finishes the sort, SORTDIR redisplays the original screen. Press Q to return to DOS. The next time you use the PROCOMM PLUS Dialing Directory screen, the dialing directory is listed in the sorted order.

Returning to Terminal Mode

When you are using the Dialing Directory screen to connect to another computer and the connection is successful, PROCOMM PLUS returns to the Terminal mode screen. When you finish the communications session with the remote computer, the screen remains in the Terminal mode. But to return to the Terminal mode screen from the Dialing Directory screen without connecting to another computer, just press Esc (Exit).

Chapter Summary

This chapter has described how to use the Dialing Directory screen to control the process of dialing and connecting to remote computers. You have learned how to build dialing directories and how to add, edit, and delete directory entries and dialing codes. The chapter has showed you how to dial another computer by using a single dialing directory entry, a dialing queue, and the Manual Dial feature, and has also explained how to print a dialing directory, how to switch between multiple dialing directories, and how to sort dialing directory entries.

Now that you know how to connect with a remote computer, turn to Chapter 4, "A Session with PROCOMM PLUS," to see what you can do next.

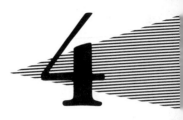

A Session with
PROCOMM PLUS

This chapter focuses on the primary commands useful during a PROCOMM PLUS communications session. You first learn the steps you must take to connect with another computer, either as a calling computer or as an answering computer. The chapter then describes how you can use PROCOMM PLUS to converse with a person or another computer program. Next, the chapter discusses how to manage information that is flowing to and from the other computer, including several ways to capture some or all of this information for later viewing. Finally, you learn what to do when you want to disconnect from the other computer.

Most of the commands explained in this chapter are keyboard commands that you can execute by pressing a keystroke combination from the Terminal mode screen, which is shown in figure 4.1. These keystroke commands are listed on the PROCOMM PLUS Command Menu, which is displayed by pressing Alt-Z from the Terminal mode screen (see fig. 4.2). Although you can invoke many of the features also through the PROCOMM PLUS menu line, this chapter emphasizes the keyboard commands because they require fewer keystrokes and are therefore the most efficient way to perform the functions.

Connecting

The first phase of each PROCOMM PLUS communications session is the *connection* phase, which is usually done through modems and telephone lines. One party's modem is set to *Auto Answer mode* so that the modem

87

PROCOMM PLUS Ready!

Fig. 4.1.

*The Terminal
mode screen.*

| Alt-Z FOR HELP| ANSI | | FDX | 2400 N81 | LOG CLOSED | PRINT OFF | OFF-LINE |

PROCOMM PLUS Ready!

Fig. 4.2.

*The PROCOMM
PLUS Command
Menu.*

```
        P R O C O M M   P L U S   C O M M A N D   M E N U

        ┌──────→ COMMUNICATIONS ◄──────┐     ┌──→ SET UP ◄──┐
        ── BEFORE ──        ── AFTER ──
        Dialing Directory Alt-D  Hang Up ......... Alt-H    Setup Facility .. Alt-S
                                 Exit ............ Alt-X    Line/Port Setup . Alt-P
        ── DURING ──                                        Translate Table . Alt-W
        Script Files ... Alt-F5  Send Files ....... PgUp    Key Mapping .... Alt-F8
        Keyboard Macros . Alt-M  Receive Files .... PgDn
        Redisplay ...... Alt-F6  Log File On/Off  Alt-F1    ──→ OTHER FUNCTIONS ◄──
        Clear Screen .... Alt-C  Log File Pause . Alt-F2
        Break Key ....... Alt-B  Screen Snapshot . Alt-G    File Directory .. Alt-F
        Elapsed Time .... Alt-T  Printer On/Off .. Alt-L    Change Directory Alt-F7
        ── OTHER ──                                         View a File ..... Alt-V
        Chat Mode ....... Alt-O  Record Mode ..... Alt-R    Editor .......... Alt-A
        Host Mode ....... Alt-Q  Duplex Toggle ... Alt-E    DOS Gateway .... Alt-F4
        Auto Answer ..... Alt-Y  CR-CR/LF Toggle  Alt-F3    Program Info .... Alt-I
        User Hot Key 1 .. Alt-J  Kermit Server Cmd Alt-K    Menu Line Key .......`
        User Hot Key 2 .. Alt-U  Screen Pause .... Alt-N
```

Press Alt-Z for extended help

answers the telephone line when the phone rings. The other party's
modem is instructed to dial the appropriate telephone number and to listen for a modem's answer. Although this scenario is typical, it is not the
only way to use PROCOMM PLUS to connect PCs for communication.
This chapter discusses the normal connection method—using modems to
place and answer a telephone call—but also describes how you can connect computers by using a telephone line that you are already using for

voice communication, as well as how you can connect PCs for communication without using a modem at all.

Initiating a Call

More often than not, when you use PROCOMM PLUS to connect to
another computer, your computer does the calling. Once the computer
you intend to call has placed its modem in Auto Answer mode, the next
step is for you to instruct your modem to place the call. (Host systems
such as bulletin boards and on-line services always set their modems to
Auto Answer. When you want to connect to a PC that is not a host system,
you may have to place a voice call to the other PC's operator to ask him
or her to set the system modem to Answer mode.)

To call another computer, press Alt-D (Dialing Directory) from the Terminal mode screen to display the Dialing Directory screen (see fig. 4.3).
Select the directory entry you want to use, and press Enter. (Refer to
Chapter 3, "Building Your Dialing Directory," for detailed instructions on
creating and using dialing directory entries.) PROCOMM PLUS sends
an instruction to the modem causing it to dial the appropriate telephone
number. When the modem on the other end answers, the modems begin
to communicate.

```
DIALING DIRECTORY: PCPLUS

        NAME                      NUMBER   BAUD P D S D  SCRIPT
 1 100+ (P. Ellison)             525-1256  2400 N-8-1 F  rbbs
 2 A Dark Night                  425-7201  1200 N-8-1 F  wwiv
 3 Abbey Road(W.Kraslavsky)      620-3271  2400 N-8-1 F  wildcat
 4 AMateur RADio(Kesteloot)      734-1387  2400 N-8-1 F  pcboard
 5 APSI (J. R. Aiello)           358-5011  2400 N-8-1 F  wwiv
 6 ARGEN (T. Cooper)             435-5949  2400 N-8-1 F  pcboard
 7 Arlington S/W Exchg-CPM       532-5568  2400 N-8-1 F  pcboard
 8 ASHS Horticult (M.Neff)       836-2418  2400 N-8-1 F  pcboard
 9 ASTROnomy (K. Riegel)         524-1837  2400 N-8-1 F  rbbs
10 Autobahn (J. Lurz)            660-8561  1200 N-8-1 F  rbbs

PgUp Scroll Up    ↑/↓ Select Entry   R Revise Entry    C Clear Marked
PgDn Scroll Dn    Space Mark Entry    E Erase Entry(s)  L Print Directory
Home First Page   Enter Dial Selected F Find Entry      P Dialing Codes
End Last Page     D Dial Entry(s)     A Find Again      X Exchange Dir
Esc Exit          M Manual Dial       G Goto Entry      T Toggle Display

Choice:

PORT: COM2  SETTINGS:  2400 N-8-1  DUPLEX: FULL  DIALING CODES:
```

Fig. 4.3.

*The Dialing
Directory screen.*

TIP | Remember that the communications parameters—transmission speed and line settings (parity, data bits, and stop bits)—must be the same for the calling computer and the answering computer. Otherwise, data transmitted will not be intelligible when received. If your modem connects with the modem of the other computer, but only gibberish (such as a stream of characters that don't spell anything) comes across your screen, you should check these parameters.

Even when communications parameters are set properly, receiving illegible information is still possible. This problem can occur because of "noisy" telephone lines, which result in loss of synchronization between the sending modem and the receiving modem. If you suspect poor line quality, terminate the communications session and try again. Sometimes the transmission quality of telephone lines significantly improves when you use them during off-peak hours (when fewer people are using the lines to place calls).

Receiving a Call

As explained in Chapter 1, one of the four categories of PC communication is PC to PC. Because you may at some time be on the receiving end of a PC-to-PC connection, you need to understand what actions are required in order to set up PROCOMM PLUS and your modem for that role.

When someone else calls you on the telephone for the purpose of voice communication, you don't do anything until you hear the phone ring. You then pick up the phone and say "Hello" to begin the conversation. You have learned that you are supposed to answer the ringing telephone. When another computer's modem is calling, however, you want your modem to answer the phone and to connect with the calling modem. Before this sequence of events can happen, you must send your modem the proper command to place the modem in Auto Answer mode. Once in Auto Answer mode, the modem will pick up a ringing line and say "Hello" with its carrier signal.

If you can dedicate a telephone line for your computer's use, you can set PROCOMM PLUS so that each time you use the program it places your modem in Auto Answer mode. (Refer to "Setting Modem Options," in Chapter 8, "Tailoring PROCOMM PLUS," for more information on how to set the initialization command to send the Auto Answer command to the modem.) Then when the phone rings, your modem answers it with the carrier signal. If the calling modem and your modem are compatible, the two modems "shake hands" and begin a communications session. You can then use PROCOMM PLUS to send and receive information to and from the remote computer.

If your modem's telephone line is sometimes used for voice communication and other times used for computer communication, however, you cannot always leave the modem in Auto Answer mode. With your modem in this mode, anyone who calls—modem or not—gets an earful of the modem's carrier signal. You should therefore place your computer's modem in Auto Answer mode only when you expect another modem to be calling your number. You usually need the operator of the calling computer to inform you so that you can place your computer's modem in Auto Answer mode. This requirement often may mean making two telephone calls: one voice call followed by a modem call. Take a look, however, at the section "Connecting during a Voice Call," in this chapter.

Although you can use the initialization command to set up your modem for Auto Answer mode, PROCOMM PLUS provides two ways to place your modem *temporarily* in Auto Answer mode. You can either invoke the Host mode or use the Alt-Y (Auto Answer) command.

To invoke PROCOMM PLUS Host mode, press Alt-Q (Host Mode) from the Terminal mode screen. PROCOMM PLUS sends the proper command to your modem to place it in Auto Answer mode. Remote users can then use their computers to call your computer and use the PROCOMM PLUS Host mode as a small computer bulletin board. PROCOMM PLUS Host mode provides file uploading and downloading as well as a rudimentary electronic mail system. Turn to Chapter 9, "Using Host Mode," for instructions on when and how to make the most of PROCOMM PLUS Host mode. When you want to return to Terminal mode, just press Esc (Exit). PROCOMM PLUS sends the initialization command again to turn off Auto Answer mode.

If you expect a call from another computer but don't want to use Host mode, place your modem in Auto Answer mode by pressing Alt-Y (Auto Answer). PROCOMM PLUS sends the appropriate command to your modem and displays the command on-screen. The Auto Answer command for Hayes-compatible modems is as follows:

ATS0 = 1

This command is preceded on your screen by + + +, which says to the modem, "Go into command mode; I'm about to send you a command." The command ATS0 = 1 then instructs the modem to go into Answer mode and be ready to answer the telephone after the first ring. You can tell that your modem has received and executed the command when the modem displays OK on the screen beneath the Auto Answer command (see fig. 4.4). Refer to "Setting Modem Options," in Chapter 8, for a discussion of how to instruct the modem to wait for more rings before answering the line.

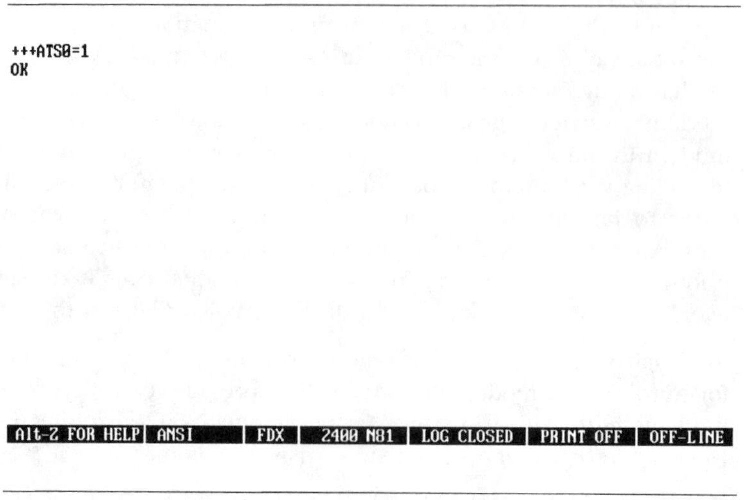

```
+++ATS0=1
OK
```

Fig. 4.4.

The command that instructs the modem to answer the next telephone call on the first ring.

| Alt-Z FOR HELP | ANSI | | FDX | 2400 N81 | LOG CLOSED | PRINT OFF | OFF-LINE |

To turn off Auto Answer mode, type the following command, starting at the left margin of the Terminal mode screen:

   ATS0 = 0

Then press Enter. The modem sends the response OK to the screen, letting you know that it understands your command and has turned off Auto Answer. The modem will no longer answer the phone.

TIP When two modems are communicating, one is in *Originate mode* and the other is in *Answer mode*. These modes refer to the technicalities of how the modems manage to send and receive data at the same time, once they are connected. *Auto Answer mode*, on the other hand, refers to the capability of the modem to answer automatically an incoming telephone call. When a modem is in Auto Answer mode, the modem is ready to answer a call but is not yet in Answer mode. Once the modem answers a call from another modem, the answering modem communicates with the calling modem and does so in Answer mode. The calling modem communicates in Originate mode. Refer to "Connecting during a Voice Call" for an explanation of how to switch your modem to Answer mode without first using Auto Answer mode.

Connecting during a Voice Call

Occasionally you may find yourself talking on the phone to an associate when the two of you decide that you want to connect your computers. You may want to send the other individual a copy of the spreadsheet file you were just discussing, for example.

Ideally, each of you would have a second telephone line, separate from the one on which you are currently talking. If so, you can use the procedures described in "Initiating a Call" and "Receiving a Call," in this chapter, to make the connection.

More typically, however, you each have your modem connected to the same line as your telephone. You can, of course, hang up and use the modems to place and answer the call again; but making the connection without hanging up would be more convenient (and more economical if the call is a toll call).

To connect computers over the same phone line as your voice call (when using Hayes-compatible modems), each of you should first start PROCOMM PLUS and display the Terminal mode screen. Make sure that the line settings (transmission speed, parity, and so on) are the same at both ends. Decide which one of your computers will act as the *originate* computer. The other will act as the *answer* computer (it doesn't matter which one you choose). The operator of the answer computer then should type the following command, starting at the left side of the Terminal mode screen and pressing Enter after the command:

ATA

This command places the modem in Answer mode and forces the modem to begin transmitting the carrier signal.

In the meantime, the operator of the originate computer should type the following command, starting at the left side of the Terminal mode screen and pressing Enter after the command:

ATD

This command places the modem in Originate mode. The modem then "listens" for a carrier signal from an Answer mode modem.

After each operator enters the proper command, and as soon as the answer modem emits its carrier signal, both operators should immediately hang up the telephones. The modems keep the lines open and begin to communicate.

Using a Direct Connection

When two computers are physically close enough to be connected by a 50-foot (or shorter) cable, you don't have to use a modem at all. Instead, connect the computers by a *null modem cable*. (Refer to Appendix B for information on building a null modem cable.) Connect the cable to the serial port you designated when you installed PROCOMM PLUS. If you need to use a different serial port, refer to "Assigning COM Port and Line Settings," in Chapter 8.

With a null modem cable, the communications connection is established as soon as you start PROCOMM PLUS. You can tell that the two computers are properly connected if you can type in the Terminal mode screen of one computer and see the results in the other. If you see nothing or you see characters different from what you are typing, make sure that you have all communications parameters properly set.

Conversing

Once you are connected to another computer, you have a number of ways to send and receive information. At some point in nearly every communications session, you send information by typing it at the keyboard, and you receive information by reading on your screen information typed by the operator or program at the other end. In essence, you are carrying on the computer equivalent of a conversation. When you are using PROCOMM PLUS to "converse" in this manner, you need to understand three PROCOMM PLUS features: *Duplex mode toggle*, *carriage return/ line feed toggle*, and *Chat mode*. The discussions that follow describe these features.

Toggling Duplex

Duplex is one of those computer terms that through continual usage has acquired a meaning that is a bit different from its original definition. When used most accurately, duplex refers to the capability of a modem to send and receive information at the same time. A modem that is capable of sending data and receiving data simultaneously is referred to as a *full-duplex* modem. *Half-duplex* modems, on the other hand, operate like a one-lane bridge. When receiving data, a half-duplex modem cannot send data. When sending data, the modem cannot receive data. The modem continually switches from send to receive and from receive to send in order to communicate with another modem—much as a CB radio does.

When your computer is connected to a host computer (such as an on-line service or a bulletin board) through a full-duplex modem, typically the computer at the other end "echoes" back to your screen any character you type on your keyboard. This echo provides a crude but effective way to confirm that the other computer received the character you typed. Consequently, when the host computer echoes characters to your screen, PROCOMM PLUS refers to the mode as *Full-duplex mode.*

Sometimes the computer on the other end does not echo characters to your screen. PROCOMM PLUS refers to this mode as *Half-duplex* mode. When your computer is connected to another computer that is using Half-duplex mode, PROCOMM PLUS must provide the "echo" in order for you to see what you are typing.

PROCOMM PLUS shows the current duplex mode setting in the status line at the bottom of the Terminal mode screen. The characters FDX in the status line indicate that PROCOMM PLUS is in Full-duplex mode, as shown in figure 4.5. The characters HDX in the status line mean that PROCOMM PLUS is in Half-duplex mode. As with other communications settings in PROCOMM PLUS, the Installation Utility (see Appendix A) establishes a default duplex mode setting that you can change with the Setup Utility (see Chapter 8). The default setting is overridden by a different setting in a dialing directory entry (see Chapter 3).

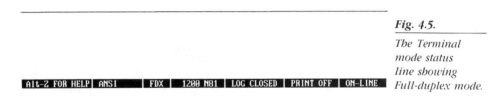

Fig. 4.5.

The Terminal mode status line showing Full-duplex mode.

Just as other communications settings must match at both ends of the PC communications line, the duplex mode of your computer's modem must match the mode of the computer on the other end. Otherwise, one of two problems occurs:

- When the computer on the other end is echoing the characters you send back to your screen, but you have set PROCOMM PLUS to half duplex, you see double characters like:

    HHeelllloo

    To remedy this problem, you must switch PROCOMM PLUS to Full-duplex mode by pressing Alt-E (Duplex Toggle).

- If the other computer is not echoing characters you transmit back to your screen, but you have set PROCOMM PLUS to Full-duplex mode, you cannot see any characters at all as you type. Again, the remedy is to press Alt-E (Duplex Toggle), to switch your modem to Half-duplex mode.

Toggling Carriage Return/Line Feed

Most people are familiar with the way an electric typewriter works. When you finish typing a line of text, you press the Return key. The typewriter advances the paper one line and moves the typing element to the beginning of this new line. Although you press a single key, the results are two distinct operations. Advancing the paper by one line is often referred to as *line feed*, and repositioning the typing element to the beginning of the line is called *carriage return* (a term carried over from mechanical typewriters where the carriage moves and the typing element remains stationary).

For you to be able to read more than one line of text on your computer's screen, something analogous to the typewriter's line feed and carriage return must occur at the end of each line. Indeed, the same terms are used to describe these operations in the context of your computer's screen.

If PROCOMM PLUS receives characters in a continuous stream with no carriage return at all (either from you or from the remote computer), the program inserts line feeds and carriage returns on its own. If you type a line of characters at the Terminal mode screen or receive a line of characters transmitted by another computer, PROCOMM PLUS displays the characters one-by-one across the screen, starting in the uppermost line in the Terminal mode screen. When PROCOMM PLUS receives the 80th character or space, which fills up the line (the screen is 80 characters wide), the program takes care of the chore of moving the cursor down one line— line feed—and returning the cursor to the left side of the screen— carriage return—before placing the next character or space on the screen.

For example, someone might type a message to you and fail to press Enter before reaching the edge of the screen. Figure 4.6 shows the Terminal mode screen after a remote user typed a message without pressing the Enter key. The word Hope is split over two lines, but PROCOMM PLUS does automatically insert a line feed and carriage return.

More typically, however, the other user typing a message does press the Enter key at the end of each line of text, much as if the individual were using an electric typewriter. Although this method prevents words from

```
Hi, I've just loaded my new copy of PROCOMM PLUS and wanted to try it out. I Hop
e you don't mind giving up your lunch hour to help me.
```

Fig. 4.6.

*A message typed
by a remote
user who didn't
press Enter.*

```
Alt-Z FOR HELP  TTY        FDX    1200 N81  LOG CLOSED  PRINT OFF  ON-LINE
```

being broken, it raises another issue: When PROCOMM PLUS receives (from the keyboard or from a remote computer) a carriage-return character, should the program translate the character as a carriage return only (CR) or as a carriage return and line feed (CR/LF)?

As with the Duplex mode setting described in the preceding section, your answer to this question depends on how the computer on the other end is operating. The discussions that follow will help you decide whether PROCOMM PLUS should add a line feed and show you how to get the program to do so.

Most host computers, such as on-line services and bulletin boards (including PROCOMM PLUS Host mode), add a line feed to each carriage return. If PROCOMM PLUS also adds a line feed, everything on your screen is double-spaced. For this reason, when you first install PROCOMM PLUS, the default CR Translation setting is CR rather than CR/LF, as shown in figure 4.7. (Refer to "Setting Terminal Options," in Chapter 8, "Tailoring PROCOMM PLUS," for information on changing this setting with the Setup Utility.)

When you plan to connect to a PC that is not operating as a bulletin board, set CR Translation to CR/LF. You can change the setting by using the Setup Utility, but because you will probably connect to on-line services and bulletin boards more often than to a PC, an easier method is to leave the default CR Translation set to CR, and just press Alt-F3 (CR-CR/LF Toggle). You can easily tell when you need to change the setting. When-

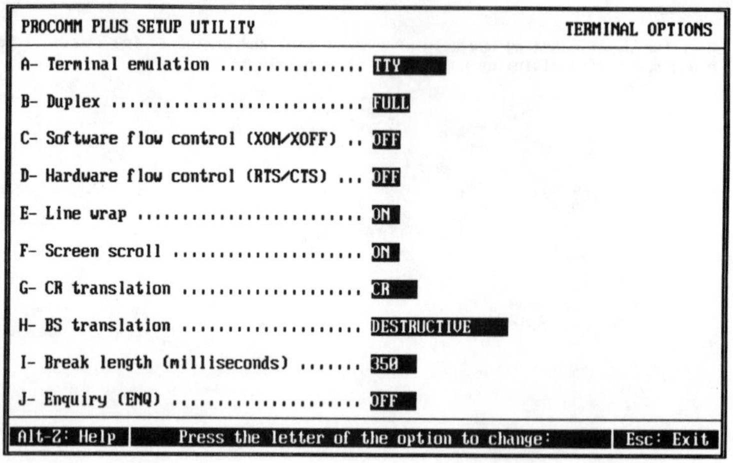

```
PROCOMM PLUS SETUP UTILITY                              TERMINAL OPTIONS

A- Terminal emulation ................ TTY

B- Duplex .......................... FULL

C- Software flow control (XON/XOFF) .. OFF

D- Hardware flow control (RTS/CTS) ... OFF

E- Line wrap ......................... ON

F- Screen scroll ..................... ON

G- CR translation .................... CR

H- BS translation .................... DESTRUCTIVE

I- Break length (milliseconds) ....... 350

J- Enquiry (ENQ) ..................... OFF

Alt-Z: Help  |  Press the letter of the option to change:  |  Esc: Exit
```

Fig. 4.7.

The Terminal Options screen in the PROCOMM PLUS Setup Utility with CR Translation set to CR.

ever all lines of characters you type or receive from another computer overwrite the preceding lines, you need to toggle CR translation to CR/LF.

If for some reason you change the default CR Translation setting to CR/LF and notice that all text on the screen is double-spaced, press Alt-F3 (CR-CR/LF Toggle). PROCOMM PLUS briefly displays the message NO LF AFTER CR in the status line and then translates carriage-return characters as just CR and not as CR/LF.

When you use the PROCOMM PLUS Chat mode, the CR Translation setting doesn't matter. PROCOMM PLUS displays incoming information and text you type without extra lines and without overwriting lines.

Using Chat Mode

When you carry on a two-way conversation with the operator of another computer, you may have difficulty separating what you are typing from what the person at the other end is typing. If you both type at the same time, the letters all run together on your screen, resulting in unintelligible alphabet soup. To help you solve this problem, PROCOMM PLUS provides its *Chat mode*.

To start Chat mode, press Alt-O (Chat Mode) from the Terminal mode screen. PROCOMM PLUS clears the screen of all data and displays the Chat mode screen, shown in figure 4.8. The Chat mode screen is split vertically into two sections. The top section is labeled REMOTE, and the bottom portion is labeled LOCAL.

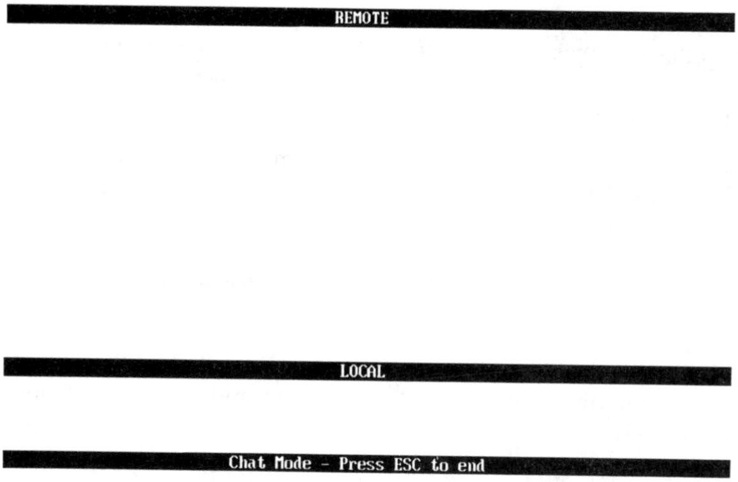

```
████████████████████████      REMOTE      ████████████████████████
```

Fig. 4.8.

The Chat mode screen.

```
████████████████████████       LOCAL       ████████████████████████
```

```
████████████████████   Chat Mode - Press ESC to end   ████████████████████
```

The REMOTE section of the Chat mode screen contains information received from the computer at the other end of the line, the *remote* computer. You see only the remote user's side of the conversation in this portion of the screen, which can display up to 18 lines of information at a time.

Incoming text starts at the bottom of the REMOTE section and, as more text comes in, scrolls up the screen. When the section fills and still another line is received, the information at the top scrolls off the screen and out of sight. (Refer to "Redisplaying Information," in this chapter, for an explanation of how to review the information after it scrolls off the screen.)

Your side of the conversation appears in the LOCAL section of the Chat mode screen.

In addition to the convenience of being able to distinguish easily your input from the remote user's input, you will note two other significant differences between Chat mode and typing at the normal Terminal mode screen:

- While you are using Chat mode, you do not have to worry about whether PROCOMM PLUS is in Full-duplex or Half-duplex mode or whether the remote computer is or is not in Echoplex mode. PROCOMM PLUS always displays exactly what you type. The CR Translation setting (CR or CR/LF) doesn't matter either.

- PROCOMM PLUS sends a line of text only after you press Enter. Up until the time you press Enter, you can read and edit the line without the remote user ever seeing it. This feature eliminates the embarrassment of knowing that someone else is watching as you make typing errors. You have an opportunity to review each line before irretrievably sending it on to the remote computer.

To edit a line in the LOCAL section of the Chat mode screen (when you have not yet pressed Enter), use the Backspace key to erase characters through the error. Each time you press Backspace, PROCOMM PLUS erases the character to the left of the cursor. When you have erased through the error, type the correct information, and press Enter to send the line.

When you have finished using Chat mode, press Esc (End). PROCOMM PLUS scrolls everything off the top of the screen and displays a blank Terminal mode screen.

You may at first be tempted to use Chat mode all the time, but you will quickly discover that few of the keyboard commands available from the Terminal mode screen have any effect from the Chat mode screen. Only Alt-F6 (Redisplay) and keyboard macros (Alt-0 to Alt-9) operate in Chat mode. The Chat mode screen is best reserved for use when you're conversing directly with someone who is typing messages to you on the other end of the line. For example, several on-line services offer a capability to chat with other users who are connected to the system. These on-line service features are sometimes called *CB simulators*.

Managing Incoming Information

When you consider using PROCOMM PLUS and PC communications to gather information, you need to be aware of three basic properties of Terminal mode:

- *The screen size is limited.* Although the amount of information you can receive in Terminal mode is unlimited, you never can see more than 80 columns by 24 rows at once. This area amounts to 1,920 characters and spaces, a little more than can fit on a normal sheet of paper double-spaced with one-inch margins.

- *Information scrolls by quickly.* If you are using a modem that operates at a transmission speed of 1200 bps, information can flow in at a rate of approximately 1,000 six-character words per minute. Only speed readers can read that fast. A 2400-bps modem can receive twice as many words per minute.

Obviously, at that rate you will never be able to read all the information as it scrolls across your screen.

- *Information is not saved automatically.* PROCOMM PLUS does not automatically save information that another computer sends to your Terminal mode screen. Once the text flows off the top of your screen, the information is gone unless you take some affirmative step to save it.

These properties are not unique to PROCOMM PLUS. They are true of any PC communications program. PROCOMM PLUS, however, provides several features that enable you to overcome these limitations easily. The discussions that follow explain how best to use these features to manage incoming information.

Pausing the Screen

Most bulletin boards and on-line services pause at the end of each screenful of information and require you to press Enter or some other key to indicate that you are ready to see the next screen. But you may occasionally run into a situation in which you begin to view a lengthy bulletin or file that scrolls so fast up your screen that you cannot read it. In this situation, you can use the Screen Pause feature to stop the scrolling long enough to read the pertinent information.

When in Terminal mode, the best way to stop the scrolling of information up your screen temporarily is to press Alt-N (Screen Pause). PROCOMM PLUS displays in the status line the following message:

```
Screen paused.  Press any key to continue...
```

Until you press a key, PROCOMM PLUS prevents any more text from displaying. If the remote computer continues to send information, however, PROCOMM PLUS continues to receive it in the background. When you press a key, PROCOMM PLUS displays any information that was received while the screen was paused, followed by any new information that continues to flow in.

Many remote systems, including most on-line services and bulletin boards, support a flow-control method known as *XON/XOFF*. These systems stop transmitting information when they receive an ASCII (American Standard Code for Information Interchange) character referred to as *XOFF*. This character is the same one that you generate by pressing Ctrl-S on your keyboard. The remote system continues its transmission after receiving another ASCII character known as *XON*. The XON character is the same character generated when you press Ctrl-Q.

When you pause the scrolling of incoming information by pressing Alt-N (Screen Pause), PROCOMM PLUS also sends the XOFF character to the other system. If the remote system supports XON/XOFF flow control, the system suspends transmission of data. Once you press a key to resume the display of incoming information, PROCOMM PLUS also sends XON to the remote system so that it resumes transmission.

TIP One of the Terminal Options screen settings that you can modify by using the Setup Utility (see Chapter 8) is software flow control (XON/XOFF) (again see fig. 4.7). But PROCOMM PLUS sends XON and XOFF when you use the Alt-N (Screen Pause) feature described in this section, regardless of this Terminal Options setting. The XON/XOFF setting determines whether PROCOMM PLUS and PROCOMM PLUS Host mode automatically use XON/XOFF flow control when connected to another computer that supports it.

The Screen Pause command does not work in the Chat mode screen. But when you are using Chat mode while communicating with a system that supports XON/XOFF, you can press Ctrl-S to suspend incoming data and Ctrl-Q to resume incoming data.

Redisplaying Information

The preceding section explains how to pause the screen to read information that is still displayed on-screen. But PROCOMM PLUS's Redisplay feature enables you to review data even after it has scrolled off the screen.

Throughout each PROCOMM PLUS session, PROCOMM PLUS continually saves the information that scrolls across your screen. The data is saved into a portion of memory referred to as the *redisplay buffer*. This buffer is a temporary storage area that is emptied every time you exit PROCOMM PLUS and return to DOS. The redisplay buffer is of limited size, capable of holding up to 10,000 characters. When the redisplay buffer fills, PROCOMM PLUS discards the oldest data, in favor of data more recently displayed on your screen.

To see the data stored in the redisplay buffer, press Alt-F6 (Redisplay). PROCOMM PLUS displays the last 22 lines that scrolled off the Terminal mode (or Chat mode) screen. Figure 4.9 shows an example.

All 10,000 characters potentially stored in the redisplay buffer obviously cannot be shown at once. In fact, more than 5 filled screens can be stored in the buffer. And because many lines may contain only a few characters, the redisplay buffer often holds as many as the last 16 or 17 screens that

```
City            State Net   AC    Access #
-----------------------------------------------
Burlington       NC   CS    919   584-3762
Davidson         NC   CS    919   725-1550
Durham           NC   CS    919   682-6239
Durham           NC   CS M  919   683-1866
Greensboro       NC   CS M  919   272-4994
Greensboro       NC   CS    919   373-1635
Raleigh          NC   CS M  919   876-8150
Raleigh          NC   CS    919   876-9095
Research TriPrk  NC   CS    919   682-6239
Wilmington       NC   CS M  919   392-4700
Winston-Salem    NC   CS M  919   723-9471
Winston-Salem    NC   CS    919   725-1550Last Page !off

Thank you for using CompuServe!

Off at 18:42 EDT 21-Aug-89
Connect time = 0:05
+E⁄3↓2u(0;&z6~JE<D!!$"<9sd&jo-I=dX0
NO CARRIER
```

```
PgUp/PgDn:Pages  ↑⁄↓:Scrolls  Home:Top  End:Bottom  F:Find  W:Write  ESC:Exit
```

Fig. 4.9.

The Redisplay Buffer screen.

scrolled off the top of the Terminal mode screen. Several commands, however, can help you move quickly around the buffer. The last line of the Redisplay Buffer screen contains a list of available keyboard commands. These commands are described in table 4.1.

Table 4.1
Redisplay Buffer Screen Keyboard Commands

Command	Name	Effect
PgUp	Pages	Scrolls screen up one page/screen (22 lines)
PgDn	Pages	Scrolls screen down one page/screen (22 lines)
↑	Scrolls	Scrolls screen up one line
↓	Scrolls	Scrolls screen down one line
Home	Top	Jumps screen to top of buffer
End	Bottom	Jumps screen to bottom of buffer
F or /	Find	Finds specific character string
W	Write	Writes buffer information to file or printer
Esc	Exit	Returns to Terminal mode screen

Searching for a Specific String

You may sometimes be looking for a particular word or phrase that you recall having seen on a previous screen. Before you begin searching for a particular character string, press Home (Top) to move the screen to the top of the redisplay buffer. Then press **F** (Find) or press the slash key (/). PROCOMM PLUS displays the following prompt in the bottom line of the Redisplay Buffer screen:

    String to look for:

Type the string of characters you want to find. You can include any displayable characters as well as spaces and can use either upper- or lower-case letters. After typing the string, press Enter. PROCOMM PLUS searches down through the buffer, looking for the string in any portion of a word. For example, searching for the string *com* finds all the following words:

    becoming
    Welcome
    compiled
    PROCOMM PLUS
    COMPOSE
    CompuServe

When PROCOMM PLUS locates the text you requested, the program displays the line that contains the matching word or phrase as the top line on the screen (unless the match is found in the last 22 lines at the bottom of the redisplay buffer, in which case the program simply displays the last 22 lines of the buffer).

Saving the Buffer to a File

Occasionally you may decide that you need to keep a copy of the redisplay buffer. For example, the buffer might contain a list of telephone numbers you want to keep. To save the redisplay buffer to disk, press **W** (Write). PROCOMM PLUS displays the following message in the last line of the Redisplay Buffer screen:

    File name to save to:

Type a valid DOS file name, such as *phone#.txt*, and press Enter (see fig. 4.10). PROCOMM PLUS saves the contents of the entire redisplay buffer to disk, using the file name you specified. The result is a plain ASCII file, which you can read and edit with any editor or word processor that can handle ASCII files. You can also press Alt-V (View a File) directly from the Terminal mode screen to read the saved redisplay buffer.

```
City           State Net  AC   Access #
-----------------------------------------------
Burlington      NC   CS   919  584-3762
Davidson        NC   CS   919  725-1550
Durham          NC   CS   919  682-6239
Durham          NC   CS M 919  683-1866
Greensboro      NC   CS M 919  272-4994
Greensboro      NC   CS   919  373-1635
Raleigh         NC   CS M 919  876-8150
Raleigh         NC   CS   919  876-9095
Research TriPrk NC   CS   919  682-6239
Wilmington      NC   CS M 919  392-4700
Winston-Salem   NC   CS M 919  723-9471
Winston-Salem   NC   CS   919  725-1550Last Page !off

Thank you for using CompuServe!

Off at 18:42 EDT 21-Aug-89
Connect time = 0:05
+E×/3↓2υ(0:&z6~JE<D!!$"<9sd&Jo-I=dX0
NO CARRIER
```

File name to save to: phone#.txt

Fig. 4.10.

Saving the redisplay buffer to a disk file named PHONE#.TXT.

Printing the Redisplay Buffer

You can also easily print the contents of the redisplay buffer. Again, use the **W** (Write) command, but type *prn* as the file name. This special name is recognized by DOS to mean the first parallel port. Make sure that your printer is turned on, loaded with paper, and connected to the parallel port. Press Enter, and PROCOMM PLUS sends the contents of the buffer to your printer.

TIP | To print the redisplay buffer, you can also use the name *lpt1* in place of *prn*. If your printer is connected to the second parallel port on your computer, use the name *lpt2*.

To return to the Terminal mode screen (or to the Chat mode screen if you were in Chat mode when you pressed Alt-F6), press Esc (Exit).

Taking a Snapshot

From time to time during a communications session, you may decide that you want a copy of a single screen. As described elsewhere in this chapter, PROCOMM PLUS offers several ways to save much more than one screen, but many times one screen is all you need. The PROCOMM PLUS Screen Snapshot feature enables you to save a single screen at a time.

To save to disk the data currently displayed on the Terminal mode screen, including the contents of the status line, press Alt-G (Screen Snapshot).

PROCOMM PLUS briefly displays the message SNAPSHOT in the leftmost section of the status line and saves the screen to disk as an ASCII file, using a file name specified in the File/Path Options screen of the Setup Utility. The File/Path Options screen is shown in figure 4.11. As you can see, the default name for screen snapshot files is PCPLUS.SCR. (See Chapter 8, "Tailoring PROCOMM PLUS," for more information.)

Fig. 4.11.

The File/Path Options screen of the Setup Utility.

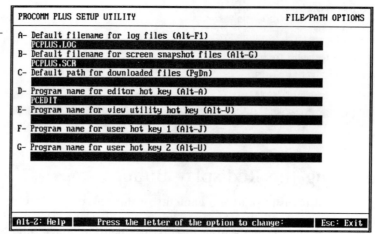

Each time you save a screen in this fashion, PROCOMM PLUS adds the screen's contents to the end of the snapshot file.

TIP | The Screen Snapshot feature is not available when you're viewing the redisplay buffer. To save the redisplay buffer to a file, use the **W** (Write) command instead (see "Saving the Buffer to a File").

Capturing the Session to Disk

Using the Alt-G (Screen Snapshot) command, you can save a screenful of information; and using the Alt-F6 (Redisplay) command, you can save up to 10,000 characters. PROCOMM PLUS, however, also provides the Log File feature, which enables you to capture any or all of a session directly to disk, without having to use either the Screen Snapshot or the Redisplay command.

Any time you want to begin capturing to a disk file everything that scrolls across the Terminal mode screen, press Alt-F1 (Log File On/Off).

PROCOMM PLUS opens a narrow window across the Terminal mode screen that contains the following prompt:

    Enter log filename, or CR for default:

This prompt means that you can specify a disk file to receive the captured data by typing a valid DOS file name and pressing Enter. (If you don't specify a different DOS directory, PROCOMM PLUS places the log file in the current directory.) Alternatively, you can simply press Enter without typing a file name, and PROCOMM PLUS uses the default log file name, which is established in the File/Path Options screen of the Setup Utility (again see fig. 4.11). When you first install PROCOMM PLUS, the default log file name is PCPLUS.LOG.

Once you press Enter at the log file name prompt, PROCOMM PLUS begins capturing to the specified disk file all characters that display on your Terminal mode screen. If the log file you are using already contains data, PROCOMM PLUS adds the new data to the end. To remind you that PROCOMM PLUS is capturing the session to disk, the program displays the message LOG OPEN in the status line.

TIP | The Log File feature does not capture information that is already on-screen when you press Alt-F1 (Log File On/Off). If you want to save the current screen of information to disk, press Alt-G (Screen Snapshot), as explained in the preceding section of this chapter.

Using the Log File feature is much like turning on your VCR to record a ball game or movie for viewing later. Even if you miss something the first time through, you have the chance to look at it again as often as you like.

Even while you're using the Log File feature, you can skip information that you don't want saved to the log file. Perhaps, for example, you have been reading your electronic mail and saving a copy to disk, but you are about to read a particularly sensitive piece of correspondence that you don't want saved.

To suspend the Log File feature temporarily, press Alt-F2 (Log File Pause). PROCOMM PLUS has already logged to disk any information that is currently on-screen, but the program does not save anything else that is added to the Terminal mode screen until you reactivate the log file. PROCOMM PLUS displays the message LOG ON HOLD in the status line to remind you that the Log File feature is suspended. When you want to resume saving the session to disk, press Alt-F2 (Log File Pause) again. PROCOMM PLUS redisplays the message LOG OPEN in the status line, indicating that all incoming and outgoing data is again being saved to the log file.

To toggle off the Log File feature, press Alt-F1 (Log File On/Off). The message LOG CLOSED appears in the status line.

TIP | During Chat mode, you cannot activate the Log File feature. You can, however, turn on the feature in the Terminal mode screen before entering Chat mode, and PROCOMM PLUS continues to save incoming and outgoing information to disk while you "chat." You must return to the Terminal mode screen before you can turn off the Log File feature. The Alt-F2 (Log File Pause) command has no effect in Chat mode.

Viewing a File

With PROCOMM PLUS's File View feature, you can display the information that you have saved to a disk file with the Redisplay feature, the Screen Snapshot feature, the Log File feature, or the call-logging file (PCPLUS.FON). You can also read any file that contains only ASCII characters.

To display the contents of a file created by PROCOMM PLUS or to display any other ASCII file on the disk, press Alt-V (View a File) from the Terminal mode screen. PROCOMM PLUS opens a window containing the prompt Please enter filename. Type the name of the disk file you want to view, and press Enter. Unless you specify a directory, PROCOMM PLUS looks for the file in the current DOS directory. If you mistype the file name or don't specify the correct DOS directory, PROCOMM PLUS beeps, briefly displays the message FILE NOT FOUND..., and then redisplays the Please enter filename prompt. Once PROCOMM PLUS locates the file you specify, the file displays in the File View screen, as shown in figure 4.12.

Whenever PROCOMM PLUS displays a file in the File View screen, you can scroll the screen, using the commands shown at the bottom of the screen. These commands and their effects are listed in table 4.2.

Table 4.2
File View Screen Keyboard Commands

Command	Name	Effect
Home	Top of File	Moves screen to top of file
PgUp	Previous Page	Scrolls screen up one page/screen
PgDn	Next Page	Scrolls screen down one page/screen
Esc	Exit	Returns to Terminal mode screen

```
    *** No mail waiting ***

 2 COMPOSE a new message
 3 UPLOAD a message
 4 USE a file from PER area

 5 ADDRESS Book
 6 SET options

 9 Send a CONGRESSgram ($)Enter choice !F█ █GO PHONE

CompuServe (FREE)PHONES

COMPUSERVE NUMBERS
 1 Search by area code
 2 Search by City and State
 3 List all CompuServe Numbers

ALL NETWORK NUMBERS
 4 Search by area code
 5 Search by City and State
 6 List all network numbers
```

Home: Top of file	PgUp: Previous page	PgDn: Next page	ESC: Exit

Fig. 4.12.

An ASCII file displayed in the File View screen.

When you have finished viewing the file, press Esc (Exit) to return to the Terminal mode screen.

Although the File View feature does give you the capability to view ASCII files, other file-viewing programs have more powerful features, such as line-by-line scrolling and text-searching capabilities. Refer to "Setting File and Path Options," in Chapter 8, for an explanation of how to access other such programs from PROCOMM PLUS.

TIP With the File View feature, you can *read* ASCII files but not *edit* them. PROCOMM PLUS does, however, enable you to access a text editor whenever you want to edit an ASCII file. For example, you may want to edit a short ASCII message you intend to send through an electronic mail service. Refer to Chapter 6, "Using the PROCOMM PLUS Editor," for an explanation of how to use the rudimentary text-editor program, PCEDIT, which is included with PROCOMM PLUS. That chapter also explains how to access other text editor programs from PROCOMM PLUS.

Sending Information to Your Printer

In addition to saving to disk any information displayed on-screen during a communications session, you can send the information to your printer.

Press Alt-L (Printer On/Off) to toggle on this feature. PROCOMM PLUS indicates that it is sending incoming and outgoing characters to the printer by displaying the message PRINT ON in the Terminal mode screen's status line. All characters that subsequently display on your screen are also printed on your printer.

TIP When you press Alt-L, PROCOMM PLUS sends information to the DOS device named PRN, which is the first connected parallel port on your computer (usually LPT1). If your printer is connected to a different parallel port, you cannot use this feature.

To turn off printing, press Alt-L (Printer On/Off) again. PROCOMM PLUS changes the message in the status line to PRINT OFF and stops sending information to the printer.

TIP PROCOMM PLUS's capability of printing a communications session is limited by the fact that no "hand shaking" or "flow control" occurs between PROCOMM PLUS and the printer. PROCOMM PLUS simply sends data out to the printer port, whether or not the printer is functioning or is ready to receive more data. If you want to use this feature, you should also use a *print-spooler* program, such as the DOS PRINT command. A number of other commercial and public-domain print-spooler programs are available. Without such a spooler program, your printer may not be able to keep up with the speed of incoming data. Make sure to load your print spooler before going into PROCOMM PLUS, not from the DOS Gateway.

Another method of printing your session is to use the PROCOMM PLUS Log File feature to save the session to disk, and then use the DOS COPY command to send the file to your printer. For example, if you save your on-line session to the default log file PCPLUS.LOG, use the following DOS command to print the file:

    COPY PCPLUS.LOG PRN

Clearing the Screen

During a communications session, you may occasionally want to clear everything off your screen. Perhaps you are about to receive important information from the remote computer, and you want to remove extraneous data from the screen. Or maybe you have just completed a session with one remote computer and want to clear the screen before calling the next computer.

To clear the Terminal mode screen, press Alt-C (Clear Screen). PROCOMM PLUS removes all information from the Terminal mode screen.

TIP | Even though pressing Alt-C (Clear Screen) erases all information from the screen, the command does not erase the screen contents from the redisplay buffer. Refer to "Redisplaying Information," in this chapter, for a description of how to review information that is no longer displayed on-screen.

When you use the Alt-C (Clear Screen) command, PROCOMM PLUS also resets screen colors to those established in the Terminal Color Options screen of the Setup Utility (refer to "Setting Terminal Options," in Chapter 8). Some remote computers may send codes to your computer that have the effect of temporarily changing screen colors. If you encounter this problem, you can use the Alt-C (Clear Screen) command to return the colors to normal.

Viewing Elapsed Time

Whether you are connected to an on-line service, over a toll telephone line, or both, you undoubtedly want to keep close tabs on the connection time. Most on-line information services—as well as the telephone company—typically charge you by the minute, so you need to be aware of just how long you have been connected. PROCOMM PLUS provides this information through its Elapsed Time command.

To determine the amount of time that has elapsed since you began the current communications session, press Alt-T (Elapsed Time). PROCOMM PLUS opens a window in the center of the Terminal mode screen, showing the current date, time, and the following message:

```
ELAPSED TIME THIS CALL: HH:MM:SS
```

HH is hours, *MM* is minutes, and *SS* is seconds. As long as the window is displayed, it continues to tick off the seconds. Press any key to remove this window from the screen.

Ending a Call

The last step in any communications session is to hang up the line. When you are connected to an on-line service or a bulletin board, you should issue the sign-off command that is appropriate for that system. The host computer usually drops the line first. You then see a few meaningless characters on your screen followed by the message NO CARRIER. This message from your modem means that it recognizes that the modem at the other end has stopped transmitting and tells you that therefore your modem is going to hang up the telephone line.

Sometimes, however, you have to disconnect the line from your end. For example, you may have connected to an on-line service like CompuServe through a public data network like Telenet. After you log off the on-line service, you are still connected to the network. From the Terminal mode screen, press Alt-H (Hang Up). PROCOMM PLUS briefly displays the message DISCONNECTING in the status line and sends the signal to your modem to hang up the telephone line.

TIP | You should make sure that the modem did in fact drop the telephone connection. Pick up a telephone that is connected to the same line as the modem and listen for a dial tone. If the line is still open, the modem didn't hang up the line. Refer to Appendix B for an explanation of how to install and set up a modem properly for use with PROCOMM PLUS to ensure that the modem hangs up the telephone line when instructed to do so.

Chapter Summary

This chapter has discussed the major PROCOMM PLUS commands used during a communications session. You have learned how to connect to another computer, how to converse with a person or computer program on the other end, and how to manage incoming and outgoing information. Finally, the chapter has explained what you need to do when you want to end the communications session. Turn now to Chapter 5, "Transferring Files," for a discussion of how to use PROCOMM PLUS's feature-rich file-transfer capability.

Transferring Files

This chapter explains how to use PROCOMM PLUS to send and receive electronic files to and from another computer—an operation referred to as a *file transfer*. In concept, transferring a file from one computer to another is nearly the same as copying a file from one disk drive to another. You are making an electronic duplicate, or copy, of the original file and then placing the copy in a different location. From the user's perspective, however, file transfer between computers is significantly more complicated than file transfer between disk drives. This chapter shows you how to perform this operation easily by using PROCOMM PLUS.

First, you learn the basic steps to take when sending a file from your computer to a remote computer (the computer on the other end of the telephone line). Then you learn what to do to receive a file that is transmitted by a remote computer. Next, the chapter discusses when and how to use the file-transfer protocols available in PROCOMM PLUS.

This chapter also discusses the PROCOMM PLUS commands that enable you to work with disk files, including commands to view a DOS directory, change the working directory, and use the PROCOMM PLUS DOS Gateway feature to access the operating system.

Sending Files

File transfer is a two-way street. PROCOMM PLUS enables you to send computer files to the remote computer and to receive files from the remote computer. The steps you take to send a file are similar to but not exactly the same as the steps necessary to receive a file. This section describes how to send one or more files to a remote computer.

TIP In PROCOMM PLUS, the phrases *send a file*, *upload a file*, and *transmit a file* mean the same thing. They refer to the act of transferring a computer file from your computer to a remote computer. Sometimes new users are confused by the term *upload*, thinking that it refers to uploading a file to your computer—which would be receiving a file. When used in PROCOMM PLUS, however, the phrase *upload a file* means to send a file to another computer. This usage is easiest to understand in the context of sending files to a host computer, such as a bulletin board. When you send a file to the bulletin board, you are uploading the file to the host. Receiving a file from the host is called *downloading a file*.

Before you can send a file by using PROCOMM PLUS, you must be connected to the remote computer, and you must be at the Terminal mode screen. If the remote computer is a PC that is not operating a bulletin board or running PROCOMM PLUS Host mode, you must inform the remote computer's operator that you are about to upload a file. For file transfer to occur, this operator must begin the appropriate download procedure soon after you execute your upload procedure.

When you are connected to a host system (an on-line service, such as CompuServe, or a PC bulletin board, for example), you inform the host program that you intend to upload a file by selecting an appropriate menu option or command. (The exact command depends on the host program with which you are communicating.) You usually have to type the name of the file you are going to upload to the host and indicate a file-transfer protocol. (File-transfer protocols are explained in the paragraphs that follow.) When the host instructs you to "Begin your transfer procedure," or words to that effect (see the last line of the Terminal mode screen in fig. 5.1), you begin the upload procedure.

To upload a file, press PgUp (Send Files). PROCOMM PLUS displays a small window containing a list of 16 *upload* protocols available for use when you upload a file, as shown in figure 5.1. This chapter refers to this window as the *Upload menu*.

A *protocol* is a mutually agreed-on set of rules. A *file-transfer protocol* is an agreed-on set of rules that control the flow of data between the two computers. These rules also often screen transmitted data for errors introduced by the transmission process. This error-screening process is usually referred to as *error checking*, or *error control*. Because not all telephone systems have been optimized for sending computer data, you cannot assume that the signal between your modem and the receiving modem will remain clear enough to achieve an error-free transmission. Even one lost bit can change critically the content of the information you are send-

```
Password: *****
    ┌─┤ Upload Protocols ├──────────────────────────────────────┐
    │                                                            │
    │  1) XMODEM      5) TELINK      9) WXMODEM   13) YMODEM-G BATCH │
    │  2) KERMIT      6) MODEM7     10) IMODEM    14) EXTERN 1    │
    │  3) YMODEM      7) SEALINK    11) YMODEM-G  15) EXTERN 2    │
    │  4) ASCII       8) COMPUSERVE B 12) YMODEM BATCH 16) EXTERN 3 │
  F)│                                                            │
  H)│ Your Selection:      (or press ENTER for XMODEM)           │
  R)└────────────────────────────────────────────────────────────┘

Your choice? U

A)scii  K)ermit  S)ealink  X)modem  Y)modem Batch
I)modem  T)elink  W)xmodem  G)Ymodem-G Batch
Your choice? X


File name? sales889.wk1
           !...................!...................!
Description: Sales figures for August, 1989.

Begin your  XMODEM  transfer procedure...
▐Alt-Z FOR HELP▐ VT102 ▐  FDX ▐ 2400 N81 ▐ LOG CLOSED ▐ PRINT OFF ▐ ON-LINE ▐
```

Fig. 5.1.

The Upload menu.

ing. Using a file-transfer protocol that performs error-checking ensures that the data received by the remote computer is the same as the data your computer sends. Refer to "File-Transfer Protocols," later in this chapter, for more information on each of the protocols listed in figure 5.1 and for details on how to decide which one to use.

An *upload protocol* is a file-transfer protocol that is available for your use when uploading files to another computer. Like most other communication parameters, the file-transfer protocol you choose must match that used by the remote computer.

When you press PgUp (Send Files) to display the Upload menu, PRO-COMM PLUS lists the *current* file-transfer protocol in the bottom line of this menu:

(or press ENTER for *current_protocol*)

The term *current_protocol* is the name of the current protocol. XMODEM is the current protocol when you start PROCOMM PLUS. After you call a remote system by using the dialing directory, the dialing directory entry establishes the current protocol (refer to Chapter 3, "Building Your Dialing Directory"). Otherwise, each time you select a file-transfer protocol during an upload or download operation, the selected protocol becomes the current protocol. When you want to use the current file-transfer protocol, press Enter. To use a different protocol, type its number (listed in the Upload menu), and press Enter.

> **TIP** Several file-transfer protocols available in PROCOMM PLUS let you specify multiple files for uploading. Refer to the discussions later in this chapter on the protocols KERMIT, YMODEM BATCH, TELINK, MODEM7, and YMODEM-G BATCH for more information on transmitting multiple files.

After you select an upload (file-transfer) protocol, PROCOMM PLUS removes the Upload menu from the screen and displays a window containing the following prompt:

    Please enter filename:

Type the name of the file you want to transmit to the remote computer, and press Enter. Include the full DOS path if the file is not in the current DOS directory. This chapter refers to this file as the *transfer file*.

Often, you type the file name of the transfer file on the screen just before you press PgUp (Send Files). For example, when you are uploading a file to a host computer, such as a bulletin board system, the host requests that you enter the file name and choose a file-transfer protocol. Then the host instructs you to begin your upload. When you press PgUp (Send Files), PROCOMM PLUS searches for the latest file name on-screen. If the program finds a valid file name (from one to eight characters followed by a period and then followed by from one to three characters, no spaces), PROCOMM PLUS places this name to the right of the prompt Please enter filename, as shown in figure 5.2. This feature of PROCOMM PLUS relieves you from having to type the file name twice. (Note: This feature can be turned off. See "Adjusting General Options," in Chapter 8.) If this name is not the name of the file you want to send, you can use the keys listed in table 5.1 to edit the transfer-file name. When the file name is correct, press Enter.

Table 5.1
Transfer-File Name Editing Keys

Key	Function
←	Moves the cursor one space to the left
→	Moves the cursor one space to the right
Home	Moves the cursor to the left end of name
End	Moves the cursor to the right end of name
Insert	Toggles Insert/Overtype modes
Del	Deletes the character at cursor
Backspace	Deletes the character to the left of cursor
Tab	Deletes the entire line
Ctrl-End	Deletes the characters from cursor to right end of file name

Password: *****

Fig. 5.2.

PROCOMM PLUS
reading the
file name from
the screen.

```
F)iles  U)pload  D)ownload
H)elp  T)ine  C)hat  G)oodbye
R)ead ┌─┤ Send XMODEM ├──────────────────────────────────┐
      │                                                  │
Your  │ Please enter filename: sales889.wk1              │
      │                                                  │
A)sci └──────────────────────────────────────────────────┘
I)modem  T)elink  W)xmodem  G)Ymodem-G Batch
Your choice? X

File name? sales889.wk1
         !.................!.................!
Description: Sales figures for August, 1989.

Begin your  XMODEM  transfer procedure...
Alt-Z FOR HELP│ VT102      │ FDX │ 2400 N81 │ LOG CLOSED │ PRINT OFF │ ON-LINE
```

TIP | When you are connected to another PC that is using PROCOMM PLUS, file transfer is easiest if your computer or the other PC uses the Host mode. Refer to Chapter 9, "Using Host Mode" for details on how Host mode handles file transfer.

After you start the upload procedure and the remote computer begins its download procedure, PROCOMM PLUS displays the following message at the bottom of the screen:

    File transfer in progress... Press ESC to abort

PROCOMM PLUS also displays a window on the right side of the screen, as shown in figure 5.3 (except when you are using the ASCII protocol—see "ASCII," later in this chapter). This chapter refers to this window as the *Progress* window.

The Progress window tracks a number of parameters that are of interest during file transfer:

- The first line in the Progress window is labeled PROTOCOL. It indicates the file-transfer protocol that PROCOMM PLUS is using.

- The line labeled FILE NAME shows the name of the transfer file.

- The third line, labeled FILE SIZE, lists the size of the transfer file in bytes.

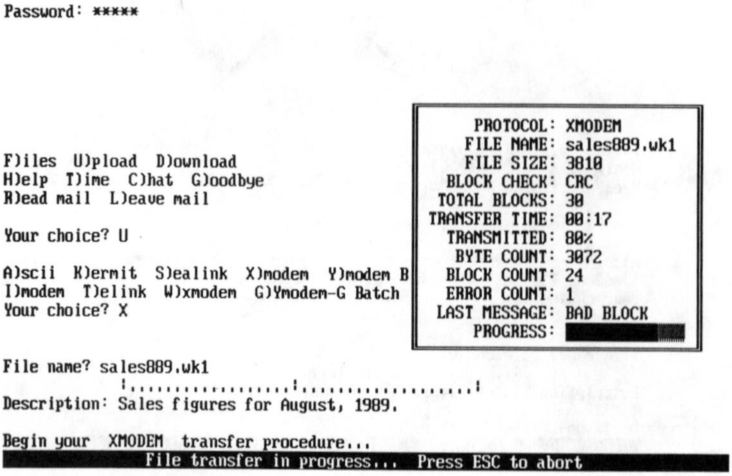

Password: *****

Fig. 5.3.

*The Progress
window during
an upload.*

```
                                        ┌─────────────────────────────┐
F)iles  U)pload  D)ownload              │ PROTOCOL: XMODEM            │
H)elp  T)ime  C)hat  G)oodbye           │ FILE NAME: sales889.wk1     │
R)ead mail  L)eave mail                 │ FILE SIZE: 3810             │
                                        │ BLOCK CHECK: CRC            │
Your choice? U                          │ TOTAL BLOCKS: 30            │
                                        │ TRANSFER TIME: 00:17        │
A)scii  K)ermit  S)ealink  X)modem  Y)modem B│ TRANSMITTED: 80%        │
I)modem  T)elink  W)xmodem  G)Ymodem-G Batch │ BYTE COUNT: 3072        │
Your choice? X                          │ BLOCK COUNT: 24            │
                                        │ ERROR COUNT: 1             │
File name? sales889.wk1                 │ LAST MESSAGE: BAD BLOCK    │
                                        │ PROGRESS: ▓▓▓▓▓▓▓▓░         │
                                        └─────────────────────────────┘
      !......................!....................!
Description: Sales figures for August, 1989.

Begin your  XMODEM  transfer procedure...
           File transfer in progress...  Press ESC to abort
```

- The fourth line, labeled BLOCK CHECK, refers to the method the file-transfer protocol uses to detect errors in the data. Figure 5.3 indicates that XMODEM is using CRC, which stands for Cyclical Redundancy Check (also known as CRC-16; see "XMODEM," later in this chapter). Some other protocols and some versions of XMODEM use another, less reliable, method called checksum.

- The fifth line of the Progress window shows the TOTAL BLOCKS of data you must send for a successful transfer of the file. As a part of the file-transfer procedure, each upload protocol except the ASCII protocol sends a set number of bytes of data, known as a block, and waits. When the remote computer receives the block, the computer checks for errors in the data. When an error is detected, the remote computer requests that the block be retransmitted. If no errors are found, the remote computer requests that the next block be sent. This process continues until PROCOMM PLUS has sent the entire file.

Not all protocols, however, send blocks of the same size. XMODEM, for example, sends 128 bytes of data in each block. Figure 5.3 shows the Progress window for a file transfer using the XMODEM protocol. A total of 30 blocks are required to send the file. (This number is calculated by taking the file size in bytes (3,810), and dividing it by the number of bytes in a block (128). Then you round the result up from 29.77 to 30.

The remote computer receives 30 blocks at 128 bytes each for a total of 3,840 bytes. XMODEM fills the extra 30 bytes at the end of the file with ^Z characters, which don't affect the meaning of the file.

- The sixth line, labeled TRANSFER TIME, indicates an approximate amount of time that the file transfer should take using the chosen upload protocol.

- The line labeled TRANSMITTED indicates as a percentage the portion of the file that has been successfully transmitted to the remote computer.

- The BYTE COUNT line denotes the number of bytes of data that have been transmitted.

- The BLOCK COUNT line shows the number of blocks of data that have been transmitted successfully. The block count is equal to the byte count divided by the number of bytes in a block. In figure 5.3, for example, the block count, 24, equals 3,072 divided by 128.

- The ERROR COUNT is incremented by one each time the file-transfer protocol detects an error. When XMODEM detects an error, the protocol requests that the block be re-sent. If XMODEM detects an error in the same block 15 times, the protocol aborts the transfer. Otherwise, if the block arrives without an error, XMODEM acknowledges that it has received the block correctly, and the next block is sent. The occurrence of many errors usually means that the modems are not communicating clearly, probably as a result of poor telephone transmission.

- When the file-transfer protocol detects an error, the protocol sends a message to your computer indicating the type of error. The most recent message generated by the file-transfer protocol is displayed in the LAST MESSAGE line. The error shown in figure 5.3, BAD BLOCK, indicates that a block of data did not pass the CRC test. Your computer therefore sends that block again.

- The last line in the Progress window, labeled PROGRESS, shows graphically how much of the file has been transferred. The horizontal progress bar grows from left to right as the file is transferred. When the TRANSMITTED line shows 100%, the progress bar extends to the right side of the Progress window.

As soon as the remote computer receives the entire file, your computer beeps, and the word COMPLETED flashes in place of the word PROTOCOL in the top line of the Progress window. After several seconds, PROCOMM PLUS removes the Progress window from the menu and returns to the normal Terminal mode screen.

When you have a lengthy file to send, you typically want to use the fastest transmission method possible. But because of the need for error checking, information sent as a file is transmitted more slowly than the same amount of information transferred as straight ASCII characters without error checking. Some file-transfer methods are faster than others. The pros and cons of the various protocols are covered in "File-Transfer Protocols," later in this chapter.

Since 1987, many modem manufacturers have offered relatively expensive modems that perform error-checking in hardware. These hardware-based error-checking, or error-control, modems not only do a better job of detecting errors caused by telephone line problems, but these modems are much faster than the error-checking methods used by software-based file-transfer protocols. The de facto standard as of 1989 was a series of protocols developed by Microcom, Inc., and known as MNP (Microcom Networking Protocol) Classes 1 through 4. However, an internationally recognized standard known as CCITT V.42 (refer to "New Standards," in Chapter 1) has become the official industry standard. To maintain compatibility with existing modems, V.42 incorporates MNP Classes 1 through 4 as a fall-back standard. For example, when a V.42 connects with a modem that doesn't support the V.42 error-control method but does support MNP Classes 1 through 4, the V.42 modem will use MNP.

In addition to the file-transfer protocol, a couple of other factors have a bearing on just how fast you can send a file to a remote computer. Not surprisingly, transmission speed of the modems at each end of the connection most directly affects the speed at which you can transmit files. Modems that have a maximum transmission speed of 2400 bps can potentially transmit files at twice the speed of a 1200-bps modem, and these faster modems are widely available at reasonable prices.

Another way you can dramatically increase effective file-transfer speed is through either of the two related techniques known as *file compression* and *data compression*. File compression is analogous to sitting on a loaf of bread. A file-compression program squeezes the "air" from a file so that it takes up less space. The significant information—the "bread"—is still there, but in a compressed form. After you transmit the compressed file, someone at the other end runs a program that decompresses the file, putting it back in the form that is expected by the word processing program or other type of program that created the file. File-compression programs can squeeze more "air" out of some files than others. The most popular programs of this type often can shrink a file by more than half and sometimes by as much as 75 percent or more.

TIP You can obtain excellent file-compression software from most PC user groups and PC bulletin boards. These shareware programs not only compress files, but also can create a *library*, or *archive*, file from multiple normal DOS files. For example, you can use one of these archiving programs to bundle four spreadsheet files and three word processing files into a single archive file. This archive file will be compressed so that its size in bytes is perhaps half the total size of the seven files that comprise the archive file. You then can transmit one file to a colleague rather than having to send seven separate files.

File compression and file archiving are particularly useful when you want to upload or download software that requires a number of different files in order to operate. For example, the user-supported (often referred to as shareware) program ProComm 2.4.3 normally is found on PC bulletin boards archived into two or three large files. After you download the files, you use an archiving program to decompress and unarchive the many files needed to run ProComm. Be aware, however, that at least two competing programs of this type are used widely to compress and archive files stored on bulletin boards. You have to use the correct program to be able to unarchive and decompress a file you download. You usually can tell by the file name extension of the bulletin board file which program was used to compress and archive the file (for example, .ARC versus .ZIP).

Data-compression schemes (as opposed to file-compression programs) do not require that you run a program to compress the file before you send it. Instead, the data is compressed by the modem *as* you send the file and decompressed at the other end by the receiving modem. The net effect is that the file is sent faster. Many modem manufacturers produce modems that support a data-compression scheme known as MNP Class 5, developed by Microcom, Inc. When this data-compression scheme is used with the MNP Classes 1 through 4 error-control protocols, a 2400-bps modem can achieve an effective transmission rate of more than 4800 bps. Hayes V-series modems achieve similar results using a different (incompatible) data-compression scheme. The CCITT V.42bis standard includes a third data-compression scheme for modems, which achieves even faster transmission speed results. Ratified in September 1989, CCITT V.42bis is likely to become the industry standard.

Receiving Files

PROCOMM PLUS enables you to send computer files to a remote computer and to receive computer files from a remote computer. The steps you take to receive a file are similar to but not exactly the same as the steps necessary to send a file. This section describes how to receive a file.

TIP | In PROCOMM PLUS, the phrases *receive a file* and *download a file* mean the same thing. They refer to the act of receiving a computer file sent to your computer by a remote computer. The term *download* is easiest to understand when used in the context of obtaining a file from a host computer, such as a bulletin board. Receiving a file from the host is called *downloading a file* from the host. Conversely, when you send a file to the bulletin board, you are uploading the file to the host.

Before you can receive a file by using PROCOMM PLUS, you must be connected to the remote computer, and you must be at the Terminal mode screen. If the remote computer is a PC that is not operating a bulletin board or running PROCOMM PLUS Host mode, you must inform the remote computer's operator that you are about to start downloading a file. The operator should begin the upload procedure first, and then you should immediately begin your download procedure. Most protocols are receiver-driven; that is, the sender will not actually begin uploading data until it receives a signal to start from the receiver. The only exception is the ASCII file-transfer protocol. The receiver should begin the download procedure before the sender begins uploading. Otherwise, the first few characters of the file will be lost.

When you are connected to a host system, such as an on-line service like CompuServe or a PC bulletin board, you usually inform the host program that you intend to download a file by selecting an appropriate menu option or command. You usually have to type the name of the file you are going to download from the host and select a file-transfer protocol (refer to "File-Transfer Protocols," later in this chapter). When the host instructs you to "Begin your transfer procedure," or words to that effect (see the last line of the Terminal mode screen in fig 5.4), you begin the download procedure.

To download a file, press PgDn (Receive Files). PROCOMM PLUS displays a small window containing a list of 16 download protocols available for use when you download a file, as shown in figure 5.4. This chapter refers to this window as the *Download menu*. PROCOMM PLUS also indicates in the top edge of the window around the Download menu the free space available on your working disk.

```
Is this correct (Y/N)? Y

Pa┌┤ Download Protocols - 1998848 bytes free ├─────────────────┐
  │  1) XMODEM      5) TELINK      9) WXMODEM      13) YMODEM-G BATCH
  │  2) KERMIT      6) MODEM7     10) IMODEM       14) EXTERN 1
  │  3) YMODEM      7) SEALINK    11) YMODEM-G     15) EXTERN 2
  │  4) ASCII       8) COMPUSERVE B 12) YMODEM BATCH 16) EXTERN 3
  │
  │  Your Selection:    (or press ENTER for XMODEM)
F)└─────────────────────────────────────────────────────────────┘
H)elp  T)ime  C)hat  G)oodbye
R)ead mail  L)eave mail

Your choice? D

A)scii  K)ermit  S)ealink  X)modem  Y)modem Batch
I)modem  T)elink  W)xmodem  G)Ymodem-G Batch
Your choice? Y


File name? modems.arc

Begin your  YMODEM BATCH  transfer procedure...
 Alt-Z FOR HELP  VT102    FDX   2400 N81   LOG CLOSED   PRINT OFF   ON-LINE
```

Fig. 5.4.

The Download menu.

A *download protocol* is a file-transfer protocol that is available for your use when you are downloading files from another computer. The file-transfer protocol you choose must match that used by the remote computer (the computer from which you are downloading the file). Refer to "File-Transfer Protocols," later in this chapter, for more information on each of the protocols listed in figure 5.4.

When you press PgDn (Receive Files) to display the Download menu, PROCOMM PLUS lists the current file-transfer protocol in the bottom line:

   (or press ENTER for *current_protocol*)

In this line, *current_protocol* is the name of the current protocol. XMODEM is the current protocol when you start PROCOMM PLUS. After you call a remote system by using the dialing directory, the dialing directory entry establishes the current protocol (see Chapter 3). Otherwise, each time you select a file-transfer protocol during an upload or download operation, the selected protocol becomes the current protocol. When you want to use the current file-transfer protocol, press Enter. To use a different protocol, type its number (listed in the Download menu), and press Enter.

When you select a download (file-transfer) protocol, PROCOMM PLUS removes the Download menu from the screen. If you select XMODEM, YMODEM, ASCII, WXMODEM, IMODEM, or YMODEM-G protocol, PROCOMM PLUS displays a window containing the following prompt:

   Please enter filename:

Type the name of the file you want to download from the remote computer (or any name you want the file to have on your system), and press Enter. This chapter refers to this file as the *transfer file*.

TIP | Several of the file-transfer protocols available in PROCOMM PLUS enable you to specify multiple files for downloading. Refer to the discussions later in this chapter on the protocols KERMIT, YMODEM BATCH, TELINK, MODEM7, and YMODEM-G BATCH for more information on downloading multiple files.

Often, you type the file name of the transfer file on-screen just before you press PgDn (Receive Files). For example, when you are downloading a file from a host computer, such as a bulletin board system, the host requests you to enter the file name and file-transfer protocol and then instructs you to begin your download. When you press PgDn (Receive Files), PROCOMM PLUS searches for the latest file name on-screen. If the program finds a valid file name, PROCOMM PLUS places this name to the right of the prompt Please enter filename. This feature of PROCOMM PLUS relieves you from having to type the file name twice. (Note: This feature can be turned off. See "Adjusting General Options," in Chapter 8.) If this name is not the name of the file you want to send, you can use the keys listed in table 5.1 to edit the transfer-file name. When the file name is correct, press Enter.

After you enter a file name, PROCOMM PLUS searches the download directory for a file with the same name (refer to "Setting File and Path Options," in Chapter 8, for an explanation on how to assign a download directory). If a file already exists with the name you specified, PROCOMM PLUS displays the following message:

    File already exists. Overwrite it? (Y/N)

Answer Yes to continue with the download procedure and to overwrite the existing file. Answer No if you don't want to replace the existing file with the new file. Unless you have chosen the ASCII protocol, you will be prompted again to Please enter filename. At this point, you may type a different name for the file to be named on your system so that the downloaded file does not overwrite the existing file. If you decide that you don't want the transfer file, simply press Esc to abort the download procedure. If you have chosen ASCII and you get the prompt File already exists. Overwrite it? (Y/N), answer No to cause incoming data to be appended to the end of the existing file.

TIP | When you select one of the three available external protocols, PROCOMM PLUS enables you to enter a file name and any other parameters needed by the external program. PROCOMM PLUS then passes these parameters to the external program when you press Enter. Refer to Appendix C for more information.

When you are using KERMIT, TELINK, MODEM7, SEALINK, COMPUSERVE B, YMODEM BATCH, or YMODEM-G BATCH protocol, you don't specify a file name for the transfer file. PROCOMM PLUS uses the file name you or the operator of the other computer specified to the remote computer.

TIP | Several of these protocols enable you to use *wild-card* characters to download multiple files with just one download command. Refer to the discussions of protocols, later in this chapter.

After you have started the download procedure and the remote computer has begun its upload procedure, PROCOMM PLUS displays the following message at the bottom of the screen:

    File transfer in progress... Press ESC to abort

PROCOMM PLUS also displays a window on the right side of the screen (except when you are using the ASCII protocol—see "ASCII," later in this chapter).

As soon as the your computer receives the entire file, your computer beeps, and the word COMPLETED flashes in place of the word PROTOCOL in the top line of the Progress window. After several seconds, PROCOMM PLUS removes the Progress window from the menu and returns to the normal Terminal mode screen.

File-Transfer Protocols

When you type a message in the Terminal mode or Chat mode, PROCOMM PLUS sends ASCII characters to the screen of the remote computer. The operator of the remote computer can detect transmission errors simply by reading the characters on-screen. If a character looks like hieroglyphics, it reflects a transmission error. Most computer files, however, are not stored entirely as ASCII characters. Computer programs, most word processing files, spreadsheet files, database files, and other types of computer files contain data that cannot be displayed on-screen as it is transmitted. These non-ASCII files are typically referred to as *binary* files (although ASCII is technically a binary—base 2—code). To send binary files, you must use a file-transfer protocol that can send data without displaying it to the screen, and you must take steps through software or hardware to detect data errors that may be caused by degradation of the telephone signal.

Each time you begin uploading or downloading files using PROCOMM PLUS, you can choose between 16 file-transfer protocols (including 3 optional external protocols). The following guidelines will help you decide which protocol to use:

- *Use matching file-transfer protocols.* You must use the same file-transfer protocol on both the sending and receiving ends of the file-transfer process. This requirement limits the number of protocols available to you during any particular communication session to the protocols PROCOMM PLUS and the remote program have in common.

- *Use ASCII protocol only for ASCII files.* When sending or receiving a file that contains only printable ASCII characters, you have the option of choosing ASCII protocol. This protocol sends data character-by-character, just as when you type a message at the keyboard, and can optionally display the data to the screen during transmission. The ASCII protocol, however, really does not perform error checking. You cannot use the ASCII protocol for binary files.

- *Send and receive the largest blocks.* For maximum transmission speed when sending binary files, use the file-transfer protocol that sends blocks of the largest size. In PROCOMM PLUS, the YMODEM, COMPUSERVE B (when using with CompuServe Quick B), YMODEM BATCH, YMODEM-G, and YMODEM-G BATCH protocols can send blocks of 1,024 bytes.

- *Send and receive multiple files.* Some PROCOMM PLUS file-transfer protocols enable you to send or receive multiple files with one command, often called a *batch transfer*. For example, you can send the files SALES889.WK1 and SALES989.WK1 in one upload procedure. When asked for the file name, you specify

      *.wk1

  PROCOMM PLUS then sends these files one at a time. The PROCOMM PLUS protocols that can send or receive multiple files are KERMIT, YMODEM BATCH, TELINK, MODEM7, and YMODEM-G BATCH.

- *Send and receive file characteristics.* Several file-transfer protocols available in PROCOMM PLUS also send with the file such file characteristics as file name, file size, and the date and time the file was last changed. When you are downloading a file, this information enables PROCOMM PLUS to make full use of the Progress window. The protocols that transmit at least one of these file characteristics include COMPUSERVE B, KERMIT, IMODEM, MODEM7, SEALINK, TELINK, YMODEM BATCH, and YMODEM G BATCH.

- *Use sliding windows on PDNs and over long-distance lines.* When you connect to a host computer over a public data network (PDN), such as Tymnet and Telenet (or any other packet-switching network), or over long-distance telephone lines that may go through satellite relays, most file-transfer protocols can be significantly slowed. PDNs and satellite relays often cause an appreciable increase in the time needed for the receiving computer to reply to each block sent. The sending computer can spend a great deal of time just waiting. The KERMIT, SEALINK, and WXMODEM protocols, however, take advantage of the full-duplex nature of your modem (refer to "Toggling Duplex," in Chapter 4 for a discussion of *duplex* and related concepts). Instead of waiting for a reply before sending another block, the protocols send blocks and at the same time watch for the reply to previous blocks. These protocols will send several blocks before requiring any reply.

- *Use appropriate protocols with error-control modems.* If you are using a modem that performs error-control in hardware, such as a modem that supports MNP Classes 1 through 4 or CCITT V.42, use a PROCOMM PLUS protocol that sends data without doing any error-checking in software. The protocols in this group are IMODEM, YMODEM-G, and YMODEM-G BATCH.

File-transfer protocols used in PC communications programs and on host computers are not always named in the same way as in PROCOMM PLUS. For example, the protocol called XMODEM in PROCOMM PLUS is sometimes called XMODEM/CRC by other communications programs. PROCOMM PLUS's YMODEM is often referred to as 1K-XMODEM, and PROCOMM PLUS's YMODEM BATCH is called YMODEM. Purists may argue for hours about the "correct" names for these protocols, but it is more practical just to familiarize yourself with the protocols implemented in PROCOMM PLUS so that you can recognize the one you need at any particular time. Table 5.2 lists the most important properties of the available protocols so that you can compare them at a glance. Each protocol is discussed in more detail in the sections that follow.

Table 5.2
PROCOMM PLUS File-Transfer Protocols

Protocol	Error-Checking	Block Size	Multiple Files	File Characteristics			Sliding Windows
				Name	Size	Date	
XMODEM	CRC/Checksum	128	No	No	No	No	No
KERMIT	CRC/Checksum	94 max	Yes	Yes	Yes	Yes	Yes
YMODEM	CRC	128/1024	No	No	No	No	No
ASCII	None		No	No	No	No	No
TELINK	CRC/Checksum	128	Yes	Yes	Yes	Yes	No
MODEM7	CRC/Checksum	128	Yes	Yes	No	No	No
SEALINK	CRC	128	Yes	Yes	Yes	No	Yes
COMPUSERVE B	CRC	512/1024	No	Yes	No	No	Yes
WXMODEM	CRC	128	No	No	No	No	Yes
IMODEM	None	128	No	Yes	No	No	No
YMODEM-G	None	128/1024	No	No	No	No	No
YMODEM BATCH	CRC	128/1024	Yes	Yes	Yes	No	No
YMODEM-G BATCH	None	128/1024	Yes	Yes	Yes	No	No

As you peruse the remainder of this chapter, you may wonder why there are so many file-transfer protocols yet so few differences among them. The reason lies in the fact that the majority of the communications software written for PCs has been developed by hobbyists for hobbyists. Commercial success has not been a significant factor in determining the viability of these public domain or user-supported programs. Many different programs have flourished, and no one program has dominated. Everyone seems to have his or her favorite bulletin board program and terminal program (such as PROCOMM PLUS and its older sibling, ProComm). Although the XMODEM file-transfer protocol is a common thread between virtually all the popular PC communications programs, its built-in limitations have led to the development of many enhanced versions of XMODEM.

In an effort to provide you with the capability to communicate with virtually anyone's favorite bulletin board or terminal program, PROCOMM PLUS gives you a list of choices. Table 5.2 and the discussions that follow will help you sort these protocols and decide which one will become your favorite.

ASCII

ASCII—American Standard Code for Information Interchange—forms the lowest common denominator between the countless programs that run on a PC and the many different types of computers that proliferate in our high-tech society. Virtually every word processing, spreadsheet, and database program that runs on a PC can create an ASCII file. Most likely, every computer with which you will ever need to communicate can handle ASCII characters. The primary reason, however, that you will need to know how to use the ASCII file-transfer protocol is that most electronic mail systems can display only ASCII characters.

Most PC programs are character-based (with the notable exception of programs that run under such graphics-based programs as Microsoft Windows and OS/2 Presentation Manager). Each character or symbol you see displayed on-screen is represented in memory by one of the codes referred to collectively as the *ASCII character set*. This chapter calls this group of codes the *IBM character* set.

Assume that you have created an ASCII file with a word processor or PCEDIT (refer to Chapter 6), and you want to transmit the file to an electronic mail system. Before you begin sending the file, you have to cause the remote computer to begin its ASCII receive procedure. On an electronic mail system, this point occurs when the system expects you to type your message. Issue the proper command to the host or bulletin board so that it is waiting for ASCII text to be sent. The remote computer will not be able to distinguish whether PROCOMM PLUS is sending the ASCII characters from a file or whether you are typing them at the keyboard.

When the remote system is ready for you to begin transmitting the file, follow the procedure described in "Sending Files," in this chapter, selecting ASCII as the file-transfer protocol. As PROCOMM PLUS transfers the file, the program does not show the Progress window, but instead displays the message ASCII FILE UPLOAD - PRESS ESC TO ABORT in the Terminal mode screen status line, as shown in figure 5.5. During the transmission, PROCOMM PLUS displays a line count at the left end of the status line. This count indicates the number of lines of characters that PROCOMM PLUS has successfully transmitted. The receiving computer does no error-checking on the transmitted characters.

When PROCOMM PLUS finishes sending the entire file, the computer beeps five times. To complete the transfer, you must remember to inform the remote computer that you are no longer sending ASCII characters and that you want to save the file. For example, if you are sending a message to an electronic mail system, you issue the command that means to save and send the message.

Fig. 5.5.

*Sending a file
using the ASCII
file-transfer
protocol.*

```
F)iles  U)pload  D)ownload
H)elp  T)ime  C)hat  G)oodbye
R)ead mail  L)eave mail

Your choice? U

A)scii  K)ermit  S)ealink  X)modem  Y)modem Batch
I)modem  T)elink  W)xmodem  G)Ymodem-G Batch
Your choice? A


File name? menu3.scr
          !.................!.................!
Description:

Begin your  ASCII  transfer procedure...
(Press <CR> twice to end)
 LINE: 14         ASCII FILE UPLOAD  -  PRESS ESC TO ABORT
```

Although the ASCII file-transfer protocol performs no error check, PROCOMM PLUS does enable you to control several aspects of an ASCII upload through the Setup Utility ASCII Transfer Options screen shown in figure 5.6. You also have the option of enabling XON/XOFF software flow control through the Terminal Options screen of the Setup Utility. Refer to "Setting ASCII Transfer Options" and "Setting Terminal Options," in Chapter 8, for discussions of these features.

Fig. 5.6.

*The Setup Utility
ASCII Transfer
Options screen.*

```
 PROCOMM PLUS SETUP UTILITY                   ASCII TRANSFER OPTIONS

 A- Echo locally ............... NO    K- Strip 8th bit ............... NO

 B- Expand blank lines .......... YES

 C- Expand tabs ................. YES

 D- Character pacing (millisec).. 15

 E- Line pacing (1/10 sec)....... 10

 F- Pace character .............. 0

 G- CR translation (upload) ..... NONE

 H- LF translation (upload) ..... STRIP

 I- CR translation (download) ... NONE

 J- LF translation (download) ... NONE

 Alt-Z: Help  |  Press the letter of the option to change:  | Esc: Exit
```

When you want to use the ASCII protocol to receive a file, you must initiate the download procedure before the remote computer begins sending the file.

As PROCOMM PLUS downloads the ASCII file, the program does not display the Progress window, but displays the message ASCII FILE DOWNLOAD – PRESS ESC TO END in the Terminal mode screen status line, as shown in figure 5.7. During the transmission, PROCOMM PLUS displays a line count at the left end of the status line. This count indicates the number of lines of characters that PROCOMM PLUS has successfully received. PROCOMM PLUS does no error-checking on the ASCII characters received.

```
Copyright (C) 1987, 1988 DATASTORM TECHNOLOGIES, INC.
All Rights Reserved

/**************** UNAUTHORIZED DUPLICATION PROHIBITED **************/
/* This disk and the files it contains may be copied for backup    */
/* purposes only.                                                   */
/******************************************************************/

To automatically install PROCOMM PLUS: Place this diskette in drive A: and
type "A:PCINSTAL [Enter]".

This diskette should contain the following files:

PCPLUS.EXE     - The PROCOMM PLUS Executable Program
PCSETUP.EXE    - The PROCOMM PLUS Setup Facility
PCINSTAL.EXE   - The PROCOMM PLUS Installation Program
PCEDIT.EXE     - The PROCOMM PLUS Editor program

PCPLUS.HLP     - On-line help file for PROCOMM PLUS
PCPLUS.KBD     - Default keyboard mapping definitions for PROCOMM PLUS
PCPLUS.USR     - Sample user file for PROCOMM PLUS host mode

DSTOR
 LINE: 26            ASCII FILE DOWNLOAD  -  PRESS ESC TO END
```

Fig. 5.7.

Receiving a file using the ASCII download protocol.

When the file has been received, press Esc. This command causes PROCOMM PLUS to close the download file and save it to disk.

TIP | As an alternative to the ASCII download protocol, you can use the log-file feature to save incoming information to disk. Refer to "Capturing the Session to Disk," in Chapter 4, for an explanation of the log-file feature.

XMODEM

The most widely available PC-based file-transfer protocol is XMODEM. Like so many other communications terms, XMODEM can mean different things to different people, so you need to understand a little about the background of this protocol, as well as how PROCOMM PLUS uses the term.

As most broadly used, the term *XMODEM* refers to a file-transfer protocol included in the program MODEM2, written by Ward Christensen and introduced in 1979. This file-transfer protocol was originally called the MODEM protocol and was intended to transfer files between computers running the CP/M operating system. Over the years since 1979, however, the MODEM file-transfer protocol has become known as XMODEM and has been implemented in countless communications programs for use in transferring files between computers running many different operating systems. Virtually all popular communications programs for IBM PC-type computers include an implementation of XMODEM. Since its introduction, several "new and improved" versions of the protocol have appeared under various names, several of which are discussed in later sections of this chapter.

The original XMODEM used the checksum error-checking scheme. This method of detecting errors is adequate for low-speed data transmission (300 bps or less) but can miss errors that are more likely to occur when sending data at higher transmission speeds (or over noisy phone lines). One popular variation of XMODEM adds the CCITT CRC-16 error-checking scheme, which is much more reliable than the checksum method at the higher transmission rates.

The CRC-16 version of XMODEM is often called XMODEM/CRC, for obvious reasons, but PROCOMM PLUS refers to the protocol simply as XMODEM. PROCOMM PLUS's usage of the name in this manner leads to no problems, however, because the CRC-16 version of XMODEM is "backward compatible" with the checksum version. In other words, you can use PROCOMM PLUS's XMODEM to send or receive a file to or from a computer that is using the original checksum version of XMODEM or the newer CRC-16 version.

Several other modified versions of XMODEM seem to have overtaken XMODEM in popularity. To varying degrees, these other protocols overcome XMODEM's recognized weaknesses: 128K blocks, no multiple file transfers, no file characteristics transferred, no sliding windows. Protocols that are available in PROCOMM PLUS and are essentially improved versions of XMODEM include YMODEM (often called 1K-XMODEM), TELINK, SEALINK, WXMODEM, YMODEM-G, YMODEM BATCH (often called YMODEM), and YMODEM-G BATCH.

TIP | Never use XMODEM to transfer files to or from the on-line service CompuServe. The delays caused by packet-switching result in excessive time-out errors, aborted transfers, and slow transmissions. Use the COMPUSERVE B protocol instead (discussed later in this chapter).

YMODEM, YMODEM BATCH, YMODEM-G, and YMODEM-G BATCH

The YMODEM protocol was introduced in 1985 by Chuck Forsberg, Omen Technology Inc, as an extension of Ward Christensen's XMODEM protocol. As originally implemented, Forsberg's YMODEM protocol includes CRC-16 error-checking; 1,024-byte blocks; multiple file transfer; transmission of file name and file size; and, optionally, the date and time each file was last modified. Forsberg's YMODEM also includes an option called the G option, which enables you to take full advantage of modems that perform error-control.

Other implementations of YMODEM, however, do not always include all these features, so the developers of PROCOMM PLUS chose to provide YMODEM in four varieties:

- *YMODEM*. This version implements only the CRC-16 and 1,024-byte block enhancements. Other programs sometimes refer to this protocol as 1K-XMODEM.

- *YMODEM BATCH*. Closest to the Forsberg YMODEM, this protocol includes the CRC-16, 1,024-byte blocks, multiple file transfer, and transmission of file name and file size. This protocol does not, however, include the G option.

- *YMODEM-G*. This protocol is a streaming protocol. As implemented in PROCOMM PLUS, this protocol is the same as the PROCOMM PLUS YMODEM file-transfer protocol but with no error-checking. You can use YMODEM-G to transfer a single file using an error-control modem like a Hayes V-series modem, a modem that uses MNP Classes 1 through 4, or a CCITT V.42-compliant modem.

- *YMODEM-G BATCH*. This streaming protocol is closest to the Forsberg YMODEM-G protocol. YMODEM-G BATCH is the same as the PROCOMM PLUS YMODEM BATCH protocol but without error-checking.

The names PROCOMM PLUS uses for these variations of Forsberg's YMODEM match closely the names many popular PC bulletin board programs use for these same protocols.

To send or receive a single file, you can use any one of the four YMODEM protocols available in PROCOMM PLUS. When receiving a file with YMODEM BATCH or YMODEM-G BATCH, you don't specify the transfer file name, because the protocol sends the file name with the file.

When you want to send multiple files using a version of the YMODEM protocol, however, you must use YMODEM BATCH or YMODEM-G BATCH (use YMODEM-G BATCH only with an error-control modem). After you select the upload protocol, PROCOMM PLUS displays the following prompt:

```
Please enter file spec:
```

The term file spec is an abbreviation for *file specification*—the name of the file (or files) to be sent. PROCOMM PLUS searches the screen for the latest valid DOS file name and enters the name to the right of this prompt. Use the editing keys listed in table 5.1 to edit this line so that it includes the name of the file you want to transmit. You must specify a full path with the file name if the file is not in the current directory. You can use either or both of the two wild-card characters available in DOS—the question mark (?) and asterisk (*). After you type the file specification, press Enter. PROCOMM PLUS transfers all the specified files one at a time, without any further action on your part.

For example, you can use the YMODEM BATCH protocol to send the files SALES889.WK1 and SALES989.WK1 by using a single upload command. First, inform the remote computer operator or host program (for example, the bulletin board or on-line service) that you are about to upload files using the YMODEM BATCH protocol. Then, from the Terminal mode screen, press PgUp (Send Files) to display the Upload menu. Select YMODEM BATCH from this menu. PROCOMM PLUS instructs you to enter the file specification. Type *.wk1, and press Enter. PROCOMM PLUS displays the Progress window and begins transmitting SALES889.WK1. After SALES889.WK1 has been sent, PROCOMM PLUS uploads SALES989.WK1. After PROCOMM PLUS transmits SALES989.WK1, no more files remain to be sent, so PROCOMM PLUS sounds a beep five times and removes the Progress window from the screen.

In order to download multiple files with YMODEM BATCH or YMODEM-G BATCH, you must specify the file names to the remote computer. When the remote computer is ready to transmit, press PgDn (Receive Files) and select YMODEM BATCH or YMODEM-G BATCH, as appropriate (use YMODEM-G BATCH only with an error-control modem). The first file transfer begins, and PROCOMM PLUS displays the Progress window. Because YMODEM BATCH and YMODEM-G BATCH send the file name of each file transmitted, you don't have to provide the file names to PROCOMM PLUS. Also, because the sending computer transmits the size of each file, PROCOMM PLUS can make full use of the Progress window.

In figure 5.8, you see a PROCOMM PLUS Progress window indicating that the file 800R8.ARC is being transferred using the YMODEM BATCH protocol. In this case, you are downloading two files, 800R8.ARC and RS232V23.ARC, from an electronic bulletin board. The bulletin board displays the prompt Ymodem Batch send of 800R8.ARC RS232V23.ARC ready! Press PgDn (Receive Files), and select YMODEM BATCH from the Download menu. PROCOMM PLUS sends to the bulletin board a signal that begins the transmission of 800R8.ARC. The Progress window then appears, as shown in figure 5.8. As soon as PROCOMM PLUS has successfully received the first file, the program sends a signal to the bulletin board to start transmission of the second file. After both files are received, no more files remain for the bulletin board to send, so the transmission is complete. PROCOMM PLUS flashes the word COMPLETED in place of the word PROTOCOL for a few seconds in the Progress window and then removes the Progress window from the screen, returning to the full Terminal mode screen.

```
(R) Zmodem Resume (HANGUP)        (B) Ymodem Batch
(Y) Ymodem (Xmodem-1K)            (C) Slow Xmodem/CRC
(G) Qmodem-G (^X^X^X^X^X at end)  (F) Ymodem-G (registered DSZ only)

Protocol: (Enter)=B? b

Download Ymodem Batch.                    ┌──────────────────────────┐
Enter up to 9 filespecs. Wildcards are Okay,│  PROTOCOL: YMODEM BATCH │
End the list with a blank line.           │ FILE NAME: 800r8.arc     │
                                          │ FILE SIZE: 5261          │
(9764k, 28 min. left) Filespec 1: 800r8 rs232│BLOCK CHECK: CRC        │
  6k,  0.5 min.  800R8.ARC                │TOTAL BLOCKS: 6           │
  4k,  0.4 min.  RS232V23.ARC             │TRANSFER TIME: 00:25      │
                                          │TRANSMITTED: 77%          │
(9755k, 27 min. left) Filesp              │ BYTE COUNT: 4096         │
                                          │BLOCK COUNT: 4            │
Ymodem Batch Download Estimate:           │ERROR COUNT: 0            │
  2 files, 10k bytes, 0.8 minutes         │LAST MESSAGE:             │
                                          │  PROGRESS: ████████      │
LAST CHANCE!  (Enter) or (S)tart, (G)oodbye after transfer, (A)bort? s

Ymodem Batch send of 800R8.ARC RS232V23.ARC ready!
You have 60 seconds to start.  Type (Ctrl-X) several times to abort.

       File transfer in progress... Press ESC to abort
```

Fig. 5.8.

Downloading files using the YMODEM BATCH file-transfer protocol.

TIP Of the file-transfer protocols available in PROCOMM PLUS, YMODEM BATCH is probably the best to use when transferring files to and from PC bulletin boards. Its 1,024-byte block size, coupled with multiple-file transfer capability and automatic transfer of file name and size, makes YMODEM BATCH a fast protocol that also is easy to use. If you want to transfer large files over public data networks (such as Tymnet or Telenet) or other packet-switching networks, however, consider using one of the protocols that implement sliding windows: KERMIT, SEALINK, or WXMODEM.

MODEM7 and TELINK

At various times, you may want to send or receive a file from a bulletin board running on a microcomputer under the CP/M operating system. For this reason, PROCOMM PLUS supplies the MODEM7 file-transfer protocol. MODEM7 is a variation of XMODEM that is used most often on CP/M computers.

Using MODEM7, you can send multiple files with one command, including the name of each file. This protocol uses CRC-16 but can fall back to the checksum error-checking method. MODEM7 does not, however, send file size or the date and time the file was last modified. Data blocks are 128 bytes in length, so transmission is slower than with 1,024-byte block protocols like YMODEM BATCH.

The TELINK file-transfer protocol is similar to MODEM7 but also sends file size and the date and time the file was last modified. TELINK was developed by Tom Jennings and is used mainly on bulletin boards running the FIDO bulletin board software.

SEALINK and WXMODEM

A computer transmitting data using XMODEM sends a block of data and then waits. The receiving computer checks the data for errors and sends a reply requesting the next block. The sending computer is in a holding pattern for the entire time the receiving computer is checking the data and transmitting the reply. After the sending computer receives the reply, it sends the next block. Even though your modem is capable of transmitting at 2400 bps, or 240 characters per second, data may actually be strolling along at speeds as slow as 120 characters per second over satellite relays or packet-switching networks.

The SEALINK file-transfer protocol was developed by Thom Henderson of System Enhancement Associates, Inc., and released in 1986. SEALINK is an enhancement of XMODEM that uses *sliding windows*. Sliding windows refers to a file transmission technique that is capable of sending data blocks and accepting replies from the receiving computer at the same time. This feature is sometimes called full-duplex operation because it takes advantage of your modem's capability to send data and receive data simultaneously (refer to the discussion of duplex in "Toggling Duplex," in Chapter 4).

SEALINK sends up to six 128-byte blocks while "listening" for a reply from the receiving computer before the protocol has to wait. As long as SEALINK "hears" a reply from the first block by the time the sixth block is sent, the protocol doesn't have to stop to wait for a reply. When you use this technique, data is sent more efficiently.

SEALINK uses CRC-16 error-checking and is capable of sending multiple files, file name, and file size. The SEALINK protocol implemented in PROCOMM PLUS does not send the file date and time.

SEALINK is *backward compatible* with XMODEM. In other words, you can use this file-transfer protocol to send a single file to a computer that supports only XMODEM, but you will achieve no advantage over PROCOMM PLUS's normal XMODEM. To benefit from SEALINK's transmission-speed enhancement and other features, both sending and receiving computers have to be using the protocol.

The file-transfer protocol WXMODEM, sometimes called Windowed XMODEM, is similar to SEALINK in that WXMODEM uses sliding windows to increase transmission speed. WXMODEM sends up to four 128-byte blocks, while "listening" for a reply, before it has to wait. WXMODEM, however, can be used only to send a single file and does not transmit file name or file size with the transmitted data. WXMODEM was developed by Peter Boswell.

IMODEM

Like YMODEM-G, the IMODEM file-transfer protocol is a streaming protocol. That is, IMODEM does no error-checking. IMODEM was developed by John Friel for use with modems that perform the error-control function (for example, Hayes V-series modems that support MNP Classes 1 through 4 and CCITT V.42 compliant modems). When transferring files, you should use a streaming protocol like IMODEM to take full advantage of this type of modem.

Refer to Appendix B for information on properly installing and setting up an error-control modem for use with PROCOMM PLUS.

KERMIT

KERMIT is the name of a program, a file-transfer protocol, and a famous frog. In fact, the program and protocol are named after the Jim Henson muppet, Kermit the Frog. Unless specifically noted otherwise, the remainder of this section of the chapter discusses the file-transfer protocol KERMIT, rather than the Kermit program.

KERMIT was developed in 1981 at Columbia University by Frank da Cruz and Bill Catchings and released to the public domain. In contrast to XMODEM, which requires 8 data bits to operate, KERMIT can be used on computers that can handle only 7 data bits (for example, many mainframe computers) and still can manage to transmit files that contain bytes made up of 8 bits per byte. Since 1981, KERMIT has been implemented on countless brands and models of computers, from mainframe to microcomputer. Consequently, PROCOMM PLUS also provides this protocol for you to use.

Since its introduction, KERMIT has enjoyed numerous enhancements. The PROCOMM PLUS version of the KERMIT file-transfer protocol has all the desirable properties listed in table 5.2, including multiple-file transfer and transmission of file name, size, date, and time. KERMIT's data blocks, however, have a maximum length of 94 bytes (the default length in PROCOMM PLUS is 90 bytes). This small block size is, however, balanced by KERMIT's use of the sliding windows technique to increase transmission speed over satellite-relayed lines and packet-switching networks. As icing on the cake, the version of KERMIT implemented in PROCOMM PLUS also includes data compression to further increase transmission speed.

When you want to send one or more files to a remote computer that uses the KERMIT file-transfer protocol, inform the remote computer software that you are about to begin a KERMIT upload (you don't need to specify the file name). Then press PgUp (Send Files).

As with other multiple-file-transfer protocols, PROCOMM PLUS asks for the file spec. PROCOMM PLUS looks for the latest valid DOS file name on-screen and fills it in the space to the right of the prompt. Use the editing keys listed in table 5.1 to edit this name. You can also use DOS wild-card characters.

After you enter the file specification, press Enter to begin the file transfer. PROCOMM PLUS displays a special KERMIT Progress window (see fig. 5.9). PROCOMM PLUS lists Super Kermit as the file-transfer protocol, calling attention to this version's enhanced features.

Because KERMIT is continually evolving and is implemented on so many different computer systems, not all versions have exactly the same features. To maintain complete compatibility between all KERMIT implementations, the protocol performs a unique "handshake" at the beginning of the transmission. This handshake can be described as *feature negotiation*. When PROCOMM PLUS first displays the KERMIT Progress window, the program displays the following message in the LAST MESSAGE line at the bottom of the window:

Exchanging initialization parameters

```
Password: *

                        PROTOCOL: Super Kermit
                       FILE NAME: QANDA.TXT
                       FILE SIZE: 25266
                       FILE TYPE: BINARY
F)iles U)pload        FILE NUMBER: 1
H)elp T)ime C)h       COMPRESSION: YES
R)ead mail L)eav      8 BIT PREFIX: NO
                      WINDOW SIZE: 31
Your choice? U        BLOCK CHECK: 1 BYTE CHECKSUM
                     TRANSFER TIME: 02:20 (Approximate)
A)scii K)ermit         BYTE COUNT: 3489
I)modem T)elink      LAST MESSAGE: Attribute packet transferred
Your choice? K

File name?
            !...................!...................!
Description:

Begin your  KERMIT  transfer procedure...@# M3
      ESC: Abort Transfer    CTRL-B: Cancel Batch    CTRL-F: Cancel File
```

Fig. 5.9.

The KERMIT Progress window.

Each computer sends the other a block (packet) that contains a list of features implemented in its version of KERMIT. The transmission then proceeds, using the features that the two KERMITs have in common.

Next, PROCOMM PLUS displays the message File header transferred in the LAST MESSAGE line of the KERMIT Progress window, meaning that file name, file size, and modification date and time for a particular file have been transferred. PROCOMM PLUS then displays the message Attribute packet transferred (as shown in fig. 5.9), indicating that the DOS file attributes have been sent.

When a file transfer is finished, PROCOMM PLUS displays the message End of file packet transferred and then Transfer OK in the LAST MESSAGE line of the KERMIT Progress window. After all files have been successfully transferred, PROCOMM PLUS displays the message Transaction ended.

Downloading files with KERMIT is similar to uploading. First, inform the remote computer that you want it to use KERMIT to send files to your computer. Also, specify the names of the files to the remote computer. When the remote computer is ready to send a KERMIT transfer, press PgDn (Receive Files), and choose KERMIT from the Download menu. PROCOMM PLUS displays the Progress window and executes the file transfer.

When you are uploading or downloading files using any PROCOMM PLUS file-transfer protocol, you can abort the entire transfer in midstream by pressing Esc. When sending multiple files (often called a *batch*), KERMIT also enables you to cancel transmission of individual files selectively, without canceling the entire transmission. While the KERMIT Progress window is displayed, press Ctrl-F (Cancel File) to abort transmission of the file currently being transferred. No other file transfers are affected. You can cancel the whole batch transmission by pressing Ctrl-B (Cancel Batch).

When you are communicating with a system that is running the Kermit Server program, as opposed to using just the KERMIT file-transfer protocol, you execute uploads and downloads in a manner different from the method described in the preceding paragraphs. First, you must switch to the Kermit Server mode on the other system. Issue the appropriate command for the Kermit program that system is using (for example, type *server*, and press Enter).

When the remote computer is running the Kermit Server mode, press Alt-K (Kermit Server Cmd) to display a window containing four commands, as shown in figure 5.10. These commands have an affect only when you are using PROCOMM PLUS to communicate with the Kermit program in Server mode. This chapter refers to this window as the Kermit Server mode menu. From this menu, select one of the following commands:

Fig. 5.10.

The Kermit Server mode menu.

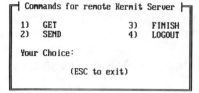

• **1) GET.** Select this command when you want to receive (download) one or more files from the remote computer. PROCOMM PLUS removes the Kermit Server mode menu from the screen and displays a smaller window that prompts you to Please enter file spec. Type the name of the file you want to download, and press Enter. The remote Kermit Server program may let you use wild-card characters to specify the files to be received. After you press Enter, PROCOMM PLUS displays the KERMIT Progress window and works with the server to execute the transfer.

- **2**) SEND. Choose this command to send one or more files to the remote computer. PROCOMM PLUS displays a window that prompts you to Please enter file spec. PROCOMM PLUS searches the screen for the latest valid DOS file name and displays it to the right of this prompt. Use the editing keys listed in table 5.1 to enter the name of the file you want to send, and press Enter. PROCOMM PLUS displays the KERMIT Progress window and works with the server to transfer the file. (Once again, you may use wild cards in file spec.)

- **3**) FINISH. Use this command to quit the Kermit Server mode without leaving the Kermit program itself. The server issues a message, such as Good-bye, and you return to the Kermit program's normal prompt.

- **4**) LOGOUT. Select this command to quit the Kermit Server mode and to log out of Kermit altogether. (Depending on how you are connected to Kermit, this command may also disconnect you from the remote computer.)

You also can customize the KERMIT file-transfer settings by using the PROCOMM PLUS Setup Utility Kermit Options screen, shown in figure 5.11. Refer to "Changing KERMIT Options," in Chapter 8, for more information.

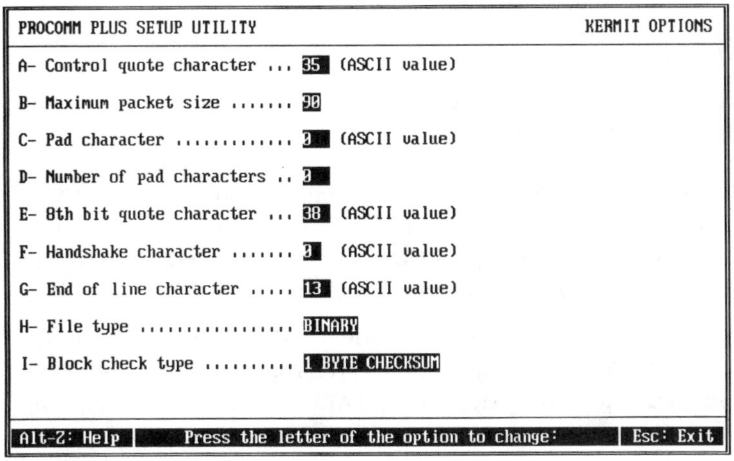

```
PROCOMM PLUS SETUP UTILITY                           KERMIT OPTIONS

A- Control quote character ... 35  (ASCII value)

B- Maximum packet size ....... 90

C- Pad character ............. 0   (ASCII value)

D- Number of pad characters .. 0

E- 8th bit quote character ... 38  (ASCII value)

F- Handshake character ....... 0   (ASCII value)

G- End of line character ..... 13  (ASCII value)

H- File type ................. BINARY

I- Block check type .......... 1 BYTE CHECKSUM


Alt-Z: Help |    Press the letter of the option to change:    | Esc: Exit
```

Fig. 5.11.

The Setup Utility Kermit Options screen.

COMPUSERVE B

Use the COMPUSERVE B file-transfer protocol with the CompuServe B or the CompuServe Quick B protocols when uploading and downloading files on the CompuServe on-line service.

TIP The primary differences between the CompuServe B and CompuServe Quick B protocols are data block size and windowing. CompuServe B uses 512-byte blocks, and CompuServe Quick B uses 1,024-byte blocks. More important, Quick B implements sliding windows. As implied by the name, CompuServe Quick B transfers files faster than CompuServe B. Therefore, you should always use Quick B.

A third version of the CompuServe B protocol, referred to as the CompuServe B+ protocol, has also been implemented on CompuServe. This protocol is not yet (as of this writing) supported in PROCOMM PLUS but can be added as an external protocol by using one of several public domain or shareware programs that can be downloaded from the IBM Communications Forum on CompuServe.

Never use XMODEM to transfer files to or from CompuServe. The delays caused by packet-switching result in excessive time-out errors, aborted transfers, and generally slow transmissions.

To take best advantage of the COMPUSERVE B protocol, set the Enquiry (ENQ) setting in the Setup Utility Terminal Options screen to CIS B, as shown in figure 5.12 (refer to "Assigning a Response to ENQ," in Chapter 8). You also can add the following line to your CompuServe log-on script file:

    SET ENQ CISB

Add the preceding line somewhere in the script after the log-on ID and password have been transmitted.

The purpose of either of these actions is to enable PROCOMM PLUS to recognize an Enquiry (ENQ) signal from CompuServe, instructing PROCOMM PLUS to begin its upload or download procedure. Refer to Chapters 7 and 11 for more information about script files.

For example, suppose that you want to download the file DTP11B.ARC from the DATASTORM library on CompuServe. You first select the name of the file you want to download, DTP11B.ARC in this case. Then issue the appropriate CompuServe command to begin a download, and choose

```
┌─────────────────────────────────────────────────────────────┐
│ PROCOMM PLUS SETUP UTILITY                    TERMINAL OPTIONS │
│                                                               │
│ A- Terminal emulation ................ VT102                  │
│                                                               │
│ B- Duplex ............................ FULL                   │
│                                                               │
│ C- Software flow control (XON/XOFF) .. OFF                    │
│                                                               │
│ D- Hardware flow control (RTS/CTS) ... OFF                    │
│                                                               │
│ E- Line wrap ......................... ON                     │
│                                                               │
│ F- Screen scroll ..................... ON                     │
│                                                               │
│ G- CR translation .................... CR                     │
│                                                               │
│ H- BS translation .................... DESTRUCTIVE            │
│                                                               │
│ I- Break length (milliseconds) ....... 350                    │
│                                                               │
│ J- Enquiry (ENQ) ..................... CIS B                   │
│                                                               │
│ Alt-Z: Help │  Press the letter of the option to change: │ Esc: Exit │
└─────────────────────────────────────────────────────────────┘
```

Fig. 5.12.

The Setup Utility Terminal Options screen with ENQ *set to CIS B.*

the CompuServe QB protocol. CompuServe prompts you to type the file name *for your computer.* Type the name you want to use for the file on your computer. After you press Enter, CompuServe sends the ENQ signal to PROCOMM PLUS, and PROCOMM PLUS begins the download. You never press PgDn (Receive Files) to begin the download.

External Protocols

Even though PROCOMM PLUS supplies 13 built-in file-transfer protocols, it also provides a means for you to invoke up to three other protocols from within PROCOMM PLUS. These non-PROCOMM PLUS protocols are referred to as *external* protocols.

For example, you may want to add the ZMODEM protocol from Chuck Forsberg and the CompuServe B+ protocol from CompuServe. Neither of these two protocols is included in PROCOMM PLUS, but both have enhanced features that warrant your attention. Refer to Appendix C for instructions on how to use PROCOMM PLUS to access an external protocol, with specific discussions on accessing ZMODEM.

Working with Disk Files

Both the upload and download capabilities of PROCOMM PLUS operate on DOS files that reside on your computer's disk. This section describes several PROCOMM PLUS commands that enable you to work with these disk files. The material explains how to list the files on the current working directory, how to change to a different working directory, and how to access DOS from within PROCOMM PLUS.

Listing the Current Working Directory

As you upload and download files, you frequently need to review the names of the files on your disk. Perhaps you cannot remember the name of the file you want to send to the other computer. Maybe you are not sure which file names you have used already.

To see a list of files in the working directory (initially, the directory from which you started PROCOMM PLUS is the working directory), press Alt-F (File Directory) from the Terminal mode screen. PROCOMM PLUS displays a window in the middle of the screen, with the following prompt in the top border line of the window:

```
Enter FILE SPEC: (Carriage Return = *.*)
```

The cursor is blinking inside the window. When you want to see a list that includes all the files in the working directory, press Enter without typing anything in the window. To see only a partial list of files, type an appropriate file specification in the window, and press Enter.

For example, to see a list of all the file names that end in .TXT, type *.*txt* as the file specification, and press Enter. After you press Enter at the file specification, PROCOMM PLUS displays a window containing a list of the files you specified, as shown in figure 5.13.

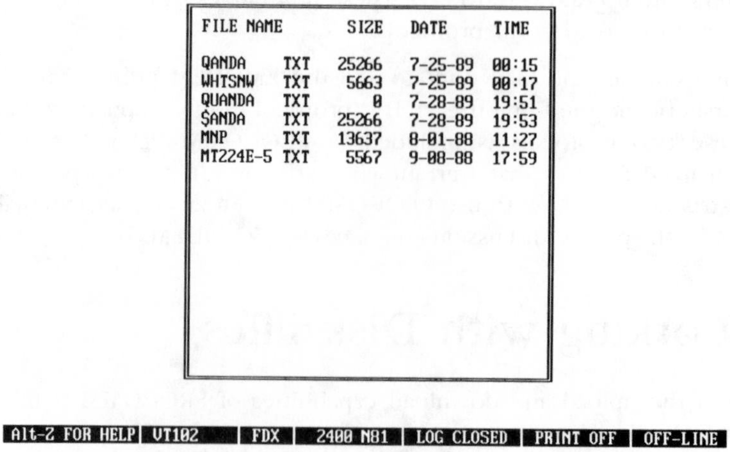

Fig. 5.13.

Displaying file names in the current working directory that meet the specification *.txt.

```
FILE NAME        SIZE    DATE     TIME

QANDA     TXT    25266   7-25-89  00:15
WHTSNW    TXT    5663    7-25-89  00:17
QUANDA    TXT    0       7-28-89  19:51
$ANDA     TXT    25266   7-28-89  19:53
MNP       TXT    13637   8-01-88  11:27
MT224E-5  TXT    5567    9-08-88  17:59
```

```
Alt-Z FOR HELP | VT102 | FDX | 2400 N81 | LOG CLOSED | PRINT OFF | OFF-LINE
```

When more than 15 file names meet the specification, PROCOMM PLUS displays the first 15 names followed by the word - MORE - at the bottom of the list. Press Esc to return to the Terminal mode screen, or press any other key to display the next 15 file names. Each time you press a key, PROCOMM PLUS displays another 15 file names. After PROCOMM PLUS displays all the file names that meet the specification, the program returns to the Terminal mode screen.

Changing the Current Working Directory

When you specify a file for PROCOMM PLUS to upload, as described in this chapter, PROCOMM PLUS looks for the file on the current working directory. By default, this directory is the directory from which you started PROCOMM PLUS. You may not always want to upload a file from the same directory, so PROCOMM PLUS enables you to switch the working directory easily.

To change to a different current working directory, from the Terminal mode screen, press Alt-F7 (Change Directory). PROCOMM PLUS displays a window prompting you for the NEW PATH. A cursor blinks to the right of this prompt. The name of the current working directory is displayed in the top border line of this window, to the right of the prompt CURRENT DIR. For example, if you started PROCOMM PLUS from the C:\PCPLUS directory, this top border of the window reads as follows:

CURRENT DIR: C:\PCPLUS

Type to the right of the prompt NEW PATH the DOS path of the new working directory, and press Enter. Type a valid DOS path (refer to your DOS manual).

You may, for example, have a subdirectory named C:\PCPLUS\UPLOAD. You already have copied into this directory all the files you want to upload. Before beginning an upload, press Alt-F7 (Change Directory). PROCOMM PLUS displays the window for the new path. Assuming that the default working directory is C:\PCPLUS, you type the path *upload* at the prompt NEW PATH, and press Enter. This command changes the current working directory to C:\PCPLUS\UPLOAD. PROCOMM PLUS now looks for upload files in this new working directory.

Accessing the Operating System

If you do much transferring of files with PROCOMM PLUS, you frequently need to perform such DOS functions as copying files, renaming files, deleting files, making directories, and so on. It would be inconvenient to leave PROCOMM PLUS every time you need to accomplish one of these DOS tasks, especially if you are on-line to another computer. Consequently, PROCOMM PLUS provides a way for you to access DOS from within PROCOMM PLUS without disconnecting from an ongoing communications session.

When you want to suspend PROCOMM PLUS temporarily and access the operating system (DOS), from the Terminal mode screen, press Alt-F4 (DOS Gateway). PROCOMM PLUS removes the Terminal mode screen and loads another copy of the operating system (PC-DOS or MS-DOS) into memory (RAM). This action does not disconnect you from an open modem connection, but any ongoing file transfer stops until you return to the Terminal mode screen. You generally should wait until a file transfer is complete before using this feature.

After PROCOMM PLUS loads the second copy of the operating system, DOS displays the following prompt:

```
Enter 'EXIT' to return to PROCOMM PLUS
C:\cur-dir›
```

In line 2, *cur-dir* is the name of your current working directory. This prompt is the *DOS prompt*. In other words, you can execute any DOS commands from this prompt.

> **TIP** For this feature to work, the DOS command-interpreter program (COMMAND.COM) must be on the root directory of your boot disk or in a directory that is specified by the SET COMSPEC command in your AUTOEXEC.BAT file. Refer to Appendix A and your DOS manual for information about the AUTOEXEC.BAT file and the SET COMSPEC command.

At the DOS prompt, for example, you may want to copy several files from a floppy disk to the current working directory on your hard disk. Use the DOS COPY command for this purpose.

While working at the DOS prompt, you can use the DOS CD (or CHDIR) command to switch to a different DOS directory. When you are ready to return to PROCOMM PLUS, however, make sure that you first return to the PROCOMM PLUS current working directory. Then, type *exit*, and press Enter. PROCOMM PLUS immediately returns to the Terminal mode screen.

Chapter Summary

This chapter explains how to use PROCOMM PLUS to send and receive electronic files. The text first described the basic steps to send a file from your computer to a remote computer, and then explained how to receive a file that is transmitted by a remote computer. Next, you learned when and how to use the file-transfer protocols available in PROCOMM PLUS. Finally, you learned how to view a DOS directory, change the working directory, and use the PROCOMM PLUS DOS Gateway feature to access the operating system.

Now that you are familiar with how to transfer files, you need some files to send. Turn to Chapter 6, "Using the PROCOMM PLUS Editor," to find out how to create text files using PROCOMM PLUS's text editor.

Using the
PROCOMM PLUS
Editor

PCEDIT is a rudimentary text editor, a simple word processor intended to create files made up entirely of characters from the ASCII character set. PCEDIT is distributed with PROCOMM PLUS but is actually a separate program. Using PCEDIT, you can create ASCII files up to 80 characters across and 500 lines long. PCEDIT has some of the characteristics of a typical word processor, such as block copy and block move, but lacks others, such as word wrap or adjustable margins. The program is intended as a quick tool and not as a replacement for full-featured word processing programs.

This chapter explains how to use PCEDIT. The text describes how to enter text and move around a file; how to make changes to the text in an existing file; how to copy, move, or delete portions of the text; how to search for character strings; and how to save ASCII text to a disk file. This chapter also lists the 20 macros available in PCEDIT that are of particular interest when you create PROCOMM PLUS script files. Refer to Chapters 7 and 11 for more information on script files.

Even though DATASTORM supplies PCEDIT as a part of the PROCOMM PLUS package, you may already have a favorite text editor you prefer to use instead. Consequently, PROCOMM PLUS's designers make it easy to substitute a different text editor for PCEDIT so that you can activate that text editor from within PROCOMM PLUS. The last portion of this chapter discusses how to add a different text editor to PROCOMM PLUS, as well as how to start the text editor from within PROCOMM PLUS.

149

Using a Text Editor with PROCOMM PLUS

One of the more popular reasons to use your PC for communications is to send and receive correspondence electronically. This procedure is usually referred to as *electronic mail*, or just *E-mail*. While E-mail systems enable you to type correspondence on-line, you may find it preferable to type your letters ahead of time, before connecting to the on-line service. This practice not only saves money by reducing connect time, but also enables you to use a full-screen editor or word processor with which you are familiar rather than the line editor provided by the E-mail system. Because E-mail systems expect only ASCII characters, text editors like PCEDIT are well suited for drafting your electronic correspondence.

Text editors like PCEDIT also are handy for creating and editing ASCII files for purposes other than E-mail. For example, PROCOMM PLUS's script files (discussed in Chapter 7, "Automating PROCOMM PLUS with Macros and Script Files," and Chapter 11, "An Overview of the ASPECT Script Language") are ASCII files you can create and edit using PCEDIT. In fact, PCEDIT has 20 keystroke commands referred to as *macros*. When executed, each of these 20 macros types a particular ASPECT command. These macros save keystrokes during the creation of a script file and reduce the potential for spelling errors (see "Using PCEDIT Macros," later in this chapter).

Starting PCEDIT

In order to give you quick access to a text editor, PROCOMM PLUS enables you to activate PCEDIT from the Terminal mode screen (refer also to "Using a Different Editor," in this chapter). Press Alt-A (Editor), and PROCOMM PLUS displays a narrow window across the middle of the screen. The top borderline of this window includes the title PCEDIT Parameters, and the window itself displays the prompt Enter parameters. In this window, type the name of the new file you want to create or the name of the existing file you want to edit, and press Enter.

Once you type a file name and press Enter, PROCOMM PLUS temporarily suspends itself. (It is still loaded in memory but is in a holding pattern.) PROCOMM PLUS then loads another copy of the DOS command processor and runs the program PCEDIT. PROCOMM PLUS also provides PCEDIT with the file name you typed.

Alternatively, you can start PCEDIT directly from DOS. After exiting PRO-COMM PLUS, at the DOS prompt type *pcedit*, and press Enter. At the top left of the screen, PCEDIT displays the prompt Enter file name. Type the name of the new file you want to create or the name of the existing file you want to edit, and press Enter.

TIP When you are creating a new file, you may want to include the extension .TXT in the file name. This extension is commonly used to denote an ASCII text file. The extension .DOC is also sometimes used for the same purpose. Whatever extension you use, decide on a file-naming convention and then stick with it. You will then be able to tell at a glance which files in a DOS directory are ASCII files and which are not.

Two types of ASCII files need particular file-name extensions: PRO-COMM PLUS script files and DOS batch files. PROCOMM PLUS script files use the script language ASPECT and must have the file-name extension .ASP. DOS batch files must have the file-name extension .BAT.

After you start PCEDIT from PROCOMM PLUS or from DOS, you briefly see the PCEDIT logo screen shown in figure 6.1 and then the PCEDIT working screen. Figure 6.2 shows a new file, LETTER1.TXT, and the screen shown in figure 6.3 contains the ASPECT script file CSERVE.ASP. (This script file is distributed with PROCOMM PLUS on the Supplemental Diskette.)

```
        ┌──────────────────┐
        │     PCEDIT       │
        │   Version 1.1B   │
        └──────────────────┘

COPYRIGHT (C) 1987, 1988 DATASTORM TECHNOLOGIES, INC.
              All Rights Reserved


        UNAUTHORIZED DUPLICATION PROHIBITED
```

Fig. 6.1.

The PCEDIT logo screen.

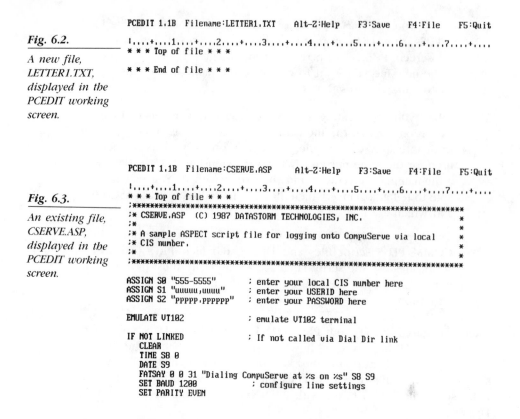

Fig. 6.2.

A new file, LETTER1.TXT, displayed in the PCEDIT working screen.

```
PCEDIT 1.1B  Filename:LETTER1.TXT     Alt-Z:Help    F3:Save    F4:File    F5:Quit
!....+....1....+....2....+....3....+....4....+....5....+....6....+....7....+....
* * * Top of file * * *

* * * End of file * * *
```

Fig. 6.3.

An existing file, CSERVE.ASP, displayed in the PCEDIT working screen.

```
PCEDIT 1.1B  Filename:CSERVE.ASP     Alt-Z:Help    F3:Save    F4:File    F5:Quit
!....+....1....+....2....+....3....+....4....+....5....+....6....+....7....+....
* * * Top of file * * *
;******************************************************************************
;* CSERVE.ASP  (C) 1987 DATASTORM TECHNOLOGIES, INC.                          *
;*                                                                            *
;* A sample ASPECT script file for logging onto CompuServe via local         *
;* CIS number.                                                                *
;*                                                                            *
;******************************************************************************

ASSIGN S0 "555-5555"        ; enter your local CIS number here
ASSIGN S1 "uuuuu,uuuu"      ; enter your USERID here
ASSIGN S2 "ppppp,pppppp"    ; enter your PASSWORD here

EMULATE VT102               ; emulate VT102 terminal

IF NOT LINKED               ; If not called via Dial Dir link
  CLEAR
  TIME S8 0
  DATE S9
  FATSAY 0 0 31 "Dialing CompuServe at %s on %s" S8 S9
  SET BAUD 1200             ; configure line settings
  SET PARITY EVEN
```

As shown in figures 6.2 and 6.3, the *top line* of the PCEDIT working screen displays the following information:

- PCEDIT 1.1B. The first section of the PCEDIT top line shows the version of the program.

- Filename:*filename*. Next, the top line displays the name of the file you are creating or editing, where *filename* is the name that you typed when you activated PCEDIT.

- Alt-Z:Help. As in PROCOMM PLUS, you can summon on-line help by pressing Alt-Z (Help). PCEDIT's help screens are discussed in more detail in "Getting Help," in this chapter.

- F3:Save F4:File F5:Quit. The last section of the PCEDIT top line lists three function key commands, each of which can be used to save the file on which you are working. Two of these commands also quit PCEDIT. Use of these and all other available PCEDIT function key commands is discussed later in this chapter.

The second line from the top of the working screen is blank. From time to time, PCEDIT displays messages in this line, so this chapter refers to the second line of the PCEDIT working screen as the *message line*. The third line from the top of the working screen contains a ruler that shows the number of spaces available across the screen, a total of 80.

When you first create a new file, PCEDIT displays the message

```
* * * Top of file * * *
```

in the fourth line of the screen, just below the ruler line. PCEDIT also displays the message

```
* * * End of file * * *
```

in line six (again see figure 6.2). (Both the top-of-file and end-of-file lines are red with yellow text on color screens.) The blank fifth line of the screen, between the top-of-file and the end-of-file markers and between the left and right edges of the screen, is the area where you can type the text for your file (this line is blue on color screens). PCEDIT enables you to expand this single line to a maximum of 500 lines, but the width remains at 80 characters.

Getting Help

You display a help screen in PCEDIT in the same manner as in PRO-COMM PLUS. Press Alt-Z (Help), and PCEDIT displays the first of two available help screens (see fig. 6.4). This screen lists the keyboard commands available in PCEDIT. The first help screen is the equivalent of PRO-COMM PLUS's command menu (refer to "Using the Command Menu as a Reminder," in Chapter 2, "Getting Around in PROCOMM PLUS").

To display the second help screen, press PgDn. This screen lists the macros available in PCEDIT (see fig. 6.5). Refer to "Using PCEDIT Macros," in this chapter, for more information.

Fig. 6.4.

PCEDIT Help screen number 1, listing PCEDIT keyboard commands.

```
PCEDIT Help (1 of 2)                            PROCOMM PLUS On-Line Help

PCEDIT is a simple ASCII text editor. It handles lines up to 80 characters
long with a maximum of 500 lines per file. The following is a list of keys
to press for available functions:

                    Ins: Switch between INSERT and OVERTYPE mode.
                    Del: DELETE CHARACTER at cursor.
              Backspace: DELETE CHARACTER at left of cursor.
           F1 or Alt-I: INSERT LINE at cursor.
           F2 or Alt-D: DELETE LINE at cursor.
               Ctrl-End: ERASE from cursor to end of line.
           F6 or Alt-S: SEARCH for a character string.
                   Home: Go to FIRST CHARACTER in the line.
                    End: Go to LAST CHARACTER in the line.
  Ctrl-Home or Ctrl-PgUp: Go to TOP OF FILE.
             Ctrl-PgDn: Go to END OF FILE.
                   PgUp: UP one screen.
                   PgDn: DOWN one screen.
                  Alt-B: MARK BLOCK for COPY, MOVE or DELETE.
                     F3: SAVE - Save file, return to editing.
                     F4: FILE - Save file, exit to DOS.
           Alt-X or F5: QUIT - Exit (with option to not save changes).

              PgDn: Next Page   Esc: Exit Help
```

Fig. 6.5.

PCEDIT Help screen number 2, listing available PCEDIT macros.

```
PCEDIT Help (2 of 2)                            PROCOMM PLUS On-Line Help

PCEDIT has several built in macros for use with the ASPECT script language.
Press one of the keys below to automatically insert the indicated macro at
the cursor position.

       KEY    MACRO               KEY     MACRO

      Alt-A: ASSIGN             Alt-F1:  IF
      Alt-C: CLEAR              Alt-F2:  ELSE
      Alt-F: FATSAY             Alt-F3:  ENDIF
      Alt-G: GET                Alt-F5:  SWITCH
      Alt-H: HANGUP             Alt-F6:  CASE
      Alt-L: LOCATE             Alt-F7:  ENDCASE
      Alt-M: MESSAGE            Alt-F8:  ENDSWITCH
      Alt-P: PAUSE              Alt-F9:  ATSAY
      Alt-R: RGET              Alt-F10:  ATGET
      Alt-T: TRANSMIT
      Alt-W: WAITFOR

              PgUp: Previous Page   Esc: Exit Help
```

Typing in PCEDIT

Text editor programs like PCEDIT are traditionally used by programmers to write lines of code rather than for true word processing. Consequently, these programs do not always provide the word processing features you may take for granted. For example, if you use PCEDIT to type correspondence, you will certainly miss a couple of features that are universally found in word processing programs: adjustable margin settings and word wrap. These features are not needed when you are drafting a program in

BASIC or writing a PROCOMM PLUS script file. Typical word processors enable you to change the left- and right-margin settings from file to file and often from line to line in the same file. Word processing programs also let you forget about the margin as you type; the program automatically wraps entire words to the next line when you type past the right margin. Neither of these two features is available in PCEDIT. As a result, you must be a bit more aware of the cursor position as you type.

A third feature that is absent in PCEDIT but found in all word processors is automatic line feed. PCEDIT does not automatically add blank lines to the end of the file as you type. Instead, you must insert each blank line manually before you can type in it. This line insertion makes typing text using PCEDIT more like typing on an electric typewriter than typing with a typical word processing package.

As shown in figure 6.2, PCEDIT initially provides only one blank line in which to type. After you type text in this line, press F1 (Insert Line). PCEDIT inserts a blank line below the cursor and moves the cursor to the left side of the screen. The effect of this command is the same as that of the Return key on an electric typewriter. The Return key performs both *line feed* and *carriage return* functions. The command F1 (Insert Line) rolls another row of "electronic paper" on to the screen and returns the cursor to the left margin. PCEDIT enables you to insert up to 500 lines in this manner.

TIP | You can also use the Enter key to perform a carriage return; but Enter does not cause PCEDIT to add another line to the file (line feed). An easy procedure gets around this problem. Each time you open a new file with PCEDIT, before you start typing, press F1 (Insert Line) 22 times and then press PgUp. This action adds 22 blank lines to the initial blank line, and places the cursor in the first line below the * * * Top of file * * * message. Now you can type 23 lines of text, pressing Enter at the end of each line, before having to add more blank lines to the file. When you reach the * * * End of file * * * message, repeat the procedure (F1 (Insert Line) 22 times followed by PgUp) to add 22 more lines for typing.

Once you have one or more blank lines in which to type, you can begin to type text in the PCEDIT work screen. All the letters and symbols represented on your keyboard are available for typing in PCEDIT. Figure 6.6 shows a letter typed in the PCEDIT work screen. You can also type the additional 128 characters from the IBM extended character set (those not found in the generic ASCII character set). Hold down the Alt key and type on the numeric keypad the decimal code for the character. After typing the number, release the Alt key; PCEDIT displays the character. For example, to type the upper left box-drawing character, hold the Alt key and type the number *169* on the numeric keypad.

Fig. 6.6.

A letter typed in PCEDIT.

```
PCEDIT 1.1B  Filename:LETTER1.TXT    Alt-Z:Help    F3:Save    F4:File    F5:Quit
!....+....1....+....2....+....3....+....4....+....5....+....6....+....7....+....
* * * Top of file * * *
                                    Wonder Widgets International
                                    121 S. Main St.
                                    Fairfax, VA  22130

                                    September 13, 1989
                                    2:45 pm

Mr. G.H. Manning
Pungo Freight Lines
P.O. Drawer Q
Santa Barabara, CA 93101

Reference: Bill of Lading Number 340-67783-98845-89

Dear George,

We received the above referenced shipment of electric motors this afternoon
minus one carton.  The missing carton is number 37 of 52.  It is urgent
that we receive this carton by Friday.  Please reply with status today!

Alston
```

Editing a File

Unless you are a professional secretary capable of error-free typing, you will be interested in PCEDIT's editing capabilities. The sections that follow discuss methods available in PCEDIT for typing over, inserting, deleting, copying, and moving text in the PCEDIT working screen. While these features are no match for full-blown word processing programs, they do enable you to correct your typos easily.

Using Insert and Overtype

PCEDIT defaults to the *Overtype mode*. In Overtype mode, you can replace an existing character in the working screen by typing on top of the character. Use the arrow keys to place the cursor at the character you want to replace and type the new character. In other words, you can correct a mistake by typing the correct version of the text on top of the original version.

Sometimes as you type, you leave one or more letters out of a word, or perhaps you forget a word altogether. You can correct this sort of error in Overtype mode, but you may have to type the entire line to do so.

Instead, PCEDIT enables you to switch to *Insert mode*. Press the Insert key and PCEDIT changes the shape of the cursor from a blinking under-score to a blinking block. This change in cursor shape is to remind you that PCEDIT is in Insert mode.

While in Insert mode, PCEDIT does not replace existing text. Rather, characters already on the screen move to the right, making room for the new text as you type. Each time you type a character, PCEDIT moves to the right any existing character at the cursor and any characters to the right of the cursor.

CAUTION: PCEDIT does not prevent you from "pushing" existing text off the right edge of the screen. If you insert text until existing characters disappear off the right side of the screen, the disappearing characters are gone. You have no way to recover this lost text short of typing it again.

Moving Around in the File

PCEDIT provides a number of ways to move the cursor around the file. The most obvious method is through use of the arrow keys ($\leftarrow$, $\rightarrow$, $\updownarrow$, $\updownarrow$). Table 6.1 lists all the cursor-movement keys available in PCEDIT.

Table 6.1
Cursor-Movement Keys

Key	Moves Cursor to
$\leftarrow$	One space to the left of the cursor
$\rightarrow$	One space to the right of the cursor
$\uparrow$	Up one line
$\downarrow$	Down one line
Tab	Four spaces to the right
Shift-Tab	Four spaces to the left
Enter	Next line, first column
Home	First character on the left end of the line
End	One space to the right of the last character on the right end of the line
PgUp	21 lines up
PgDn	21 lines down

Key	Moves Cursor to
Ctrl-PgUp	Top of file (same as Ctrl-Home)
Ctrl-Home	First screen of file (same as Ctrl-PgUp)
Ctrl-PgDn	Last screen of file
Backspace	Delete character to left of cursor

Copying, Moving, or Deleting a Block of Text

One of the more popular features of word processing programs is the capability to "cut" a block of text from one position in a file and "paste" it to a different position in the file. PCEDIT includes this cut-and-paste capability as well as the capability to copy or delete a block of text.

To begin marking a block of text that you want to copy, move, or delete, position the cursor in the first line of the target text and press Alt-B (Mark Block). PCEDIT highlights in inverse video the line of text that contains the cursor and displays the following message in the message line:

```
Move cursor to end of block then press Alt-B. Press Esc to abort
```

Use the cursor-movement keys to position the cursor in the last line of the block of text. PCEDIT highlights all the lines of text in the block, including the line that contains the cursor. Once the cursor is positioned in the last line of the block, press Alt-B (Mark Block) again. PCEDIT beeps and displays a new message in the message line:

```
Locate cursor then press Alt-C to COPY, Alt-M to MOVE,
or Alt-D to DELETE block.
```

In order to copy the marked block of text to a different position in the file, reposition the cursor to the line above where you want the copy to begin and press Alt-C (Copy Block). PCEDIT beeps and inserts a copy of the marked block of text beginning in the line below the cursor. Existing lines of text are pushed down to make room for the copied text. PCEDIT also turns off the highlighting of the marked block of text, and displays the message Block COPY complete in the message line. The marked block remains in the original location, resulting in two copies of the same lines of text in different positions in the file.

Sometimes you want to move a block of text from one position to another (the operation usually called "cut and paste"), leaving only one copy of the block. Mark the block and then position the cursor in the line above where you want the block to begin. Press Alt-M (Move Block). PCEDIT removes the marked text from the original location, inserts it again beginning in the row beneath the cursor, and displays the message Block MOVE complete.

When you want to delete the marked block, press Alt-D (Delete Block). PCEDIT beeps and asks you to confirm the deletion, displaying the question DELETE marked block? (Y/N) in the message line. PCEDIT suggests a response of Yes to this prompt. To complete the deletion, press Enter or choose Yes. PCEDIT removes the marked lines of text from the file and displays the message Block DELETE complete. If you change your mind, and decide you don't want to delete the marked text, choose No when asked to confirm the deletion. Once you have deleted the block, however, you have no way to "undelete" it, except to type it again.

The block copy, move, and delete commands discussed in this section are listed in table 6.2. The paragraphs that follow discuss the other keyboard commands available in PCEDIT and listed in table 6.2.

Table 6.2
PCEDIT Commands

Key	Name	Function
F1 or Alt-I	Insert Line	Insert a blank line at the cursor
F2 or Alt-D	Delete Line	Delete the line at the cursor
F3	Save	Save file, return to editing
F4	File	Save file, return to PROCOMM PLUS
F5 or Alt-X	Quit	Return to PROCOMM PLUS
F6 or Alt-S	Search	Search for a character string
Alt-B	Mark Block	Mark a block of text for copy, move, or delete
Alt-C	Copy Block	Copy a marked block of text
Alt-D	Delete Block	Delete a marked block of text
Alt-M	Move Block	Move a marked block of text
Alt-Z	Help	Display help screen
Insert		Toggle Insert and Overtype modes
Delete		Delete character at cursor

Deleting Text

In addition to the block-delete procedure described in the preceding section, PCEDIT enables you to delete characters from the file in several other ways. You can delete text character-by-character, line-by-line, or a portion of a line at a time.

As you type new text, the easiest way to delete a typo is to press the Backspace key. PCEDIT moves the cursor one space to the left, deleting the character or space that was in that position. For example, assume that you just typed *modme* and the cursor is resting one space to the right of the letter *e*. You really meant to type the word *modem*. To correct the mistake, you press the Backspace twice, deleting first the *e* and then the *m*. Finally, you type *em*, and the error is corrected.

You can also delete a character with the Del key. Position the cursor on the character or space that you want to delete and press Del. PCEDIT removes this character or space and "pulls" to the left any text that is on the same line and to the right of the cursor.

PCEDIT also enables you to delete a portion of a line of text, from the cursor to the right end of the line, with a single command. Position the cursor on the first character you want to delete and press Ctrl-End (Erase). PCEDIT deletes all the characters on the line from the cursor to the right end of the line.

Finally, you can delete the entire line of text at the cursor by pressing either F2 (Delete Line) or Alt-D (Delete Line).

Note: The keystroke combination Alt-D is used in two different ways in PCEDIT. When you have marked a block using Alt-B (Mark Block), Alt-D deletes the entire marked block. On the other hand, when no block is marked, Alt-D deletes the single line of text that contains the cursor.

Searching for a Character String

Because you can create up to 500 lines of text using PCEDIT, you may have a hard time locating the particular line of text you want to edit. PCEDIT provides a convenient *search* feature to help you. It can find a given group of characters anywhere in the file.

To use the PCEDIT search feature, press F6 (Search) or Alt-S (Search). In the message line, PCEDIT displays the prompt Enter search string. Type any combination of characters, including spaces, up to 58 characters in length, and press Enter. PCEDIT searches the file for matching character strings, from the position of the cursor toward the bottom of the file. If no match is found by the time the search reaches the end of the file, PCEDIT goes to the top of the file and searches down to the cursor. If PCEDIT finds no matching character string in the entire file, the program displays the message No match found.

When PCEDIT does find a match, it beeps and displays the message Found! PCEDIT also places the line that contains the target character string at the top of the screen.

For example, suppose that you are editing the CSERVE.ASP script file and want to find a line that includes the word Password. Press F6 (Search), and PCEDIT displays the message Enter search string. Type the word *Password* to the right of the prompt, as shown in figure 6.7, and press Enter. PCEDIT searches for any occurrence of the target character string.

```
PCEDIT 1.1B  Filename:CSERVE.ASP    Alt-Z:Help    F3:Save    F4:File    F5:Quit
Enter search string: Password
!....+....1....+....2....+....3....+....4....+....5....+....6....+....7....+....
* * * Top of file * * *
;******************************************************************************
;* CSERVE.ASP  (C) 1987 DATASTORM TECHNOLOGIES, INC.                         *
;*                                                                           *
;* A sample ASPECT script file for logging onto CompuServe via local        *
;* CIS number.                                                               *
;*                                                                           *
;******************************************************************************

ASSIGN S0 "555-5555"        ; enter your local CIS number here
ASSIGN S1 "uuuuu,uuuu"      ; enter your USERID here
ASSIGN S2 "ppppp,pppppp"    ; enter your PASSWORD here

EMULATE VT102               ; emulate VT102 terminal

IF NOT LINKED               ; If not called via Dial Dir link
   CLEAR
   TIME S8 0
   DATE S9
   FATSAY 0 0 31 "Dialing CompuServe at %s on %s" S8 S9
   SET BAUD 1200            ; configure line settings
   SET PARITY EVEN
```

Fig. 6.7.

Using the PCEDIT search feature.

The target character string can be in any position in a word. Also, the PCEDIT search feature is not case sensitive; PCEDIT considers upper- and lowercase letters to be the same. In the example, the PCEDIT search feature considers the words *Password*, *PASSWORD*, *password*, and *Passwords* all as matches for the target character string *Password*.

When you want to continue searching the file for the next occurrence of the same target word, press F6 (Search) or Alt-S (Search) again. This time the prompt for a search string still contains the string you typed previously, so just press Enter to continue the search.

Saving the File and Returning to PROCOMM PLUS

As you work with PCEDIT, the file you see displayed on the screen is held in your computer's memory (random-access memory, or RAM). This memory is temporary. As soon as you quit PCEDIT, the text is removed from memory. In order to preserve the file you create or edit in RAM, you must save the file to a disk file. PCEDIT provides several methods for doing this.

Any time you are working on a file in PCEDIT (or any other text editor or word processor), save the file to disk periodically (perhaps every 15 minutes) and continue working. Press F3 (Save). PCEDIT beeps and displays the message Saving file.... This practice ensures that a power outage or some other premature termination of PCEDIT will not result in complete loss of all your work. All changes made up to the time you saved the file to disk are safe.

When you are finished with the file and want to return to PROCOMM PLUS, press F4 (File). PCEDIT saves the file to disk, displays the message Saving file..., and terminates the program. Assuming that you started PCEDIT from PROCOMM PLUS (rather than from DOS), you are returned to the Terminal mode screen.

Occasionally, you may want to abandon the file or changes you have made in PCEDIT. You want to quit PCEDIT without saving the file to disk. Press F5 (Quit) or Alt-X (Quit). PCEDIT displays the following prompt:

    Save changes before exiting? (Y/N)

Respond No, the default response, to quit without saving the file. Answer Yes in order to save the file and then quit (same as F4 (File)). Assuming that you started PCEDIT from PROCOMM PLUS (rather than from DOS), you are returned to the PROCOMM PLUS Terminal mode screen.

Using PCEDIT Macros

Chapter 7, "Automating PROCOMM PLUS with Macros and Script Files," shows you several ways to make PROCOMM PLUS even easier to use than it already is, including recording your keystrokes as a script file. Chapter 11, "An Overview of the ASPECT Script Language," goes further in the discussion of script files and explains how you can design a script file from scratch by using the ASPECT script language. Whether you record a script file and simply want to make a minor change or are more ambitious and decide to create a script from the ground up, PCEDIT may be all the text editor you need.

Of particular interest for creating or editing script files are PCEDIT's 20 macros. Each PCEDIT macro command types one of 20 commands from the ASPECT command language. While these macros represent only a portion of the nearly 100 ASPECT commands, they include many of the commands that are used most often in scripts.

Table 6.3 lists the available PCEDIT macro commands. To type one of these commands, first position the cursor where you want the ASPECT command to begin and press the macro keystroke. For example, press Alt-P to insert the ASPECT command PAUSE. (When executed in a script, this command causes the script to suspend execution for a specified number of seconds.) PCEDIT types *PAUSE* beginning at the cursor. Refer to Chapters 7 and 11 in this book, as well as to Chapter 11 in the PRO-COMM PLUS documentation for a discussion of the other ASPECT commands listed in table 6.3.

Table 6.3
Macro Commands

Key	*Macro*	*Key*	*Macro*
Alt-A	ASSIGN	Alt-F1	IF
Alt-C	CLEAR	Alt-F2	ELSE
Alt-F	FATSAY	Alt-F3	ENDIF
Alt-G	GET	Alt-F5	SWITCH
Alt-H	HANGUP	Alt-F6	CASE
Alt-L	LOCATE	Alt-F7	ENDCASE
Alt-M	MESSAGE	Alt-F8	ENDSWITCH
Alt-P	PAUSE	Alt-F9	ATSAY
Alt-R	RGET	Alt-F10	ATGET
Alt-T	TRANSMIT "		
Alt-W	WAITFOR "		

Using a Different Editor

Although PCEDIT is handy, it obviously has its shortcomings. With PRO-COMM PLUS, you can easily attach one of the other excellent text editors available so that this editor can be invoked from within PROCOMM PLUS.

For example, assume that you want to use an editor named NU-Edit (a hypothetical text editor). You first install NU-Edit in the same directory as

PROCOMM PLUS. Then, you add the start-up command for NU-Edit to the Setup Utility File/Path Options screen. In this example, the start-up command is NU-EDIT, so this command must be substituted for *PCEDIT* in option **D** on the screen shown in figure 6.8. Refer to Chapter 8, "Tailoring PROCOMM PLUS," for instructions on how to make this change.

Fig. 6.8.

The Setup Utility File/Path Options screen.

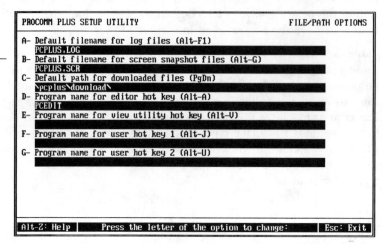

```
PROCOMM PLUS SETUP UTILITY                        FILE/PATH OPTIONS
A- Default filename for log files (Alt-F1)
   PCPLUS.LOG
B- Default filename for screen snapshot files (Alt-G)
   PCPLUS.SCR
C- Default path for downloaded files (PgDn)
   \pcplus\download\
D- Program name for editor hot key (Alt-A)
   PCEDIT
E- Program name for view utility hot key (Alt-U)

F- Program name for user hot key 1 (Alt-J)

G- Program name for user hot key 2 (Alt-U)


 Alt-Z: Help       Press the letter of the option to change:      Esc: Exit
```

After you have added the proper start-up command for your editor to the File/Path Options screen, you can activate the editor from the PROCOMM PLUS Terminal mode screen in the same manner as you formerly activated PCEDIT. Press Alt-A (Editor), and PROCOMM PLUS displays a narrow window across the Terminal mode screen prompting you to Enter parameters (see fig. 6.9). If your editor accepts a file name as a parameter (most do), type the name of a file, and press Enter. Some text editors even accept multiple file names as parameters, opening a separate window for each. Refer to your text editor's documentation for information on creating or editing files.

```
 ┤ NU-EDIT Parameters ├
  Enter parameters:
```

Fig. 6.9.

Entering parameters for the hypothetical text editor, NU-Edit.

```
 Alt-Z FOR HELP  ANSI      FDX    2400 N81   LOG CLOSED   PRINT OFF   OFF-LINE
```

When you finish with the text editor and execute the editor's command to return to DOS, you will return to PROCOMM PLUS's Terminal mode screen.

Chapter Summary

This chapter has described how to use the text editor PCEDIT, distributed with PROCOMM PLUS, in order to create and edit ASCII files. The chapter described how to enter text; move around a file; make changes to the text in an existing file; copy, move, or delete portions of the text; search for character strings; and save ASCII text to a disk file. This chapter also listed the 20 PCEDIT macros that are of particular interest when you create PROCOMM PLUS script files and explained how to start a different text editor program of your choosing from within PROCOMM PLUS.

One of the major uses for PCEDIT is to edit PROCOMM PLUS scripts. Turn now to Chapter 7 to learn how to automate your use of PROCOMM PLUS with keyboard macros and PROCOMM PLUS scripts.

Automating PROCOMM PLUS with Macros and Script Files

W hen you discover a task you perform repeatedly on your computer, look for a way to get the computer to do most of the work. Your computer can perform most tasks faster than you can, and computers don't make mistakes. PROCOMM PLUS provides two different features with which you can take advantage of the computer's capability to perform repetitive tasks quickly and correctly. These features are keyboard macros and script files.

A *keyboard macro* is a string of characters representing keystrokes. The macro is stored, along with up to 9 other macros, in a file on the disk and is available to be "played out" with one keystroke. Chapter 6, "Using the PROCOMM PLUS Editor," describes 20 built-in keyboard macros available for use in PCEDIT. This chapter explains how to create one or more sets, each containing up to 10 keyboard macros for use in PROCOMM PLUS.

A PROCOMM PLUS *script file* is a short program written in the ASPECT script (programming) language. This chapter demonstrates how to create a simple ASPECT script without programming, explains how to create a script by simply recording keystrokes, discusses several script files that DATASTORM distributes with PROCOMM PLUS, and explains how to customize them for your own use. Finally, the chapter describes the methods for activating PROCOMM PLUS scripts. This chapter is only an introduction to PROCOMM PLUS scripts. After you are comfortable with the script techniques presented in this chapter, read Chapter 11, "An Overview of the ASPECT Script Language."

Understanding Keyboard Macros

Many aspects of PC communication become routine; you type certain words or phrases frequently. For example, you log on to an on-line computer service by typing your identification number and then a password; each time you log on to a bulletin board, you type your name and a password. Typing these items is not hard work, but it does take time and holds a potential for typing errors. For example, if your ID number is *2974,ARQ*, you might type *2947,AQR* by mistake.

In order to help you save time and keystrokes and to reduce typographical errors, PROCOMM PLUS enables you to assign any string of up to 50 characters (including spaces) to an Alt-key combination referred to as a *keyboard macro*. When you press the Alt-key combination, PROCOMM PLUS types the entire set of characters in the correct order. The paragraphs that follow describe how to create, modify, save to disk, and use PROCOMM PLUS keyboard macros.

Creating Macros

The first step to create a keyboard macro is to display the Keyboard Macros screen. Press Alt-M (Keyboard Macros), and PROCOMM PLUS displays the screen shown in figure 7.1. Using this screen, you can create up to 10 macros. The left side of the screen contains a list of ten Alt-key combinations, the *macro names* (from Alt-0 to Alt-9). To the right of each macro name is a blank area, to which this chapter refers as the *macro entry*. You create a keystroke macro by typing characters in a macro entry.

To add a macro to the Keyboard Macros screen, use the up-arrow or down-arrow key to move the highlighted bar up or down in the list. Position the highlighted bar on a blank macro entry, and press **R** (Revise). A cursor begins to blink at the left end of the macro entry; you can now type up to 50 characters. You can use the cursor-movement and editing keys listed in table 7.1 while typing characters in the macro entry.

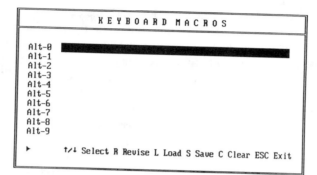

Fig. 7.1.

The Keyboard Macros screen.

Table 7.1
Keyboard Macros Screen Cursor-Movement and Editing Keys

Key	Function
←	Move the cursor one space to the left
→	Move the cursor one space to the right
Home	Move the cursor to the left end of the macro entry
End	Move the cursor one space to the right of the last character in the macro entry
Insert	Toggle Insert/Overtype modes
Delete	Delete the character at cursor
Backspace	Delete the character to the left of cursor
Tab	Delete the entire macro entry
Ctrl-End	Delete the characters from cursor to right end

When you execute a keyboard macro (discussed in "Using Keyboard Macros," in this chapter), letters and numbers translate into their corresponding keystrokes. For example, suppose that you type *BR549* in the macro entry. Later, when you execute the macro, PROCOMM PLUS types the same characters: *BR549*.

A number of special macro codes are referred to as *control codes*. These codes are called control codes because the first character in each code, the caret (^), represents the Ctrl key on the keyboard. For example, when you execute a macro that contains the code *^C*, PROCOMM PLUS does not type the caret (^) and then the letter *C*. Instead, PROCOMM PLUS transmits to the remote computer the same code that is sent if you press Ctrl-C, a keystroke combination that may have a special meaning to a particular on-line service or bulletin board. Many on-line services and bulletin boards use Ctrl-key combinations, such as Ctrl-C, Ctrl-X, Ctrl-S, and Ctrl-Q, to enable you to control the flow of information across your screen or to cancel an operation in midstream. To incorporate these keystroke combinations into a macro, you must use the codes listed in table 7.2. Table 7.2 lists only the Ctrl-key combinations most often used by on-line services and bulletin boards. In general, you can create any Ctrl-key combination by typing the caret (^) followed by a letter (A through Z, upper- or lowercase) or one of the following characters:

[ ] \ ^ _

Several of the control codes in table 7.2 translate into keystrokes that you would not normally expect. For example, suppose you type the characters ^M in a macro entry. When you execute the macro, PROCOMM PLUS sends a carriage return to the remote computer for the ^M. In other words, when you want the macro to press the Enter key, use the code ^M.

Table 7.2
Macro Control Codes

Code	Keystroke Executed by Macro
^C	Ctrl-C
^S	Ctrl-S
^Q	Ctrl-Q
^X	Ctrl-X
^M	Enter
^H	Backspace
^I	Horizontal tab (Tab key)
^J	Line feed
^K	Vertical tab
^[	Esc

You may want to create one macro to enter your CompuServe ID number and another to enter your password. Your ID is *12345,6789* and your password is *NICE DAY*. To create the ID macro, press Alt-M (Keyboard Macros) to display the Keyboard Macros screen; move the highlighted bar to an empty entry, and press **R** (Revise). Type *12345,6789^M* to the right of Alt-0, and press Enter. To create the password macro, move the highlight to another empty macro entry, and press **R** (Revise). Type *NICE DAY* ^M, and press Enter (see fig. 7.2.) Both macros end with ^M so that the macro sends the carriage return symbol. (Note: Storing a password as a macro may not be a good practice, because anyone with access to your computer and PROCOMM PLUS can display the list of macros and discover your password.)

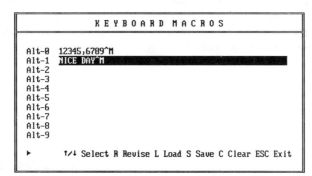

Fig. 7.2.

Keystroke macros to enter an ID number and a password.

The *pause character* has a special meaning in a PROCOMM PLUS macro entry. The pause character is one of the PROCOMM PLUS settings you can change on the Setup Utility General Options screen (refer to Chapter 8, "Tailoring PROCOMM PLUS," for information on the Setup Utility). As shown at option **F** (Pause character) in figure 7.3, the default setting for the pause character is the tilde (~). When you use the pause character

```
PROCOMM PLUS SETUP UTILITY                              GENERAL OPTIONS

A- Exploding windows ... ON        K- Menu line key ....... M

B- Sound effects ....... ON        L- Snow removal ........ OFF

C- Alarm sound ......... ON        M- Remote commands ..... OFF

D- Alarm time .......... 5   seconds   N- Enhanced kb speedup . ON

E- Translation table ... OFF       O- ANSI compatibility .. 3.x

F- Pause character ..... ~

G- Transmit pacing ..... 0   milliseconds

H- Call logging ........ ON

I- Filename lookup ..... ON

J- Menu line ........... ON

Alt-Z: Help  |  Press the letter of the option to change:  |  Esc: Exit
```

Fig. 7.3.

The Setup Utility General Options screen.

in a macro entry, PROCOMM PLUS translates this character into a half-second pause during execution of the macro. The pause character also has other uses, which are explained in "Setting Modem Options," in Chapter 8.

TIP | Don't confuse the pause character (˜) with the comma (,) that is used to pause dialing by a Hayes-compatible modem. Refer to "Adding an Entry," in Chapter 3, "Building Your Dialing Directory," for more information about using a comma to pause dialing.

Saving Macros to a Macro File

After you type a macro entry and press Enter, PROCOMM PLUS stores the macro in memory (RAM). The macro is not automatically saved permanently to a disk file. Unless you take steps to save the macro to disk, the macro can be used for the current PROCOMM PLUS session only.

To save the macro entry or entries, as they appear in the Keyboard Macros screen, choose **S** (Save). Near the top of the Keyboard Macros screen, PROCOMM PLUS displays a small window containing the prompt Save to file. PROCOMM PLUS also displays PCPLUS.KEY as a suggested response to this prompt (see fig. 7.4). Each time you start the program, PROCOMM PLUS assumes that your macros are stored in the file PCPLUS.KEY; so you normally press Enter at this prompt. PROCOMM PLUS saves the current macro entry or entries to the specified file.

Fig. 7.4.

Saving keyboard macros to the file PCPLUS.KEY.

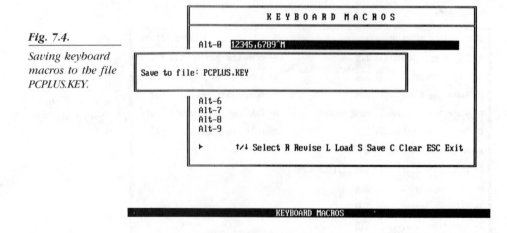

You have the option to save the current macros under a different file name. This feature enables you to create more than 10 macros. No more than 10 macros will fit on one Keyboard Macros screen; but by saving macros in separate files, you can create an unlimited number of macros.

For example, you may want to use one group of 10 macros only when connected to the CompuServe on-line service. Each macro takes you directly to a different service of CompuServe. Because these macros are of no use for any other purpose, they need to be accessible only while you are using CompuServe. To create this set of CompuServe macros, first display the Keyboard Macros screen; add the 10 macros, replacing any macros that are there (see fig. 7.5). Press **S** (Save), and type *cserve.key* as the name of the macro file; press Enter. PROCOMM PLUS saves your CompuServe macros to the disk file CSERVE.KEY. Your CompuServe macros are still current—available for use when you connect to CompuServe. The next time you start the program, however, PROCOMM PLUS again reads the set of macros found in the file PCPLUS.KEY. Refer to "Loading a Macro File," which follows, for an explanation of how to reactivate your set of CompuServe macros.

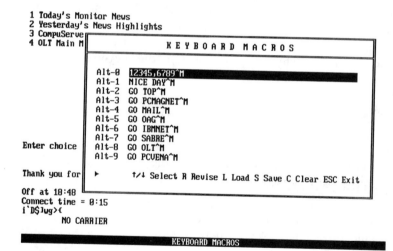

Fig. 7.5.

Keyboard macros for use while connected to the on-line service CompuServe.

Modifying Macros

The procedure for modifying a macro you have created is almost the same as the procedure for creating a macro. Press Alt-M (Keyboard Macros) to display the Keyboard Macros screen (again see fig. 7.1). Use the cursor-movement keys listed in table 7.1 to move the highlighted bar to the macro entry you want to modify. Press **R** (Revise), and PROCOMM PLUS moves the blinking cursor to the left end of the highlighted macro entry. You can then use any of the editing keys listed in table 7.1 to modify the macro entry. When you have made the needed changes, press Enter. PROCOMM PLUS saves the change to memory.

TIP | Don't forget to save the modified version of the macro entry to the disk file. Refer to "Saving Macros to a Macro File" for more information.

Loading a Macro File

Each time you start PROCOMM PLUS, it reads only the macro file named PCPLUS.KEY. When you want to use a different set of macros, you must load into memory the macro file that contains the macro set you need.

To load a macro file, press Alt-M (Keyboard Macros) to display the Keyboard Macros screen; then press **L** (Load). PROCOMM PLUS displays a small window on the Keyboard Macros screen. The window contains the prompt File to load, followed by the file name of the macro file currently in memory. Type the name of the macro file you want PROCOMM PLUS to load, and press Enter. PROCOMM PLUS loads the set of macros from the disk file and immediately displays them in the Keyboard Macros screen.

Before you dial CompuServe, you can load the set of macros you created for use with the on-line service (again see fig. 7.5). Press Alt-M (Keyboard Macros) to display the Keyboard Macros screen; then press **L** (Load). Type the file name *cserve.key* to the right of the prompt, and press Enter. PROCOMM PLUS loads the set of macros from the disk file and displays them in the Keyboard Macros screen.

When you have finished using the auxiliary set of macros, you may want to return to the set contained in the file PCPLUS.KEY. To access the PCPLUS.KEY file, press Alt-M (Keyboard Macros) to display the Keyboard Macros screen; press **L** (Load); type *pcplus.key*, and press Enter. PROCOMM PLUS reloads the macros from the disk file PCPLUS.KEY and displays them in the Keyboard Macros screen.

Using Keyboard Macros

Once you have created a macro, using it is simple. To use a macro, press the Alt-key combination listed on the left side of the Keyboard Macros screen for the macro you want to use. For example, to execute the first macro listed in figure 7.5, press Alt-0. PROCOMM PLUS types *12345,6789* and then "presses" Enter. For the second macro in figure 7.5, press Alt-1. PROCOMM PLUS types *NICE DAY*, and "presses" Enter.

Creating PROCOMM PLUS Scripts

The next step in automating PROCOMM PLUS is use of PROCOMM PLUS *scripts* (called *command files* in ProComm). A keyboard macro enables you to condense no more than 50 keystrokes into each Alt-key combination. A PROCOMM PLUS script can execute any number of keystrokes. A PROCOMM PLUS script is essentially a program. You can design the script to do much more than press keystrokes. A script can wait for a particular prompt from the remote computer before executing a set of keystrokes. You can even write a PROCOMM PLUS script to automate PROCOMM PLUS completely. Users can connect to other computers, read mail, upload and download files, and perform other functions, all by selecting options from menus of your own design. Scripts can also be set to run unattended at a predetermined time.

As your introduction to PROCOMM PLUS scripts, this portion of Chapter 7 demonstrates two easy ways to create a script: by recording your keystrokes and by customizing a script supplied by DATASTORM.

Recording a Script

When you create a macro, you are in a sense "teaching" it a sequence of keystrokes. PROCOMM PLUS enables you create simple scripts in a similar way. Instead of typing the keystrokes in a macro entry space, you turn on PROCOMM PLUS's *Record mode*, a keystroke recorder that records your keystrokes while you are on-line to a remote computer. Not only does Record mode record your keystrokes, but Record mode also records the remote system's prompts.

The Record mode records keystrokes only while you are connected to another computer. You can record the log-on sequence needed to connect to another computer by starting the recorder just before dialing the connection. The keystroke recorder is probably most often used to record a log-on sequence.

To activate Record mode, from the Terminal mode screen, press Alt-R (Record Mode). PROCOMM PLUS displays in the center of the screen a small window containing the prompt Enter script filename. Type a valid DOS file name, including the file name extension .ASP, and press Enter.

(The .ASP file-name extension designates the file as a script file written with the ASPECT script language.) If you type a file name that already exists in the current working directory, PROCOMM PLUS displays the following prompt:

    File already exists. Overwrite? (Y/N)

PROCOMM PLUS suggests Yes as the response; to continue using the same file name, just press Enter. If you don't want to replace the existing file, respond No, and PROCOMM PLUS returns to the preceding prompt. You then type a different file name, and press Enter.

After you specify a name for the script file, PROCOMM PLUS displays the message RECORDING... at the left end of the Terminal mode screen status line. This message is your clue that PROCOMM PLUS is in Record mode.

When you are connected to the remote computer, PROCOMM PLUS looks for your first keystroke. When you press a key, PROCOMM PLUS records the preceding 15 characters that were received from the remote computer. PROCOMM PLUS assumes that these 15 characters "prompted" your response. PROCOMM PLUS also records all your keystrokes. While Record mode is active and you are on-line to a remote computer, PRO-COMM PLUS continues to record all your keystrokes, as well as the remote computer's prompts. When you execute the script, PROCOMM PLUS plays back only your keystrokes when it gets the matching prompt from the remote computer. To finish recording keystrokes, press Alt-R (Record Mode) again.

For example, assume that you want to record your log-on sequence to connect to a PC that is running PROCOMM PLUS in Host mode. (Host mode is discussed in Chapter 9, "Using Host Mode.") First press Alt-R (Record Mode); type the name of the script file *pcp-host.asp*, and press Enter. PROCOMM PLUS displays the message RECORDING... in the status line, indicating that the program is in Record mode. Next, use the Dialing Directory screen to dial and connect to the remote PC. Record mode does not start recording until you connect to the remote computer.

After you connect, the remote computer displays the following message:

    Welcome to PROCOMM PLUS Host!

Then PROCOMM PLUS prompts you for your First name. In response, you type your first name, *Sam*, a semicolon (;), and your last name *Spade*, and then press Enter. (Typing first and last names on the same line, separated by a semicolon is a shortcut available in Host mode as well as in most bulletin board programs.)

After you type your name, the remote computer displays your name and asks Is this correct (Y/N)?. Assuming that you didn't misspell your name, press Y to respond Yes, and press Enter.

Finally, the remote computer asks for your password; type *falcon*, and press Enter. The log-on procedure is now completed, and the remote computer displays a menu of options. The entire sequence of prompts and responses in this example is shown in figure 7.6.

```
Welcome to PROCOMM PLUS Host!


First name: Sam Spade
SAM SPADE
Is this correct (Y/N)? Y

Password: ******
```

Fig. 7.6.

Recording a log-on sequence of prompts and responses.

```
F)iles U)pload D)ownload
H)elp T)ime C)hat G)oodbye
R)ead mail L)eave mail

Your choice?
 RECORDING...   ANSI     FDX    2400 N81   LOG CLOSED   PRINT OFF   ON-LINE
```

Now that you have finished your log-on sequence, you can turn off the recorder. Press Alt-R (Record Mode). PROCOMM PLUS saves the script to the designated file, PCP-HOST.ASP, and returns to the Terminal mode screen.

TIP | While in Record mode, make sure that you press Enter after each response to a prompt from the remote computer. Otherwise, PROCOMM PLUS does not record the next prompt from the remote computer.

You can use PROCOMM PLUS's file-viewing feature (discussed in Chapter 4, "A Session with PROCOMM PLUS") to take a look at a script file created by Record mode. Press Alt-V (View a File); type the name of the script file you want to view and press Enter. PROCOMM PLUS displays the file. Figure 7.7 shows the PCP-HOST.ASP script file generated by PROCOMM PLUS's Record mode keystroke recorder.

A PROCOMM PLUS script file consists of one or more lines of ASCII characters. Each line contains one command from the ASPECT script language.

PROCOMM PLUS executes these commands one by one, from top to bottom, unless it encounters a command that causes the execution to branch to some other portion of the script (refer to "Branching Commands," in Chapter 11, for a list of such commands).

Fig. 7.7.

The PCP-HOST.ASP script file created by the Record mode keystroke recorder.

```
;
; PROCOMM PLUS generated ASPECT script file - Editing may be required.
;
WAITFOR "^M^J^M^JFirst name: "
PAUSE 1
TRANSMIT "Sam;Spade^M"
WAITFOR " correct (Y/N)? "
PAUSE 1
TRANSMIT "y^M"
WAITFOR "^M^JPassword: "
PAUSE 1
TRANSMIT "falcon^M"
```

```
Home: Top of file      PgUp: Previous page      PgDn: Next page      ESC: Exit
```

The script shown in figure 7.7 is generated by the Record mode keystroke recorder and contains no branching commands. PROCOMM PLUS's Record mode generates only three commands of the ASPECT script command language. Figure 7.7 demonstrates all three of these ASPECT commands:

- *WAITFOR*. This command causes the script to pause (do nothing) until a specified character string is received from the remote computer. When your computer receives the anticipated character string, PROCOMM PLUS continues with the next command in the script. The Record mode places in a WAITFOR command the last 15 characters received from the remote computer before you typed a particular response. In the PCP-HOST.ASP example, the remote computer sent a carriage return (^M), a line feed (^J), another carriage return (^M), the characters *First name:*, and one space immediately before you typed *Sam;Spade*, and pressed the Enter key. The Record mode keystroke recorder generated the following command in the PCP-HOST.ASP script:

  ```
  WAITFOR "^M^J^MFirst name: "
  ```

- *PAUSE.* The PAUSE command causes the script to pause for a specified number of seconds. After the indicated time has expired, the script proceeds with the next command. The Record mode keystroke recorder always adds this command after each WAITFOR command, specifying a one-second pause. These PAUSE commands are often not necessary.

- *TRANSMIT.* This command sends a specified string of characters to the remote computer. The effect of this command is equivalent to executing a keyboard macro containing the same string of characters. While you are in Record mode, the program generates a TRANSMIT command every time you type a string of characters terminated by a carriage return (Enter). In the example, your name was the first thing you typed after connecting to the remote computer. The Record mode therefore generated the following ASPECT command line:

      TRANSMIT "Sam;Spade^M"

Any text or other characters that appear on a line to the right of a semi-colon (;) not enclosed in quotation marks are ignored when PROCOMM PLUS executes the script. This text is used to place comments, often called *internal documentation*, into the script for your future reference. The first three lines of PCP-HOST.ASP indicate that the ASPECT script file was generated by PROCOMM PLUS—Record mode—and that editing may be necessary.

Modifying a Script

When you use Record mode to record a PROCOMM PLUS script, you may need to fine-tune the script. In order to display the script for editing, you can use the text editor distributed with PROCOMM PLUS, PCEDIT, or you can use your own favorite ASCII text editor. To use PCEDIT (or any other editor that you have attached through the Setup Utility File/Path Options screen, discussed in Chapter 8, "Tailoring PROCOMM PLUS"), press Alt-A (Editor), type the name of the script file, and press Enter.

Using the example discussed in the preceding section, you may want to edit the PCP-HOST.ASP script file. Press Alt-A (Editor) to display the prompt asking you to Enter parameters. Type *pcp-host.asp*, and press Enter. PROCOMM PLUS displays the script file, as shown in figure 7.8.

The script shown in figure 7.8 will work as required, automating the procedure for logging on to a PC that is running PROCOMM PLUS in Host mode. With some editing, however, this script can be easier to follow.

```
PCEDIT 1.1B  Filename:PCP-HOST.ASP   Alt-Z:Help   F3:Save   F4:File   F5:Quit
!....+....1....+....2....+....3....+....4....+....5....+....6....+....7....+....
* * * Top of file * * *
;
; PROCOMM PLUS generated ASPECT script file - Editing may be required.
;
WAITFOR "^M^J^M^JFirst name: "
PAUSE 1
TRANSMIT "Sam;Spade^M"
WAITFOR " correct (Y/N)? "
PAUSE 1
TRANSMIT "y^M"
WAITFOR "^M^JPassword: "
PAUSE 1
TRANSMIT "falcon^M"
* * * End of file * * *
```

Fig. 7.8.

Using PCEDIT to edit a script file.

Using PCEDIT (refer to Chapter 6, "Using the PROCOMM PLUS Editor"), you can modify the script as shown in figure 7.9, deleting unnecessary text and PAUSE commands and adding comments that explain each step of the script.

```
PCEDIT 1.1B  Filename:PCP-HOST.ASP   Alt-Z:Help   F3:Save   F4:File   F5:Quit
!....+....1....+....2....+....3....+....4....+....5....+....6....+....7....+....
* * * Top of file * * *
;
; A script to sign-on to a remote PC running PROCOMM PLUS in Host mode.
;
WAITFOR "First name: "              ; Wait for "First name:" prompt.
TRANSMIT "Sam;Spade^M"              ; Send first and last names.
WAITFOR "correct (Y/N)? "           ; Wait for "correct (Y/N)?" prompt.
TRANSMIT "Y^M"                      ; Send "y"
WAITFOR "Password:"                 ; Wait for "Password:" prompt.
TRANSMIT "falcon^M"                 ; Send password.

* * * End of file * * *
```

Fig. 7.9.

An edited version of PCP-HOST.ASP.

Customizing a Predefined Script

As you can see from "Recording a Script," in this chapter, the Record mode generates scripts from only three script commands. These simple scripts are best suited for automating log-on sequences. The ASPECT script language contains more than 90 commands. To take full advantage of its robust capabilities, you must go beyond Record mode.

One alternative to using Record mode is to write a script from scratch. This method is certainly possible and may be preferable after you are familiar with the ASPECT script language. Sometimes, however, you can find examples provided by DATASTORM and others, that you can tailor to your needs. By customizing a script an experienced PROCOMM PLUS user has designed, you can save time and discover many useful ASPECT programming techniques.

The PROCOMM PLUS Program Diskette contains a sample ASPECT script file, DSTORM.ASP, which can automate logging on to the DATASTORM bulletin board. The Supplemental Diskette that is distributed in the PRO-COMM PLUS package contains 17 more predefined ASPECT script files. All 18 files are provided by DATASTORM to give you a taste of what you can do with script files. These scripts can be grouped loosely into three categories: scripts to automate dialing and logging on to bulletin boards, scripts to automate dialing and logging on to on-line services, and scripts to demonstrate useful programming techniques. Table 7.3 lists the name of each script provided by DATASTORM, along with a brief description of the script's purpose.

Table 7.3
Sample ASPECT Script Files
Contained on the PROCOMM PLUS Supplemental Diskette

File Name	Purpose
BIX.ASP	Dial and log on to BIX on-line service
COLOR.ASP	Display color attributes available in PROCOMM PLUS
CSERVE.ASP	Dial and log on to CompuServe on-line service
DELPHI.ASP	Dial and log on to Delphi on-line service
DOWJONES.ASP	Dial and log on to Dow Jones News/ Retrieval
DSTORM.ASP	Dial and log on to the DATASTORM bulletin board
FASTCOMM.ASP	Initialize Fastcomm 2496/Turbo modem
FIDO.ASP	Dial and log on to Fido bulletin boards.
GENIE.ASP	Dial and log on to GEnie on-line service
MCI.ASP	Dial and log on to MCI MAIL, sound alarm if mail waiting
NOCHANGE.ASP	Dial and log on to NOCHANGE bulletin boards
PCBOARD.ASP	Dial and log on to PC-Board bulletin boards
PURSUIT.ASP	Menu for dialing 13 PC PURSUIT cities
RBBS.ASP	Dial and log on to RBBS bulletin boards

Table 7.3—*Continued*

File Name	Purpose
RBBS13.ASP	Dial and log on to RBBS Version 13.1A and later BBSs
RTABLE.ASP	Dial and log on to DATASTORM roundtable on GEnie
DJDEMO.ASP	Demonstration of ASPECT techniques and features of the Dow Jones News/Retrieval service

Of the scripts listed in table 7.3, only COLOR.ASP, and FASTCOMM.ASP can be used without modification. (Note: Do not run FASTCOMM.ASP unless you have a Fastcomm 2496/Turbo modem.) All other scripts provided on the Supplemental Diskette require you to add telephone numbers, ID numbers, passwords, and so on, before you can actually use the scripts.

Figure 7.10 shows a portion of the CSERVE.ASP script file, distributed on the PROCOMM PLUS Supplemental Diskette. This script is intended to automate dialing and logging on to the CompuServe on-line information service. Before you can use this script, you have to use an ASCII text editor (such as PCEDIT) to replace the phone number 555-5555 (shown in fig. 7.10 to the right of the ASPECT command ASSIGN S0) with the actual local access telephone number for CompuServe (obtained when you subscribe to the service). You also must replace the characters *uuuuu,uuuu* with your assigned CompuServe User ID, and the characters *ppppp.pppppp* with your CompuServe password. If you want to use a CompuServe access number that supports connection at 2400 bps, you must also edit the SET BAUD 1200 command and change it to SET BAUD 2400.

TIP The best way to use a script provided by DATASTORM is to make a copy under a different name, and then modify the copy. For example, you can use the DOS COPY command to make a copy of the CSERVE.ASP script and give it the name CSERV24.ASP. You can then modify CSERV24.ASP by adding the correct access telephone number, your User ID number, and your password and changing the transmission rate to 2400 bps. Modifying the copy enables you to customize the script as necessary to fit your requirements without altering the original. By following this practice, you have the DATASTORM version always readily available for comparison. You should *never* modify any script or other file on the original distribution diskette. Always work from your hard disk or from a working copy of the distribution diskette.

```
;**********************************************************************
;* CSERVE.ASP  (C) 1987 DATASTORM TECHNOLOGIES, INC.                *
;*                                                                  *
;* A sample ASPECT script file for logging onto CompuServe via local *
;* CIS number.                                                      *
;*                                                                  *
;**********************************************************************

ASSIGN S0 "555-5555"          ; enter your local CIS number here
ASSIGN S1 "uuuuu,uuuu"        ; enter your USERID here
ASSIGN S2 "ppppp,pppppp"      ; enter your PASSWORD here

EMULATE VT102                 ; emulate VT102 terminal

IF NOT LINKED                 ; If not called via Dial Dir link
   CLEAR
   TIME S8 0
   DATE S9
   FATSAY 0 0 31 "Dialing CompuServe at %s on %s" S8 S9
   SET BAUD 1200
   SET PARITY EVEN            ; configure line settings
   SET DATABITS 7
   SET DUPLEX FULL
   MDIAL S0                   ; Dial CIS
```

`Home: Top of file     PgUp: Previous page     PgDn: Next page     ESC: Exit`

Fig. 7.10.

The CSERVE.ASP script supplied on the PROCOMM PLUS Supplemental Diskette.

You can also get predefined ASPECT scripts from other sources. Computer bulletin boards often contain useful ASPECT script files to automate logging on to bulletin boards within the local dialing area. Script files written for DATASTORM's ProComm 2.4.3 (and earlier versions) can also be used with PROCOMM PLUS. (Refer to "Compatibility with ProComm Command Files," in Chapter 11, "An Overview of the ASPECT Script Language," for more information.)

After you have examined the scripts produced by Record mode and those provided by DATASTORM on the Supplemental Diskette, you may be interested in trying some of the techniques demonstrated in these scripts. The only way to get the most from the ASPECT script language is to write your own scripts from scratch. Chapter 11, "An Overview of the ASPECT Script Language," at the end of this book is intended to help you get started.

Running a Script

PROCOMM PLUS provides three ways to run an ASPECT script: you can cause a script to run immediately after PROCOMM PLUS begins; you can execute a script by dialing an entry from the Dialing Directory screen; or you can run a script directly from the Terminal mode screen. The purpose of a script usually dictates the method you employ to activate it.

At Start-up

Using the ASPECT script language, you can write a script to control a PROCOMM PLUS session from start to finish. Such a script normally displays a menu from which the user selects a desired action. Because you want this type of script to control access to PROCOMM PLUS, it makes sense to activate the script with the program.

To cause a script to run immediately after PROCOMM PLUS loads, include the script name in the PROCOMM PLUS start-up command, as follows (refer to Appendix A for more information on PROCOMM PLUS start-up commands):

pcplus /fscr-name.asp

Substitute the actual script name for *scr-name.asp*.

For example, the script PURSUIT.ASP, supplied with PROCOMM PLUS on the Supplemental Diskette, is intended to automate use of the GTE Telenet's PC Pursuit service by providing a menu of 13 cities to which you can dial. In order to cause this script to execute on start-up of PROCOMM PLUS, issue the following command at the DOS prompt:

pcplus /fpursuit

When PROCOMM PLUS starts. the logo screen is displayed, then the Terminal mode screen, and then the PURSUIT.ASP script is run. Figure 7.11 shows the menu generated by the script.

Fig. 7.11.

The menu generated by the script PURSUIT.ASP.

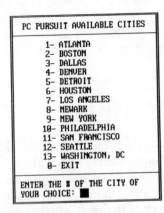

```
PC PURSUIT AVAILABLE CITIES

        1- ATLANTA
        2- BOSTON
        3- DALLAS
        4- DENVER
        5- DETROIT
        6- HOUSTON
        7- LOS ANGELES
        8- NEWARK
        9- NEW YORK
       10- PHILADELPHIA
       11- SAN FRANCISCO
       12- SEATTLE
       13- WASHINGTON, DC
        0- EXIT

ENTER THE # OF THE CITY OF
YOUR CHOICE: █
```

| PURSUIT.ASP | ANSI | FDX | 2400 N81 | LOG CLOSED | PRINT OFF | OFF-LINE |

From the Dialing Directory Screen

The most frequently used method for activating a script is through use of a dialing directory entry. As you recall from Chapter 3, "Building Your Dialing Directory," you can assign a script to each entry in a dialing directory (refer to "Adding an Entry" in Chapter 3 for more information). When you use a dialing directory entry to which you have assigned a script, PROCOMM PLUS runs the script immediately on connecting to the remote computer. Such a script is used primarily to log on to the remote computer.

Many of the sample scripts provided by DATASTORM on the PROCOMM PLUS Supplemental Diskette can be used with a dialing directory entry. For example, the RBBS13.ASP script, shown in figure 7.12, can be used with a dialing directory entry to automate logging on to a bulletin board that is running the popular RBBS-PC software (Version 13 or higher). Before you use the script, you should insert the number of the dialing directory entry in place of the number *4*, to the right of the DIAL command.

```
;**********************************************************************
;* RBBS13.ASP  (C) 1987 DATASTORM TECHNOLOGIES, INC.               *
;*                                                                  *
;* A sample ASPECT script file for logging onto RBBS version 13.1A  *
;*                                                                  *
;**********************************************************************

CLEAR
LOCATE 0 0
BOX 0 0 4 23 14
ATSAY 2 2 14 "Logging onto RBBS..."
LOCATE 6 0

IF NOT LINKED
    DIAL "4"                    ;set to your dial dir entry for RBBS
ENDIF

PAUSE 1
TRANSMIT "^M"

WAITFOR "What is your FIRST Name?"
TRANSMIT "FIRSTNAME;LASTNAME;FAKE-PASSWORD^M"

ALARM 2                        ;inform user logon complete
Home: Top of file      PgUp: Previous page      PgDn: Next page      ESC: Exit
```

Fig. 7.12.

The RBBS13.ASP script file distributed on the PROCOMM PLUS Supplemental Diskette.

> **TIP** Adding the entry number is not necessary in order to use the script from the Dialing Directory screen, because you have already specified the dialing directory entry. But by adding the entry number to the script, you have the option to run the script from the Terminal mode screen (explained in the next section) or at start-up. The portion of the script that contains the entry number executes only when the script is not run from the Dialing Directory screen (because of the IF NOT LINKED command).

You must also insert the first name, last name, and password you use on the bulletin board system (separated by semicolons) in place of FIRST-NAME;LASTNAME;FAKE-PASSWORD found to the right of the TRANSMIT command. An example of a modified version of RBBS13.ASP, renamed RBBS-PC.ASP is shown in figure 7.13.

Fig. 7.13.

A modified copy of RBBS13.ASP, named RBBS-PC.ASP.

```
;**************************************************************************
;* RBBS-PC.ASP                                                           *
;* Modified version of RBBS13.ASP  (C) 1987 DATASTORM TECHNOLOGIES, INC. *
;* A sample ASPECT script file for logging onto RBBS v. 13.1A or higher. *
;*                                                                       *
;**************************************************************************
CLEAR
LOCATE 0 0                                  ; Position cursor at top left corner.
BOX 0 0 4 23 14                             ; Draw a box on the screen, and
ATSAY 2 2 14 "Logging onto RBBS..."         ; display a message in the box.
LOCATE 6 0                                  ; Position cursor in row 6 column 0.

IF NOT LINKED                               ; If script not activated by Dial. Dir.,
    DIAL "11"                               ; dial Dial. Dir. entry number 11.
ENDIF
                                            ; Once connected,
PAUSE 1                                     ; pause 1 second, and
TRANSMIT "^M"                               ; send a carriage return.

WAITFOR "What is your FIRST Name?"          ; Wait for "Name?" prompt and then
TRANSMIT "Sam;Spade;falcon^M"               ; send name and password.

ALARM 2                                     ; Inform user log-on is complete.
```
`  Home: Top of file      PgUp: Previous page      PgDn: Next page      ESC: Exit `

To execute a script you have assigned to a dialing directory entry, press Alt-D (Dialing Directory) to display the Dialing Directory screen. Select the entry you want to use, and press Enter. (Refer to Chapter 3, "Building Your Dialing Directory," for a full discussion of several ways to select and dial a directory entry.) PROCOMM PLUS dials the entry's telephone number. When your computer connects with the remote computer, PROCOMM PLUS executes the assigned script.

For example, suppose that you assigned the RBBS-PC.ASP script to entry number 181 in the dialing directory shown in figure 7.14. (Notice that the file name extension .ASP is not shown.) To execute the script, you select this entry in the Dialing Directory screen, and press Enter. PROCOMM PLUS dials 555-1111. When your computer connects to Nickie's Place, PROCOMM PLUS executes the RBBS-PC script, logging you on to the bulletin board.

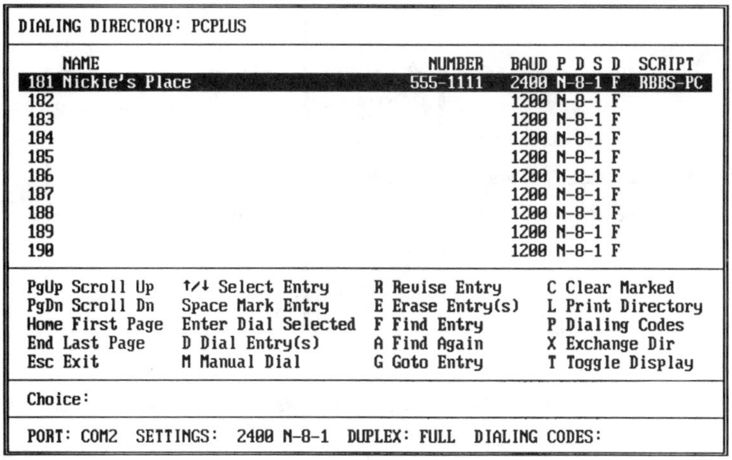

```
DIALING DIRECTORY: PCPLUS

       NAME                        NUMBER    BAUD P D S D   SCRIPT
 181 Nickie's Place              555-1111   2400 N-8-1 F   RBBS-PC
 182                                        1200 N-8-1 F
 183                                        1200 N-8-1 F
 184                                        1200 N-8-1 F
 185                                        1200 N-8-1 F
 186                                        1200 N-8-1 F
 187                                        1200 N-8-1 F
 188                                        1200 N-8-1 F
 189                                        1200 N-8-1 F
 190                                        1200 N-8-1 F

 PgUp Scroll Up    ↑/↓ Select Entry    R Revise Entry    C Clear Marked
 PgDn Scroll Dn    Space Mark Entry     E Erase Entry(s)  L Print Directory
 Home First Page   Enter Dial Selected  F Find Entry      P Dialing Codes
 End Last Page     D Dial Entry(s)      A Find Again      X Exchange Dir
 Esc Exit          M Manual Dial        G Goto Entry      T Toggle Display

 Choice:

 PORT: COM2  SETTINGS:  2400 N-8-1  DUPLEX: FULL  DIALING CODES:
```

Fig. 7.14.

A dialing directory entry showing RBBS-PC.ASP as the assigned script.

From the Terminal Mode Screen

The third way to execute an ASPECT script is available from the Terminal mode screen. Press Alt-F5 (Script Files), and PROCOMM PLUS displays a narrow window across the middle of the screen. The top border line includes the message SCRIPT SELECTION (Enter for .ASP file list), and the window itself prompts you to Please enter filename.

After PROCOMM PLUS displays this window, you have two alternatives for indicating which script you want to run:

1. Type the name of the script file and press Enter. PROCOMM PLUS assumes that the script has the file name extension .ASP, so you don't need to type it. PROCOMM PLUS executes the script.

2. If you are not sure of the script's exact spelling, press Enter to see a list of all the script files in the working directory (all files having the .ASP file-name extension). PROCOMM PLUS displays up to 15 script file names in a tall thin window, as shown in figure 7.15. The file name at the top of the list is highlighted (using high-intensity characters, not inverse video). While this second window is displayed, you can use the up arrow, the down arrow, PgDn, and PgUp to scroll the highlight to each ASPECT file name on the working directory. Position the highlight on the name of the script you want to run, and press Enter. PROCOMM PLUS runs the script.

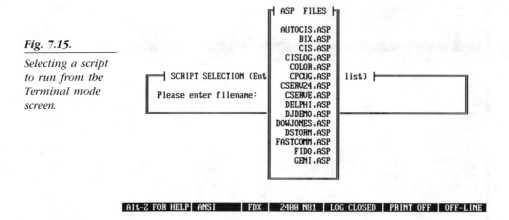

Fig. 7.15.

Selecting a script to run from the Terminal mode screen.

For example, to see a listing of all the screen colors that can be used in ASPECT display commands (refer to "Display and Printer Commands" in Chapter 11), you can execute the COLOR.ASP script, which is distributed on the Supplemental Diskette. Press Alt-F5 (Script Files), type *color*, and press Enter. The script causes your computer to beep and then display 256 different display attribute combinations.

Aborting Execution of a Script

You may decide that you didn't really want to run that script after all. To abort the script in midstream, press Esc. PROCOMM PLUS displays a small window in the upper left portion of the Terminal mode screen containing the prompt EXIT SCRIPT? (Y/N). Respond Yes, and PROCOMM PLUS terminates the script and displays the message COMMAND FILE ABORTED. Respond No, and the script resumes.

Chapter Summary

This chapter has shown you a couple of easy ways to make your computer work for you: keyboard macros and script files. The text has described how to store up to 50 keystrokes as a macro that you can play back at the touch of a single keystroke combination and has explained how to automate such tasks as logging on to an on-line service by simply recording your keystrokes in a PROCOMM PLUS script file. This chapter also discusses several script files that DATASTORM distributes with PROCOMM PLUS. This chapter is only an introduction to automation of PROCOMM

PLUS. After you are comfortable with macros and the script techniques presented in this chapter, take a look also at Chapter 11, "An Overview of the ASPECT Script Language," to discover all the capabilities of the PROCOMM PLUS command language.

This is the last chapter in Part II. You should now be well acquainted with PROCOMM PLUS, so you are ready to move on to Part III to learn how to become a PROCOMM PLUS expert.

Part III

Becoming a
PROCOMM PLUS Expert

8

Tailoring
PROCOMM PLUS

A s the first chapter in Part III, "Becoming a PROCOMM PLUS Expert," this chapter teaches you how to customize the many program settings that control the intricacies of PROCOMM PLUS's operation. PROCOMM PLUS works so well "right out of the box" that you may not believe that customizing is necessary. But as you become an experienced PROCOMM PLUS user, you may begin to think of ways you want to tailor the program's features to meet your specific needs.

Each of the four commands listed in the Set Up section of the PROCOMM PLUS Command Menu accesses a facility for making adjustments to PROCOMM PLUS (see fig. 8.1). This chapter covers only the first two of these Set Up commands: Alt-S (Setup Facility), which activates the PROCOMM PLUS Setup Utility, and Alt-P (Line/Port Setup), which displays the Line/Port Setup window. The first section in this chapter, "Assigning a COM Port and Line Settings," describes how to use the Alt-P (Line/Port Setup) command. The remainder of the chapter explains how to use the PROCOMM PLUS Setup Utility. The other two commands listed in the Set Up section of the Command Menu—Alt-W (Translate Table) and Alt-F8 (Key Mapping)—are discussed in Chapter 10, "Terminal Emulation."

When you decide to fine-tune a PROCOMM PLUS feature, with few exceptions, you can change one setting while leaving all other settings unchanged. This chapter is, therefore, arranged in an order that makes it easier to find the feature in which you are interested. Hardware-related settings, including assignment of a COM port, line settings, and modem options, are covered first. The chapter then discusses how to make adjustments to the features more closely related to the operation of PROCOMM

193

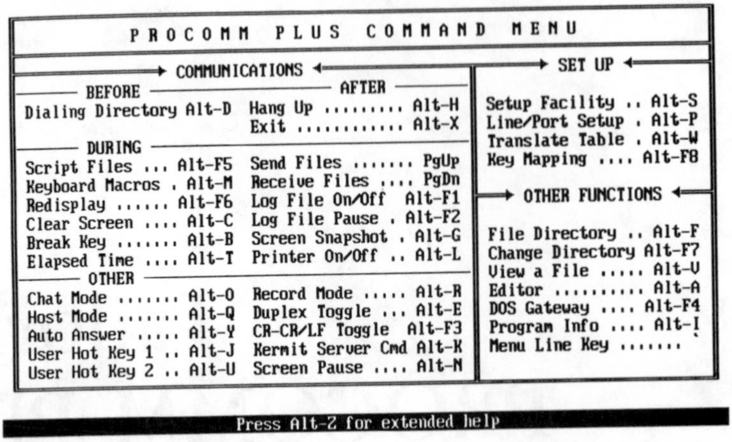

Fig. 8.1.

*The PROCOMM
PLUS Command
Menu.*

```
┌──────────────────────────────────────────────────────────────────┐
│              P R O C O M M   P L U S   C O M M A N D   M E N U     │
├─────────────────────────────────────────┬────────────────────────┤
│          ► COMMUNICATIONS ◄              │       ► SET UP ◄        │
│  ─ BEFORE ─────────────── AFTER ──────   │                        │
│  Dialing Directory Alt-D  Hang Up ........ Alt-H  Setup Facility .. Alt-S │
│                           Exit ........... Alt-X  Line/Port Setup . Alt-P │
│  ─ DURING ─                                       Translate Table . Alt-W │
│  Script Files ... Alt-F5  Send Files ....... PgUp  Key Mapping .... Alt-F8 │
│  Keyboard Macros . Alt-M  Receive Files .... PgDn                   │
│  Redisplay ...... Alt-F6  Log File On/Off  Alt-F1  ► OTHER FUNCTIONS ◄ │
│  Clear Screen .... Alt-C  Log File Pause . Alt-F2                   │
│  Break Key ....... Alt-B  Screen Snapshot . Alt-G  File Directory .. Alt-F │
│  Elapsed Time .... Alt-T  Printer On/Off .. Alt-L  Change Directory Alt-F7 │
│  ─ OTHER ─                                         View a File ..... Alt-V │
│  Chat Mode ....... Alt-O  Record Mode ..... Alt-R  Editor .......... Alt-A │
│  Host Mode ....... Alt-Q  Duplex Toggle ... Alt-E  DOS Gateway .... Alt-F4 │
│  Auto Answer ..... Alt-Y  CR-CR/LF Toggle Alt-F3  Program Info .... Alt-I │
│  User Hot Key 1 .. Alt-J  Kermit Server Cmd Alt-K  Menu Line Key ....... │
│  User Hot Key 2 .. Alt-U  Screen Pause .... Alt-N                   │
├──────────────────────────────────────────────────────────────────┤
│              Press Alt-Z for extended help                         │
└──────────────────────────────────────────────────────────────────┘
```

PLUS itself, such as the menu-line key, the pause character, terminal options, and screen color. Finally, the chapter describes how to fine-tune file-related features, including several file-transfer protocol options.

Assigning a COM Port and Line Settings

The first time you start PROCOMM PLUS, it uses the COM port assignment and line settings (transmission rate, parity, data bits, and stop bits) you specified during installation. (See Appendix A, "Installing and Starting PROCOMM PLUS," for more information about the Installation Utility.) As explained in Chapter 3, "Building Your Dialing Directory," you can also change line settings through a dialing directory entry. Each time you dial another computer by using a dialing directory entry, PROCOMM PLUS uses line settings specified in the dialing directory entry. You cannot, however, assign a different COM port through a dialing directory entry.

From time to time, you may need to reassign the COM port, perhaps because you add a mouse to COM1 and move your modem to COM2. You also may at some time need to change one of the line settings without using a dialing directory entry, perhaps after you are already on-line. Consequently, PROCOMM PLUS lets you perform adjustments to the COM port and to line settings through the Line/Port Setup window.

To display the Line/Port Setup window, from the Terminal mode screen, press Alt-P (Line/Port Setup). PROCOMM PLUS displays the window shown in figure 8.2.

```
┌─────────────────────────────────────────────────────────┐
│         CURRENT SETTINGS:  2400,N,8,1,COM2              │
│                                                         │
│  BAUD RATE    PARITY      DATA BITS    STOP BITS   PORT │
│                                                         │
│  1)    300    N) NONE     Alt-7) 7     Alt-1) 1   F1) COM1 │
│  2)   1200    E) EVEN     Alt-8) 8     Alt-2) 2   F2) COM2 │
│  3)   2400    O) ODD                              F3) COM3 │
│  4)   4800    M) MARK                             F4) COM4 │
│  5)   9600    S) SPACE                            F5) COM5 │
│  6)  19200                                        F6) COM6 │
│  7)  38400                                        F7) COM7 │
│  8)  57600    Alt-N) N/8/1                        F8) COM8 │
│  9) 115200    Alt-E) E/7/1                               │
│                                                         │
│  Esc) Exit    Alt-S) Save and Exit   YOUR CHOICE:       │
└─────────────────────────────────────────────────────────┘
```

Fig. 8.2.

The Line/Port Setup window.

LINE/PORT SETUP

The Line/Port Setup window is divided into two sections: the current settings line and the command section. At the top of the window, the current settings line shows the assigned COM port and the active line settings. In figure 8.2, this top line contains the following message:

CURRENT SETTINGS: 2400,N,8,1,COM2

This message means that the computer's line settings are currently 2400-bps baud (transmission) rate, no parity, 8 data bits, and 1 stop bit. COM2 is assigned as the serial port through which PROCOMM PLUS will attempt to communicate.

As you can see from figure 8.2, the remainder of the Line/Port Setup window contains a list of available commands. These commands are separated into five columns entitled BAUD RATE, PARITY, DATA BITS, STOP BITS, and PORT. Four commands near the bottom of this window don't really fit into any of these columns.

Selecting a command in the Line/Port Setup window is so quick and easy that you almost don't notice that anything happens. In order to make a change to the COM port assignment or to a line setting, choose the appropriate command from the Line/Port Setup window. In this window, you choose a command by pressing the key or keystroke combination listed to the left of the desired setting. For example, press F1 to change the COM port to COM1. These commands and their meanings are listed in table 8.1. As soon as you execute a command, PROCOMM PLUS changes the setting and updates the current settings line at the top of the window.

Table 8.1
Line/Port Setup Window Commands

Line/Port Setting	Command	Meaning
Baud Rate	1	300 bps
	2	1200 bps
	3	2400 bps
	4	4800 bps
	5	9600 bps
	6	19200 bps
	7	38400 bps
	8	57600 bps
	9	115200 bps
Parity	N	No parity bit
	E	Even parity
	O	Odd parity
	M	Mark parity—parity bit always 1
	S	Space parity—parity bit always 0
Data bits	Alt-7	7 data bits
	Alt-8	8 data bits
Stop bits	Alt-1	1 stop bit
	Alt-2	2 stop bits
COM port	F1	COM1
	F2	COM2
	F3	COM3
	F4	COM4
	F5	COM5
	F6	COM6
	F7	COM7
	F8	COM8
Miscellaneous	Alt-N	No parity bit, 8 data bits, 1 stop bit
	Alt-E	Even parity, 7 data bits, 1 stop bit
	Esc	Exit, return to Terminal mode
	Alt-S	Save the new settings, return to Terminal mode

PROCOMM PLUS also provides two short-cut commands. By far, the two most common line configurations are N81 (no parity bit, 8 data bits, and 1 stop bit) and E71 (even parity, 7 data bits, and 1 stop bit). To switch line settings immediately to N81, press Alt-N. When you want to change the line configuration to E71, press Alt-E.

For example, you may be connected to another computer, but the characters you see are unintelligible—made up mostly of foreign language symbols and box-drawing characters. You realize that the line settings should have been E71 (even parity, 7 data bits, and 1 stop bit); but your dialing directory entry contained the settings N81 (no parity, 8 data bits, and 1 stop bit). You could hang up, fix the dialing entry, and redial; instead, you press Alt-P (Line/Port Setup) to display the Line/Port Setup window and then press Alt-E to switch line settings to even parity, 7 data bits, and 1 stop bit.

TIP | A simple rule of thumb is to use N81 for PC-to-PC or PC-to-bulletin board connections. All PCs need 8-bit bytes to represent the full IBM character set and to transmit program files; and with 8 data bits, no room is left for a parity bit. On the other hand, when calling an on-line service, such as CompuServe, or another mainframe system, use E71. Most such systems are run on computers that can handle only 7 data bits per byte, always leaving 1 bit for a parity bit. Mainframe systems usually use even parity.

Occasionally, you may need to reassign the serial port through which you want to communicate. Perhaps you have moved your modem to a different port, or maybe you want to use PROCOMM PLUS to transfer files to another PC connected to one of your computer's serial ports by a null modem cable (refer to Appendix B for information about attaching a modem and using a null modem cable). To assign a different COM port, display the Line/Port Setup window, and press one of the function keys F1 through F8. PROCOMM PLUS immediately switches to the new COM port.

CAUTION: Do not switch COM ports while on-line to another computer. PROCOMM PLUS will drop the connection.

After you have made the necessary COM port and line setting adjustments, you still must decide whether you want the change(s) to be temporary or permanent. In other words, do you want the adjustments you have just made to last only through the current session with PROCOMM PLUS, or do you want them to be used every time you start PROCOMM PLUS until you make another change?

When you want adjustments made through the Line/Port Setup window to be temporary, return to the Terminal mode screen by pressing Esc (Exit). PROCOMM PLUS maintains the changes you have made only for the current PROCOMM PLUS session. The next time you use PROCOMM PLUS, the COM port and line settings return to their previous state. In order to make a lasting change to COM port and line settings, press Alt-S (Save and Exit) instead of Esc (Exit). PROCOMM PLUS saves the revised settings to disk and uses this new group of settings for the remainder of the current session as well as the next time you run the program.

Starting the Setup Utility and Saving Changes

The PROCOMM PLUS Setup Utility is an auxiliary program that can be activated from within PROCOMM PLUS. The Setup Utility enables you to make adjustments to nearly every PROCOMM PLUS feature, including modem options. To display the main Setup Utility screen, from the Terminal mode screen, press Alt-S (Setup Facility). PROCOMM PLUS displays the Setup Utility Main Menu (see fig. 8.3).

Fig. 8.3.

The Setup Utility Main Menu.

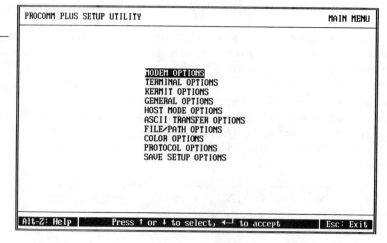

```
PROCOMM PLUS SETUP UTILITY                                      MAIN MENU




                         MODEM OPTIONS
                         TERMINAL OPTIONS
                         KERMIT OPTIONS
                         GENERAL OPTIONS
                         HOST MODE OPTIONS
                         ASCII TRANSFER OPTIONS
                         FILE/PATH OPTIONS
                         COLOR OPTIONS
                         PROTOCOL OPTIONS
                         SAVE SETUP OPTIONS





 Alt-Z: Help       Press ↑ or ↓ to select, ◄┘ to accept       Esc: Exit
```

TIP The PROCOMM PLUS Setup Utility is actually a separate program named PCSETUP.EXE. When you attempt to activate this program from within PROCOMM PLUS by pressing Alt-S (Setup Facility), you may get the following error message:

EXTERNAL PROGRAM ERROR: File or path not found.

This message means simply that PROCOMM PLUS cannot find PCSETUP.EXE. If this error occurs, review the installation instructions in Appendix A, "Installing and Starting PROCOMM PLUS," to make sure that you have installed PROCOMM PLUS correctly.

Several Setup Utility screens consist of vertical menus, as in figure 8.3. You can select options from these menus in a manner similar to that used for the PROCOMM PLUS menu line (refer to "Using the Menu Line," in Chapter 2). You can select an option in two different, equally effective, methods:

1. You can type the first letter of the option.

2. You can use the up arrow or down arrow to move a highlighted (reverse video) menu-selection bar to the option you want to use, and then press Enter.

The latter method is sometimes referred to as the "point and shoot" method. Typing the first letter of the option is usually the quicker of the two methods. The typing method, in most cases, requires fewer keystrokes on your part.

When you are finished making adjustments in the Setup Utility, press Esc (Exit) until you return to the Setup Utility Main Menu (again see fig. 8.3). To save the changes, select SAVE SETUP OPTIONS from the Setup Utility Main Menu. The Setup Utility saves the modified settings to a disk file named PCPLUS.PRM. Press Esc (Exit) again to return to PROCOMM PLUS's Terminal mode screen.

Occasionally, you may want to use the Setup Utility to make only a temporary change to one of PROCOMM PLUS's settings. After you make the desired modification, press Esc (Exit) until you return to the Setup Utility Main Menu, and then press Esc (Exit) once more. The Setup Utility displays a small window near the center of the screen containing the message

    MAKE CHANGES PERMANENT? (Y/N)

Respond No. The Setup Utility returns you to PROCOMM PLUS's Terminal mode screen without saving the setup change(s) you made. The modification is effective only for the current PROCOMM PLUS session. The next time you start PROCOMM PLUS, settings return to those saved in the PCPLUS.PRM file. Responding Yes to this prompt has the same effect as selecting the SAVE SETUP OPTIONS choice from the Setup Utility Main Menu.

Setting Modem Options

In order to make adjustments to modem-related settings, select MODEM OPTIONS from the Setup Utility Main Menu. The Setup Utility displays the Modem Options menu shown in figure 8.4. This menu lists three choices: GENERAL OPTIONS, RESULT MESSAGES, and PORT ASSIGNMENTS.

Fig. 8.4.

The Modem Options menu.

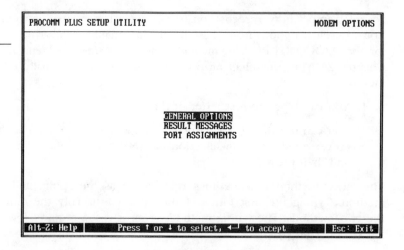

Adjusting General Modem Options

From time to time as you use PROCOMM PLUS, the program sends certain software commands to your modem. These commands are listed in the Modem Options screen.

To display the Modem Options screen, select **GENERAL OPTIONS** from the Modem Options menu. The Setup Utility displays the screen shown in figure 8.5. Each option in this screen is preceded by a letter, A through J. To select an option for modification, press the corresponding letter. For example, to choose the first option, **A** (Initialization command), press the letter **A**. The following paragraphs discuss the options listed on the Modem Options screen.

When you have made the desired changes to the settings displayed in the Modem Options screen, press Esc (Exit) twice to return to the Modem Options menu and then to the Setup Utility Main Menu.

Editing the Initialization Command

Each time you start PROCOMM PLUS, it sends a software "wake-up call," referred to as an *initialization command*, to your modem (but see "Sending the Initialization Command If CD Is High," later in this chapter). The first line in the Modem Options screen enables you to customize this command.

```
PROCOMM PLUS SETUP UTILITY                              MODEM OPTIONS

A- Initialization command ..  AT E1 V1 X4 Q0 &C1 &D2 S7=255 S0=0^M

B- Dialing command .........  ATDT

C- Dialing command suffix ..  ^M

D- Hangup command ..........  ~~~+++~~~ATH0^M

E- Auto answer command .....  ~~~+++~~~ATS0=1^M

F- Wait for connection .....  45  seconds

G- Pause between calls .....  1   seconds

H- Auto baud detect ........  ON

I- Drop DTR to hangup ......  YES

J- Send init if CD high ....  YES

Alt-Z: Help |        Press ◄┘ when complete        | Esc: Exit
```

Fig. 8.5.

*The Modem
Options screen.*

The initialization command activates a number of the modem's built-in features. For example, Hayes-compatible modems are capable of sending back to your screen verbal status codes, usually referred to as *result codes*. These result codes inform you of telephone-line conditions. The result code RING means that the telephone line to which the modem is connected is ringing. The result code CONNECT 2400 means that your modem has connected to another modem at a speed of 2400 bps. A portion of PROCOMM PLUS's initialization command activates the modem's verbal result code feature (refer also to "Setting Result Messages," later in this chapter).

The initialization command is established first by the Installation Utility (refer to Appendix A, "Installing and Starting PROCOMM PLUS"). During the installation procedure, you choose from a list of several well-known modems. When you select a modem, the Installation Utility inserts the appropriate initialization command for that modem into the Modem Options screen. The initialization command shown in figure 8.5 is intended for use with a Hayes 2400 modem. For a Hayes 1200 modem, the installation program inserts the following initialization command:

    AT E1 V1 X1 Q0 S7 = 255 S11 = 55 S0 = 0^M

The parts of the initialization command in figure 8.5 are explained later in this section of the chapter.

At times, you may want to customize the initialization command. To make changes to the initialization command, select **A** (Initialization command) from the Modem Options screen. The Setup Utility moves the cursor into the entry space to the right of the `Initialization command` label. This entry space is 45 characters in length. Use the cursor-movement keys and editing keys listed in table 8.2 to move the cursor around the entry and to make necessary modifications. Press Enter to accept the new initialization command and to return the cursor to the bottom line of the screen.

Table 8.2
Setup Utility Cursor-Movement and Editing Keys

Key	Function
←	Move the cursor one space to the left
→	Move the cursor one space to the right
Home	Move the cursor to the left end of the entry
End	Move the cursor one space to the right of the last character in the entry
Insert	Toggle Insert/Overtype modes
Delete	Delete the character at cursor
Backspace	Delete the character to the left of cursor
Tab	Delete the entire entry
Ctrl-End	Delete the characters from cursor to right end

The initialization command is actually a series of commands. For example, the initialization command shown in figure 8.5 includes the following commands:

- **AT**, the first two letters of the initialization command, stand for "attention." These letters alert the modem that a command follows. All Hayes-compatible modems recognize this command. Most modems enable you to list several commands after AT.

- **E1** causes your modem to echo the characters back to the screen as you type, but only while the modem is in *Command mode*. In Command mode, the modem recognizes and responds to commands typed at the keyboard. The E1 command has no effect while your modem is on-line and communicating with another modem. This command is not equivalent to the PROCOMM PLUS Alt-E (Duplex Toggle) command.

- **V1** causes the modem to use verbal result codes rather than numeric result codes. For example, the verbal result code CONNECT 1200 is equivalent to the numeric result code 5. PROCOMM PLUS uses the verbal result codes as it monitors the dialing process.

- **X4** selects a result code set. Each result code set monitors a different group of line conditions. For example, with set 4 chosen, the modem waits for a dial tone before dialing. If no dial tone sounds, the modem returns the code NO DIAL TONE and aborts the dialing. PROCOMM PLUS then waits for a designated period of time before trying again. This feature is desirable in most situations, but on some telephone systems, the dial tone may be so faint that the modem cannot detect the tone. In this case, check your modem manual for a result code group that ignores the dial tone (sometimes called *blind dialing*). For many Hayes-compatible modems, the result code groups that enable blind dialing are groups 1 and 3, so you change this portion of the initialization command to X1 or X3.

- **Q0** controls the modem's display of result codes. If the command is set to Q1, the modem sends no result codes, verbal or numerical. PROCOMM PLUS uses result codes, so this portion of the initialization command should always be Q0.

- **&C1** causes the modem to detect the reception of a carrier signal from another modem. When your modem detects a carrier, the modem sends to PROCOMM PLUS a signal indicating that a carrier signal is present (referred to as *asserting CD*, or *raising CD high*). PROCOMM PLUS works properly only if this feature is enabled. Many modems, however, are set at the factory so that they always assert the CD (also called DCD for Data Carrier Detect) signal, whether or not a carrier signal is actually present. If your status line displays the message ON-LINE at all times, even when you are not connected to another computer, your modem is always asserting the CD signal. The &C1 command changes this setting. If your modem does not recognize the &C1 command (many modems do not), refer to your modem manual and to Appendix B, "Installing a Modem," to determine the proper DIP switch setting that enables the modem to detect and report to PROCOMM PLUS the presence of another modem's carrier signal. Refer also to "Sending the Initialization Command If CD Is High," later in this chapter.

- **&D2** causes the modem to "watch" the DTR (Data Terminal Ready) line (one of the circuits defined by the RS-232 standard). While you are on-line to another computer, your computer continuously asserts (sends a voltage over) the DTR line to the modem. When you execute the Alt-H (Hang Up) command, PROCOMM PLUS turns off this DTR signal. (This setting is the default condition, but refer to "Dropping the DTR Signal," later in this chapter.) If the modem is watching the DTR line, the modem "hangs up" the telephone line when the DTR signal goes from on to off. If your modem does not recognize the &D2 command, refer to your modem's manual and to Appendix B to determine the appropriate DIP switch setting that achieves the same result. Even if you don't enable this feature, PROCOMM PLUS still can hang up the telephone by sending a Hangup command to the modem; but that method is not as quick as the DTR method. Refer also to "Understanding the Hangup Command," later in this chapter.

- **S7=255** establishes the number of seconds your modem will wait for a carrier signal before hanging up. The value 255 sets the wait time at 255 seconds (4 minutes 15 seconds), the maximum time allowed for most Hayes-compatible modems.

- **S0=0** disables the auto-answer feature of your modem. This command ensures that your modem will not answer the phone except when PROCOMM PLUS is in Host mode or when you execute the Alt-Y (Auto Answer) command.

- **^M** is a control code, which causes PROCOMM PLUS to send the ASCII carriage-return character. (Refer to table 7.2 of Chapter 7, "Automating PROCOMM PLUS with Macros and Script Files," for other PROCOMM PLUS control codes.) The carriage-return character causes the modem to process the initialization command by letting the modem know that the command is complete.

Refer to the manual that was distributed with your modem to find further explanation of these commands and to determine whether you need to modify the initialization command inserted by PROCOMM PLUS's Installation Utility.

Understanding the Dialing Command and Dialing Command Suffix

Below the **A** (Initialization command) option on the Modem Options screen, the Setup Utility lists two options that determine the command PROCOMM PLUS sends to your modem when you use PROCOMM PLUS to dial a telephone number. The options are **B** (Dialing command) and **C** (Dialing command suffix).

Select **B** (Dialing command) to specify the command that instructs your modem to dial a telephone number—the *dialing command*. The dialing command shown in figure 8.5 is ATDT. This command is recognized by all Hayes-compatible modems to mean "dial the telephone using touch-tone." When the telephone service in your area supports only pulse (rotary) dialing, select the **B** (Dialing command) option from the Modem Options screen and edit the entry to change the second *T* to the letter *P* (that is, change the command to ATDP).

When you issue a PROCOMM PLUS command to dial a specific telephone number, PROCOMM PLUS sends to the modem the dialing command listed in the **B** (Dialing command) entry of the Modem Options screen, followed by the telephone number you specify in the Dialing Directory screen. For example, if you issue a PROCOMM PLUS command to dial the number 555-4391 (assuming touch-tone service), PROCOMM PLUS sends the following command to your modem:

    ATDT5554391

In order to specify characters to be added to the end of each dialing command, select the **C** (Dialing command suffix) option from the Modem Options screen. PROCOMM PLUS then adds to the end of the dialing command the characters you specify in this entry. Hayes-compatible modems expect all commands to end with a carriage return; so the **C** (Dialing command suffix) entry must be a carriage-return character. As explained in "Editing the Initialization Command," earlier in this chapter, the control code ^M causes PROCOMM PLUS to send a carriage-return character; so ^M is normally the dialing command suffix, as shown in figure 8.5.

Understanding the Hangup Command

Occasionally, you have to execute the Alt-H (Hang Up) command to PROCOMM PLUS in order to disconnect from a communications session. When you issue this command, PROCOMM PLUS first tries to hang up the line by dropping the DTR signal. (This setting is the default condition, but refer also to "Editing the Initialization Command" and "Dropping the DTR

Signal," both in this chapter.) If this attempt is not successful, PROCOMM PLUS sends to the modem the Hangup command listed in the **D** (Hangup command) entry space on the Modem Options screen.

The default Hangup command is

    ~~~+++~~~ATH0^M

The first character in this modem command, repeated three times, is the tilde (~). The tilde is the default *pause character*. PROCOMM PLUS translates this character into a pause of one-half second. (Refer to "Adjusting General Options," later in this chapter, for an explanation of how to change the pause character.) The three tildes, therefore, result in a pause of one and one-half seconds. In other words, no characters are sent or received by the modem for one and one-half seconds.

Following the three tildes in the Hangup command are three plus signs (+++) followed by three more tildes (~~~). Three pluses together preceded and followed by at least one character-free second cause a Hayes-compatible modem to go from On-line mode to Command mode. (These three pluses are sometimes called an *escape code*.) In On-line mode, the modem ignores commands from PROCOMM PLUS. When the modem is in Command mode, the modem recognizes the AT commands PROCOMM PLUS sends.

The actual command that causes the modem to hang up the telephone (go *on book*) is ATH0. This modem command follows the second trio of tildes. A carriage return must also be sent to cause the modem to process the command, so the control code ^M appears at the end of the Hangup command.

Adjusting the Number of Rings before Auto Answer

"Receiving a Call," in Chapter 4, "A Session with PROCOMM PLUS," explains how to use the Alt-Y (Auto Answer) command to place your modem in Answer mode. When you press Alt-Y, PROCOMM PLUS sends to your modem the command that is listed in the **E** (Auto answer command) entry space on the Modem Options screen. The default Auto answer command is

    ~~~+++~~~ATS0=1^M

As explained in the preceding section, the first portion of the command (~~~+++~~~) places your modem in Command mode so that it is expecting a command from PROCOMM PLUS. The remainder of the Auto answer command (ATS0=1) instructs the modem to switch to Answer mode and to answer any in-coming call on the first ring.

In some situations, you may prefer that the modem not answer on the first ring. Select the **E** (Auto answer command) option from the Modem Options screen. Use the keys listed in table 8.2 to change the number *1* to another positive integer. The value of this number can be any integer between 0 and 255. Setting this number to 0 disables Answer mode. As in the other commands, the ^M code sends a carriage-return character to the modem, causing it to process the command.

Adjusting the Wait Time and the Pause Time

Chapter 3, "Building Your Dialing Directory," explains how to use PROCOMM PLUS's Dialing Directory screen to connect to a remote computer. You may recall that when your modem is dialing a remote computer's telephone line, PROCOMM PLUS displays the Dialing window (see fig. 8.6). After PROCOMM PLUS sends to the modem the instruction to dial the telephone number, the program waits the length of time listed in the WAIT FOR CONNECTION line of the Dialing window. This time is referred to in this chapter as the *wait time*. If the modem does not complete a connection in the specified wait time, either because the line is busy or because there is no answer at the other end, PROCOMM PLUS automatically puts the phone line back on hook and pauses the dialing.

```
DIALING DIRECTORY: PCPLUS

      NAME                            NUMBER     BAUD P D S D  SCRIPT
  141 Scorpio Rising                 620-2827   2400 N-8-1 F  opus
  142 Scotland the Brave             768-8637   9600 N-8-1 F  pcboardr
  143 ShanErin (D. Page)             941-8291   9600 N-8-1 F  opusr
  144 Soft Sale (G.Hendershot)       569-6876   2400 N-8-1 F  major
  145 Software AG                    391-6917   9600 N-8-1 F  pcboardr
  146 Software Lib.(Schinnell)     1-301-949-8848  2400 N-8-1 F
  147 Software Link                  734-7860   2400 N-8-1 F
  148 Source Data Corp               359-0993   2400 N-8-1 F  pcboard
  149 Space Party (S. Hawley)        385-9698   2400 N-8-1 F  opus
  150 Split Infinity                 841-1859   2400 N-8-1 F

     DIALING: Software Lib.(Schinnell)   LAST CONNECTED ON: 08/14/89
      NUMBER: 1-301-949-8848        TOTAL COMPLETED CALLS: 3
 SCRIPT FILE:                        WAIT FOR CONNECTION: 45  SECS
   LAST CALL:                        PAUSE BETWEEN CALLS: 4   SECS
 PASS NUMBER: 1                     TIME AT START OF DIAL: 12:55:58AM
ELAPSED TIME: 3               TIME AT START OF THIS CALL: 12:55:58AM

 Choice:    Space Recycle  Del Remove from list  End Change wait  Esc Abort
```

Fig. 8.6.

The Dialing window.

At times, you may decide that you want to increase or decrease the wait time. This setting can be changed temporarily, for the current PROCOMM PLUS session only, from within the Dialing window (refer to "Monitoring Call Progress," in Chapter 3). You can make a more permanent change to the wait time through the Setup Utility Modem Options screen.

The value in the **F** (Wait for connection) entry space on the Modem Options screen determines the default wait time. To change the wait time value, from the Modem Options screen, select **F** (Wait for connection). Use the cursor-movement and editing keys listed in table 8.2 to make the desired change, and press Enter.

When PROCOMM PLUS terminates dialing because the line is busy or because the wait time expires, PROCOMM PLUS does not give up. Rather, the program pauses the length of time specified in the PAUSE BETWEEN CALLS line of the Dialing window and then either retries the same entry or dials the next number in the dialing queue (refer to Chapter 3 for more information). This chapter refers to the period of time that PROCOMM PLUS pauses between dialings as the *pause time*. To change the pause time, from the Modem Options screen, select **G** (Pause between calls). Use the keys listed in table 8.2 to edit the pause time, and press Enter.

Activating Auto Baud Detect

Sometimes when you use PROCOMM PLUS to dial a remote computer, you aren't sure of the remote modem's transmission speed. At other times, you think that you know the other modem's speed, but you are mistaken. PROCOMM PLUS can handle either situation through its *auto baud detect* capability.

To activate the auto baud detect feature, the **H** (Auto baud detect) entry in the Modem Options screen must be set to ON, as shown in figure 8.5. If this entry is set to OFF, select **H** (Auto baud detect). The Setup Utility moves the cursor into the entry space to the right of the Auto baud detect prompt. Press any key to change the value of the entry to ON. Then press Enter to return the cursor to the last line of the screen.

TIP | The auto baud detect feature works properly only if the entries on the Modem Result Messages screen are correct. See "Setting Result Messages," later in this chapter.

With auto baud detect enabled, PROCOMM PLUS detects the transmission speed at which your modem connects to the remote modem. If the connect speed is different from your computer's current transmission speed setting—set by the dialing directory entry or through the Alt-P (Line/Port Setup) command—PROCOMM PLUS adjusts your computer's transmission speed setting automatically. If you have not activated the auto baud detect feature, even though the modems communicate, your computer displays only gibberish on the screen. To remedy the situation, you have to change the computer's line setting by using the Alt-P (Line/Port Setup) command (discussed earlier in this chapter).

TIP | You can easily tell whether the auto baud detect feature has worked. When PROCOMM PLUS first connects to the other computer, PROCOMM PLUS displays the connect speed in the LAST CALL line of the Dialing window (see fig. 8.6). PROCOMM PLUS also displays the connect speed at the top of the Terminal mode screen, just after the program removes the Dialing window from the screen. Compare these connect-speed numbers with the transmission speed listed in the center of the Terminal mode screen status line. All three numbers should match. If they don't, auto baud detect did not work. Check the Setup Utility Modem Options screen to make sure that the feature is enabled. Then check the Modem Result Messages screen to be sure that the result message matches the connect message you saw flashing in the LAST CALL line of the Dialing window (for example, CONNECT 1200 or CONNECT 2400).

TIP | When you are using a 2400-bps modem to call a remote computer and you have enabled the auto baud detect feature, set your modem to its highest transmission speed—2400 bps. Use this setting even if you think that the other computer is probably using a 1200-bps modem. When two modems connect and attempt to communicate, they go through a "negotiation" process to determine a transmission speed at which both can operate. During this negotiation process, one modem may have to decrease its transmission speed, but neither modem will increase its transmission speed.

For example, you may have a 2400-bps modem. You set it to 1200 bps to dial a computer you think is operating at 1200 bps. When the modems connect, the modems will never negotiate a connect speed higher than 1200 bps, even if the remote modem is actually set to 2400 bps. On the other hand, if you call the remote computer with your modem set to 2400 bps, your modem's fastest speed, the modems will negotiate a connect speed of 2400 bps or 1200 bps, depending on the speed of the remote modem.

Dropping the DTR Signal

The DTR (Data Terminal Ready) line is one of the circuits defined by the RS-232 standard. (Refer to Appendix B, "Installing a Modem," for a description of the RS-232 standard.) While you are on-line to another computer, your computer continuously asserts (sends a voltage over) the DTR line to the modem. When you execute the Alt-H (Hang Up) command, by default PROCOMM PLUS turns off this DTR signal, thus informing the modem that the computer is no longer going to send data. In response, the modem hangs up the phone line (assuming that your modem is watching the status of the DTR line—refer to "Editing the Initialization Command," in this chapter, and to Appendix B for more information).

If you cannot set your modem so that it monitors the DTR line, you can turn off this PROCOMM PLUS feature. Select **I** (Drop DTR to hangup) from the Setup Utility Modem Options screen. The cursor moves to the entry space to the right of the Drop DTR to hangup prompt. Press any key, and the Setup Utility changes the value in this entry from YES to NO. Press Enter to accept the change and return the cursor to the bottom line of the screen.

TIP | Whether or not you turn off the DTR-hangup feature, PROCOMM PLUS can still cause your modem to hang up the telephone line by using the Hangup command discussed in "Understanding the Hangup Command," in this chapter. With the DTR-hangup feature enabled, PROCOMM PLUS first tries dropping DTR. If the modem doesn't hang up, PROCOMM PLUS sends the Hangup command to the modem. Disabling the DTR-hangup feature causes PROCOMM PLUS to send only the Hangup command.

Sending the Initialization Command If CD Is High

PROCOMM PLUS normally sends the initialization command to the modem every time you start the program. In most cases, this procedure is desirable. Occasionally, when your modem is already connected to another modem, you may not want PROCOMM PLUS to try to initialize the modem.

For example, suppose that you have just managed to connect to your company's busy electronic bulletin board for the purpose of forwarding this week's sales figures. After connecting, you realize that you forgot to add one figure to the spreadsheet you are about to upload. You need to run the spreadsheet program quickly, add the missing number to the spreadsheet, return to PROCOMM PLUS, and send the file; but you don't want to lose your connection and have to dial in again.

You could try to use PROCOMM PLUS's DOS gateway, but the spreadsheet you want to upload is too big to fit into memory along with PROCOMM PLUS (see "Accessing the Operating System," in Chapter 5, for a discussion of the DOS Gateway feature). Instead, from the Terminal mode screen, you press Alt-X (Exit). PROCOMM PLUS first asks, EXIT TO

DOS? (Y/N). You respond Yes. PROCOMM PLUS then beeps and asks HANGUP LINE? (Y/N). This time you respond No. PROCOMM PLUS exits to the operating system without disconnecting your modem from the other computer.

Now you can run your spreadsheet program, add the missing number, save the spreadsheet, and return to DOS. Remember that the connection to your company's electronic bulletin board is still active. You again start PROCOMM PLUS. This time, because your modem is already connected to another modem, you don't want PROCOMM PLUS to send the initialization command. If PROCOMM PLUS sends the command, your modem will forward the command to the bulletin board with unpredictable results.

In order to prevent PROCOMM PLUS from sending the initialization command when your modem is already on-line, set the last option on the Setup Utility Modem Options screen to NO. From the Modem Options screen, select **J** (Send init if CD high). The cursor moves into the last entry space on the screen. Press any key. The Setup Utility changes the entry value to NO. Press Enter to accept the change and to return the cursor to the bottom line of the screen. Now when you start PROCOMM PLUS, the initialization command will not be sent to the modem if CD is high. The command will still be sent when CD is not high.

But when is CD high? The DCD (Data Carrier Detect) line is one of the circuits defined by the RS-232 standard. (Refer to Appendix B, "Installing a Modem," for a description of the RS-232 standard.) The DCD line is often referred to as just CD (Carrier Detect). While you are on-line to another computer, each modem sends a carrier signal to the other modem. When your modem is set up properly, your modem continuously *asserts*, or *raises high* (sends a voltage over), the CD line to the other computer as long as your modem detects the other modem's carrier signal. (Refer to Appendix B and to "Editing the Initialization Command," in this chapter, for information on how to cause the modem to use the DCD line to report the existence of a carrier signal.)

PROCOMM PLUS continuously monitors the status of the CD line from the modem. Whenever CD is high, PROCOMM PLUS "knows" that the modem is on-line to another modem and displays the message ON-LINE at the right end of the Terminal mode screen status line. If you instruct PROCOMM PLUS not to send the initialization command if CD is high, the program will not send the initialization command when your modem is already on-line.

TIP When you start PROCOMM PLUS and your modem is *not* communicating with another modem, the message in the status line should be OFF-LINE. If you notice that the message says ON-LINE, your modem is not set up properly. Your modem is holding CD high all the time, whether or not a carrier signal is actually present. Refer to Appendix B, "Installing a Modem," for help in remedying this problem.

Setting Result Messages

The second option on the Modem Options menu is **RESULT MESSAGES**. Select this option to display the Modem Result Messages screen, shown in figure 8.7. This screen displays the messages your modem sends to PROCOMM PLUS to indicate the status of a call.

Fig. 8.7.

The Modem Result Messages screen.

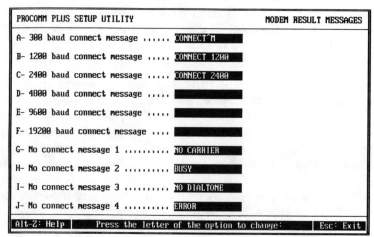

```
PROCOMM PLUS SETUP UTILITY                        MODEM RESULT MESSAGES

A- 300 baud connect message ......  CONNECT^M

B- 1200 baud connect message .....  CONNECT 1200

C- 2400 baud connect message .....  CONNECT 2400

D- 4800 baud connect message .....

E- 9600 baud connect message .....

F- 19200 baud connect message ....

G- No connect message 1 ..........  NO CARRIER

H- No connect message 2 ..........  BUSY

I- No connect message 3 ..........  NO DIALTONE

J- No connect message 4 ..........  ERROR

Alt-Z: Help  |    Press the letter of the option to change:    | Esc: Exit
```

When you use the Installation Utility to install PROCOMM PLUS on your computer, the utility asks the brand and model of the modem you are using. Based on your response to this question, the installation program inserts values into the entry spaces on the Modem Result Messages screen. Figure 8.7 shows the messages the Installation Utility inserts for a Hayes Smartmodem 2400.

Hayes-compatible modems can usually generate more than one set of result messages. The modem command ATXn, where n is a positive integer, determines which result message set the modem will use. Refer to the your modem manual and to "Editing the Initialization Command," in this chapter, for information on how to select a result message set.

The messages displayed in the Modem Result Messages screen must match the result message set you are actually using. Otherwise, some of PROCOMM PLUS's features, such as auto baud detect, may not work properly. Check your modem manual to make sure that the messages listed in the Modem Result Messages screen actually match the messages generated by your modem. If a result message is incorrect, select the corresponding option from the Modem Result Messages screen by pressing the appropriate letter. Use the cursor-movement and editing keys listed in table 8.2 to edit the entry. Press Enter to accept the change.

For example, to change the message your modem sends when it connects to another modem at 2400 bps, select **C** (2400 baud connect message). Make the necessary change(s) and press Enter.

TIP | In order for the auto baud detect feature to work effectively, the result messages that begin with the word CONNECT (referred to here as *connect messages*) must be set correctly. PROCOMM PLUS can adjust the transmission rate to match a particular connect speed only if the corresponding connect message on the Modem Result Messages screen is spelled exactly the same way the modem sends the connect message. For a Hayes-compatible modem, make sure that the 300-baud connect message reads CONNECT^M, not just CONNECT. The other messages don't need the ^M.

After you have made the desired changes to the messages displayed in the Modem Result Messages screen, press Esc (Exit) twice to return to the Modem Options menu and then to the Setup Utility Main Menu.

Altering Port Assignments

PROCOMM PLUS is one of the few communications programs that enables you to use serial ports above COM4, not withstanding the fact that no version of DOS explicitly supports serial ports beyond COM4 (and many versions of DOS don't support COM3 or COM4). Using PROCOMM PLUS, you can address up to a total of eight serial ports, no matter what version of DOS you are using.

To accomplish this feat, PROCOMM PLUS addresses the hardware directly, going around the operating system (DOS). For this method to work, however, PROCOMM PLUS needs to know the exact hardware location of each port. The location of each port is specified by two parameters: the *base address* and the *IRQ line* (interrupt request line).

The hardware locations used by PROCOMM PLUS for each of the eight available serial ports are listed on the Setup Utility Modem Port Assignments screen. To display this screen, select **P**ORT ASSIGNMENTS from the Modem Options menu. The screen shown in figure 8.8 lists the default base addresses and IRQ lines used by PROCOMM PLUS for the eight ports.

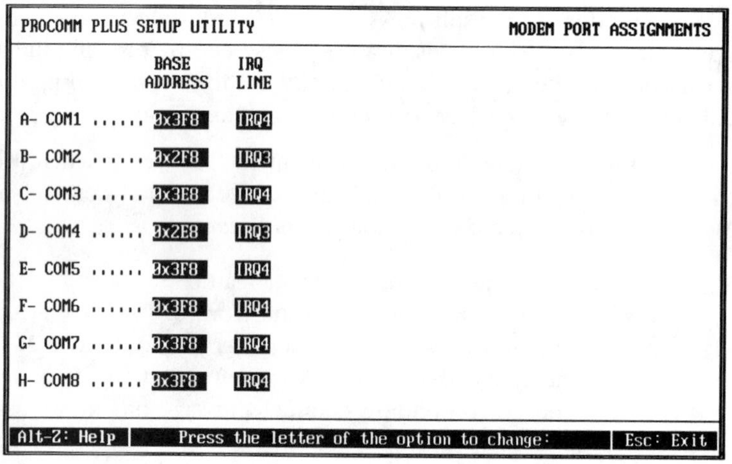

Fig. 8.8.

The Modem Port Assignments screen.

```
┌─────────────────────────────────────────────────────────────────────┐
│ PROCOMM PLUS SETUP UTILITY                        MODEM PORT ASSIGNMENTS │
│                                                                       │
│                        BASE      IRQ                                  │
│                      ADDRESS     LINE                                 │
│                                                                       │
│  A- COM1  ......  0x3F8    IRQ4                                       │
│                                                                       │
│  B- COM2  ......  0x2F8    IRQ3                                       │
│                                                                       │
│  C- COM3  ......  0x3E8    IRQ4                                       │
│                                                                       │
│  D- COM4  ......  0x2E8    IRQ3                                       │
│                                                                       │
│  E- COM5  ......  0x3F8    IRQ4                                       │
│                                                                       │
│  F- COM6  ......  0x3F8    IRQ4                                       │
│                                                                       │
│  G- COM7  ......  0x3F8    IRQ4                                       │
│                                                                       │
│  H- COM8  ......  0x3F8    IRQ4                                       │
│                                                                       │
│ ┌─────────┬──────────────────────────────────────────┬───────────┐   │
│ │ Alt-Z: Help │   Press the letter of the option to change:  │ Esc: Exit │ │
│ └─────────┴──────────────────────────────────────────┴───────────┘   │
└─────────────────────────────────────────────────────────────────────┘
```

COM1 and COM2 are supported by all versions of DOS, so the base addresses and IRQ lines of these ports are standard on all PCs and compatibles. These base addresses and IRQ lines should never have to be changed. If you install a serial port or internal modem as COM3 or COM4, the port's hardware location will most likely match the values listed in figure 8.8. Refer to the port's and modem's documentation to be sure.

In the unlikely event that you want to use PROCOMM PLUS through a serial port other than COM1 through COM4, determine the port's hardware location by consulting your computer's documentation and edit one of the lines in the Modem Port Assignments screen to match.

After you have made the desired changes to the port assignments displayed in the Modem Port Assignments screen, press Esc (Exit) twice to return to the Modem Options menu and then to the Setup Utility Main Menu.

TIP Don't install two serial devices on the same IRQ line. You may experience problems using some software when you are using two devices on the same IRQ line. For example, you may have a mouse attached to COM1 and an internal modem installed as COM3. As shown in figure 8.8, COM1 and COM3 are both normally assigned to IRQ4 (although each is assigned to a different base address). Some software may not permit you to use the mouse and the modem together when they are attached to COM1 and COM3, respectively. To resolve this conflict, install one device on IRQ3 and the other on IRQ4. In this example, you can install the modem as COM2 and avoid any potential problem.

Setting Terminal Options

As explained in Chapter 4, "A Session with PROCOMM PLUS," a number of PROCOMM PLUS settings affect the way data is displayed in the Terminal mode screen. For example, the Duplex setting determines whether PROCOMM PLUS echoes to the screen characters typed at your keyboard; and the CR-CR/LF Toggle setting determines whether PROCOMM PLUS adds a line feed to each carriage-return character PROCOMM PLUS receives. Chapter 4 describes how to make temporary changes to Duplex and CR-CR/LF settings, in effect, overriding the default settings; but the Terminal Options screen from the Setup Utility enables you to change the default settings.

First, select **TERMINAL OPTIONS** from the Setup Utility Main Menu (shown in fig. 8.3) to display the Terminal Options menu (see fig. 8.9). This menu lists two options: **GENERAL OPTIONS** and **COLOR OPTIONS**. These two groups of options are discussed in this portion of the chapter.

Adjusting Terminal General Options

Select **GENERAL OPTIONS** to display the Terminal Options screen, shown in figure 8.10. Most of the features listed on this screen control the way data is displayed on the Terminal mode screen.

After you have made all the needed changes to the settings on the Terminal Options screen, press Esc (Exit) twice to return to the Terminal Options menu and then to the Setup Utility Main Menu.

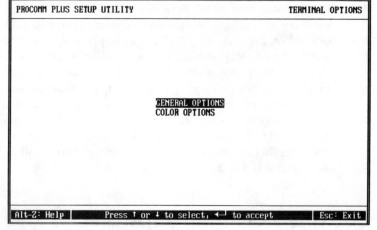

Fig. 8.9.

The Terminal Options menu.

Fig. 8.10.

The Terminal Options screen.

Setting the Default Terminal Emulation

The Installation Utility, discussed in Appendix A, sets the initial default terminal emulation. Choose the **A** (Terminal emulation) option on the Setup Utility Terminal Options screen to change the default terminal emulation setting.

The Setup Utility opens a window that lists the 16 available emulations. Use the up- and down-arrow keys to move a highlighted bar to the desired emulation, and press Enter. You can override the default terminal emulation by specifying a different terminal emulation in a dialing directory entry (see Chapter 3, "Building your Dialing Directory"). Refer to Chapter 10, "Terminal Emulation," for more information about specific emulations.

Changing the Default Duplex

Choose **B** (Duplex) from the Terminal Options screen when you want to change the default Duplex setting. The cursor moves to the entry space on this line. Each time you press any key except Esc or Enter, the value in the Duplex entry space toggles between FULL and HALF. Press Enter to accept the change.

Dialing directory entries and the Alt-E (Duplex Toggle) command override this setting. Refer to "Toggling Duplex," in Chapter 4, for a full discussion of this topic.

Controlling the Flow

Many remote systems, including most on-line services and bulletin boards, support a flow-control method known as XON/XOFF. These systems stop transmitting information when they receive an ASCII character referred to as XOFF. This character is the same as the character you generate by pressing Ctrl-S on your keyboard. When the remote system receives the ASCII character known as XON, the remote system continues its transmission. The XON character is the same as the character you generate when you press Ctrl-Q.

Set the **C** (Software flow control (XON/XOFF)) setting on the Terminal Options screen to ON to cause PROCOMM PLUS and PROCOMM PLUS Host mode to suspend the *transmission* of characters to the remote computer when XOFF is received from the remote computer and to resume transmission when XON is received. PROCOMM PLUS sends XON and XOFF when you use the Alt-N (Screen Pause) feature described in "Pausing the Screen," in Chapter 4, regardless of this Terminal Options setting.

Enable the **D** (Hardware flow control (RTS/CTS)) feature when you are using a modem that performs error-control or data compression (such as MNP-capable modems, Hayes V-series modems, and CCITT V.42-compliant modems). This feature is disabled by default.

The RTS (Request to Send) and CTS (Clear to Send) lines are two of the circuits defined by the RS-232 standard. (Refer to Appendix B, "Installing a Modem," for a description of the RS-232 standard.) When these circuits are used, the computer asserts RTS along with DTR (Data Terminal Ready) to inform the modem that the computer is ready to send data. The modem then responds by asserting CTS along with DSR (Data Set Ready) to grant permission to send the data. If the modem wants to stop the flow of data from the computer, the modem lowers RTS, which lowers the computer's CTS. This RTS/CTS method of controlling the flow of data from the computer to the modem is referred to as a *hardware flow-control* method and is necessary for some of the newer generation of error-control and data-compression modems to work properly. (Refer to Appendix B, "Installing a Modem," for additional information on this topic.)

Modems that don't perform error control or data compression generally also don't use this RTS/CTS flow-control method. The transmission rate between computer and modem is set at a pace that both can handle, and the flow of data is regulated by the software running on the two computers—*software flow control*.

To make a change to either flow-control option, select the option from the Terminal Options screen. Press any key except Esc or Enter. Each time you press a key, the value in the entry toggles to ON or OFF. Press Enter to accept the new value.

Setting Line Wrap

While the **E** (Line wrap) feature is enabled (the default setting), PROCOMM PLUS inserts line feeds and carriage returns on its own when the program receives characters in a continuous stream with no carriage returns (either from you or from the remote computer). The typical PC's screen is 80 characters wide, so when PROCOMM PLUS receives the 80th character or space, the line is filled. At this point, PROCOMM PLUS moves the cursor down one line—line feed—and returns the cursor to the left side of the screen—carriage return—before the program places the next character or space on the screen. With this option set to OFF, the 81st character and all subsequent characters overwrite the 80th character. Refer to "Toggling Carriage Return/Line Feed," in Chapter 4, for a full discussion of this topic. (Note: This setting does not affect the operation of Chat mode.)

To make a change to this option, select **E** (Line wrap) from the Terminal Options screen. Press any key except Esc or Enter. Each time you press a key, the value in the entry toggles between ON and OFF. Press Enter to accept the new value.

Understanding Screen Scroll

With the **F** (Screen scroll) option set ON (the default setting), PROCOMM PLUS scrolls the contents of the Terminal mode screen up one line when the cursor is in the 24th line on the screen, and PROCOMM PLUS receives a 25th line of characters. Any characters on line 1 of the screen scroll off the top. Setting this option to OFF causes the 25th and subsequent lines of characters to overwrite characters on the 24th line of the screen. (Note: This setting does not affect the operation of Chat mode.)

Setting CR Translation

The **G** (CR translation) option has two possible settings: CR for carriage return only and CR/LF for carriage return and line feed. The default setting is CR. (Note: This setting does not affect the operation of Chat mode.)

Most host computers, such as on-line services and bulletin boards (including PROCOMM PLUS Host mode), add a line feed to each carriage return. If PROCOMM PLUS also adds a line feed, everything on your screen will be double-spaced. Therefore, you should normally leave CR translation set to CR.

When you plan to connect to a PC that is not operating as a bulletin board (including PROCOMM PLUS Host mode) or if you are connecting to any system that sends CRs without line feeds, you can set CR translation to CR/LF. Select **G** (CR translation) from the Terminal Options screen. Press any key except Esc or Enter. Each time you press a key, the value in the entry toggles between CR and CR/LF. Press Enter to accept the new value.

TIP | Because you will probably connect to on-line services and bulletin boards more often than to a PC, the easier choice is to leave the default CR translation set to CR and just press Alt-F3 (CR-CR/LF Toggle). PROCOMM PLUS briefly displays the message LF AFTER CR at the left end of the status line. You can easily tell when you need to make this change. Any time all lines of characters you type or receive from another computer continually overwrite the preceding lines, you need to toggle CR translation to CR/LF.

Translating the Backspace Character

The **H** (BS translation) option enables you to specify how PROCOMM PLUS translates the Backspace character on the Terminal mode screen. The choices are DESTRUCTIVE (the default) or NON-DESTRUCTIVE.

When this option is set to DESTRUCTIVE, the Backspace moves the cursor one space to the left and deletes any character that is there. Most host systems (on-line services and bulletin boards) use the Backspace character this way. Occasionally, however, you may connect to a host system where the Backspace character is not intended to delete the character to the left of the cursor. If so, change the **H** (BS translation) option to NON-DESTRUCTIVE. (Note: This setting does not affect the operation of Chat mode.)

Select **H** (BS translation) from the Terminal Options screen. Press any key except Esc or Enter to toggle the entry between DESTRUCTIVE and NON-DESTRUCTIVE. Press Enter to accept the new setting.

Setting the Break Duration

When you press Alt-B (Break Key), PROCOMM PLUS temporarily interrupts the data transmission with a timed delay. The default length of the delay is 350 milliseconds. The **I** (Break length (milliseconds)) option enables you to change the duration of the delay.

To change the length of the delay, choose **I** (Break length (milliseconds)) from the Terminal Options screen. The Setup Utility places the cursor in the entry space. Use the cursor-movement and editing keys listed in table 8.2 to make the desired modification. Press Enter to accept the new break length.

Assigning a Response to ENQ

The **J** (Enquiry (ENQ)) option on the Terminal Options screen specifies how PROCOMM PLUS responds to an incoming ENQ (ASCII decimal 5) character. By default, this feature is set to OFF, and PROCOMM PLUS doesn't respond at all to the ENQ character. You do, however, have two other choices.

If you set this option to ON, PROCOMM PLUS transmits the keyboard macro Alt-0 when PROCOMM PLUS receives the ENQ character (refer to Chapter 7, "Automating PROCOMM PLUS with Macros and Script Files").

When you set the **J** (Enquiry (ENQ)) option to CIS B, PROCOMM PLUS begins file-transfer procedure by using the COMPUSERVE B file-transfer protocol when PROCOMM PLUS receives the ENQ character (see "COMPUSERVE B" in Chapter 5, "Transferring Files").

TIP If you set the **J** (Enquiry (ENQ)) option to CIS B, you may occasionally have a problem with the file-transfer Progress window popping up when you log on to CompuServe or at other times while you are connected to CompuServe. This problem results from CompuServe's sending the ENQ character. To prevent this problem, change your CompuServe Online Terminal Settings. While on-line to CompuServe, type *go terminal*. This command takes you to the TERMINAL/SERVICE OPTIONS area. Select Change permanent settings and then Terminal type/parameters. Change the Micro inquiry sequence at logon setting to NO. Finally, select Make session settings permanent.

To change the value of this option, select **J** (Enquiry (ENQ)) from the Terminal Options screen. Press any key to cycle through the three options: OFF, ON, and CIS B. Press Enter to accept the new entry.

TIP The following ASPECT script command has the same effect as setting the **J** (Enquiry (ENQ)) option to CIS B:

SET ENQ CISB

The CSERVE.ASP script, provided on the PROCOMM PLUS Supplemental Diskette, includes this command. Using this method lets you leave **J** (Enquiry (ENQ)) set to OFF so that errant ENQ characters won't cause the file-transfer Progress window to pop up.

Adjusting Terminal Color

The Setup Utility enables you to customize the color displayed in the Terminal mode screen independent of how PROCOMM PLUS displays color in other modes. (Refer to "Customizing Color," in this chapter, for instructions on tailoring the color of PROCOMM PLUS's other screens, windows, and menus.) From the Setup Utility Main Menu, select TERMINAL OPTIONS to display the Terminal Options menu (again see fig. 8.9). Then choose COLOR OPTIONS. The Setup Utility displays the Terminal Color Options screen shown in figure 8.11. (Note: The settings listed in figure 8.11 are optimized for a monochrome monitor attached to a color graphics display adapter in order to create a more legible screen reproduction. The default settings on your screen will be different if you are using a color screen.)

Use the Terminal Color Options screen to tailor the colors used in the Terminal mode screen. Select one of the five attributes: **A** (Normal), **B** (High intensity), **C** (Low intensity), **D** (Reverse video), or **E** (Underline).

Fig. 8.11.

Fig. 8.11.

The Terminal Color Options screen.

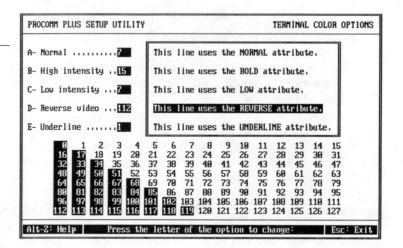

The Setup Utility positions the cursor in the appropriate entry space. Choose a new color scheme from among the 128 sample color schemes displayed in the bottom half of the screen. Press Enter to accept the new value.

As soon as you change the value of an attribute, the Setup Utility shows you the result by altering the colors of the box and 5 lines of text in the upper right quadrant of the screen. Once you are satisfied with the new colors, press Esc (Exit) twice to return to the Terminal Options menu and then to the Setup Utility Main Menu.

Adjusting General Options

Another group of default settings might be called miscellaneous special effects. These settings, grouped on the General Options screen, control such features as exploding windows, sound effects, alarms, the menu line, snow removal, and keyboard speedup.

To display the General Options screen, select **GENERAL OPTIONS** from the Setup Utility Main Menu (not to be confused with the **GENERAL OPTIONS** choice on the Terminal Options menu, which displays the Terminal Options screen). The Setup Utility displays the screen shown in figure 8.12.

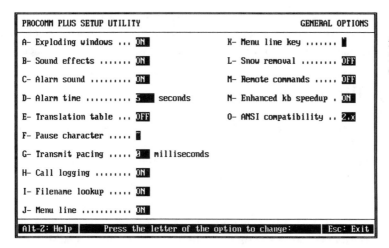

Fig. 8.12.

The General Options screen.

The features discussed in the following sections are listed on the General Options screen. Ten of these fifteen features can be set either ON or OFF. Select the option you want to change by pressing the corresponding letter. The Setup Utility positions the cursor in the option's entry space. Press any key to toggle the setting from ON to OFF or from OFF to ON, and press Enter to confirm your selection. Alteration of the other five options is explained in the corresponding paragraphs. When you are satisfied with the new settings, press Esc (Exit) to return to the Setup Utility Main Menu.

Suppressing Exploding Windows and Sound Effects

By default PROCOMM PLUS opens virtually all screens and windows—for example, the Command Menu, the Dialing Directory screen, the file-transfer Progress window, and the Dialing window—by displaying first a very small box and then rapidly expanding the box into the final window. This striking special effect is known as *exploding windows*. If you prefer that the screen windows appear instantly, without the simulated explosion, set the **A** (Exploding windows) option to OFF.

Another feature that gives PROCOMM PLUS its unique personality is its novel use of sound effects. From beginning to end, PROCOMM PLUS announces the opening and closing of each window or prompt with a friendly "tweet" or pleasant "zip." This setting may sound too much like fun, however, to suit your boss; so set the **B** (Sound effects) option to OFF if you decide that you need to turn off the PROCOMM PLUS sound effects.

Controlling the Alarm

In addition to the sound effects described in the preceding paragraph, PROCOMM PLUS also uses a beep-beep alarm to alert you to certain events, for instance, when your modem successfully connects to another modem or when a file-transfer is completed or aborted. These alarms are normally a convenient feature. You may, for example, start downloading from a bulletin board a large file, which you know will take at least 20 minutes to receive. Instead of dozing off at your keyboard, you can go about your business around the office or house because PROCOMM PLUS will sound the alarm as soon as the transfer is complete. Your spouse may not appreciate this feature, however, if you are downloading a file at 3:00 a.m. Set the **C** (Alarm sound) option to OFF to prevent PROCOMM PLUS from sounding the alarm.

Use **D** (Alarm time) option to modify the length of time that PROCOMM PLUS sounds an alarm signal. By default, PROCOMM PLUS sounds the alarm, discussed in the preceding paragraph, for five seconds. The program causes your computer to generate a beep-beep once each second for five seconds. To change the duration of an alarm, select **D** (Alarm time) from the General Options screen. When the cursor moves into the entry space, type an integer, and press Enter. For example, if you change the value to 2, the next time you download a file, PROCOMM PLUS sounds the alarm only twice.

Activating the Translation Table

Use the **E** (Translation table) option on the General Options screen to turn on use of the translation table. Refer to Chapter 10, "Terminal Emulation," for a full discussion of the translation table.

Changing the Pause Character and Setting Transmit Pacing

The pause character is used in keyboard macros and in the Hangup and Auto answer commands, both found on the Setup Utility Modem Options screen. This character causes PROCOMM PLUS to pause processing for one-half second. As shown in figure 8.12, the default pause character is the tilde (~). The tilde is used as the pause character because the tilde is seldom used for any other purpose. It is possible, however, that for some special application you might need to use the tilde in a keyboard macro to represent an actual tilde. In that case, change the pause character. Select **F** (Pause character) from the General Options screen. When the cursor moves to the entry space, type the character you want for a pause character. Press Enter to accept the change. If you change the pause character, don't forget to edit the Hangup and Auto answer commands in the Modem Options screen to replace the tildes in those commands with the new pause character.

PROCOMM PLUS normally does not insert any time delay between transmitted characters. Some host computers, however, may have difficulty processing character strings and terminal-control sequences sent by a macro or by terminal-emulation keyboard mapping without at least a slight delay between characters. If you think that you may be having this problem, try increasing to 15 the value in the **G** (Transmit pacing) option on the General Options screen. Press Enter to accept the change.

Disabling Call Logging and Filename Lookup

The **H** (Call logging) option on the General Options screen enables you to turn off the PROCOMM PLUS call-logging feature. By default, PROCOMM PLUS keeps a log of the completed calls you make with your modem. For each successful connection, PROCOMM PLUS records the name of the remote system; the telephone number; the connect date; connect time; disconnect time; and the duration of the connection in hours, minutes, and seconds. The program stores this information in a disk file named PCPLUS.FON. If you have no need for a call log, you can turn off this feature by changing the General Options **H** (Call logging) entry to OFF.

When you transfer a file to or from a remote computer, at some point PROCOMM PLUS prompts you to Please enter filename. Often, you have just typed the name of the transfer file before you pressed PgUp (Send Files) or PgDn (Receive Files). After you choose a file-transfer protocol, PROCOMM PLUS searches for the latest file name on the screen. If the program finds a valid DOS file name (from one to eight characters followed by a period and from one to three characters, no spaces), PROCOMM PLUS places this name to the right of the prompt Please enter filename. This PROCOMM PLUS feature relieves you from having to type the file name twice. On the other hand, if you find that you seldom use the file name PROCOMM PLUS finds on the screen, you may want to turn off this "lookup" feature. Select **I** (Filename lookup) on the General Options screen, and press any key to toggle the entry to OFF. Press Enter to accept the change.

Controlling the Menu Line

In addition to the commands listed in the Command Menu, you can invoke most PROCOMM PLUS features by selecting options from a Lotus 1-2-3-style top-line menu. This feature, referred to as the *menu line*, is an entirely new feature in PROCOMM PLUS, not found in ProComm. "Using the Menu Line," in Chapter 2, discusses how to execute PROCOMM PLUS commands through the menu line. You may, however, want to disable the built-in PROCOMM PLUS menus. To turn off the menu-line feature, select **J** (Menu line) on the General Options screen, and press any key. The Setup Utility changes the entry value to OFF. Press Enter to accept the change.

PROCOMM PLUS doesn't automatically display a menu at the top of the Terminal mode screen. To access the menu line, referred to in the preceding paragraph, you must be at the Terminal mode screen and then press the menu-line key. When you first install PROCOMM PLUS, the menu-line key is the left apostrophe key (` ` `), as shown in figure 8.12. The left apostrophe is on the same key as the tilde. This key is used as the menu-line key because it is seldom used for anything else while you are on-line to another computer. You may want to use a different key as the menu-line key. To reassign the menu-line key to a different key, select **K** (Menu line key) on the General Options screen. Type the new menu-line key, and press Enter.

Removing Snow and Speeding Keyboard Reaction

The original Color Graphics Adapter (CGA) found in IBM PCs and IBM PC/XTs, as well as many compatible graphics adapters, sometimes exhibits an annoying screen flicker known as "snow" (because the flicker looks a little like a light snow flurry on your screen). You see this snow when information is written to the adapter's screen buffer. Newer types of graphics adapters (such as the Enhanced Graphics Adapter—EGA), however, do not have this problem. If your screen does suffer from snow, change **L** (Snow removal) option on the General Options screen to ON.

Figure 2.3 in Chapter 2, "Getting Around in PROCOMM PLUS," shows the three most widely used styles of PC keyboards: the IBM PC keyboard, the IBM Personal Computer AT keyboard, and the IBM Enhanced Keyboard. The **N** (Enhanced kb speedup) command on the General Options screen is effective only if you are using the IBM Enhanced Keyboard or equivalent. This feature causes PROCOMM PLUS to react more quickly to each key press and also increases the character repeat rate when a key is held down.

Accepting Remote Commands and Ensuring ANSI Compatibility

By default, PROCOMM PLUS does not recognize script commands sent by a remote computer. If you decide that you want to allow script commands from a remote computer to run on your PC, set the **M** (Remote commands) option on the General Options screen to ON and refer to "Running a Script," in Chapter 11, "An Overview of the ASPECT Script Language," for further instructions. Scripts can be run remotely only from a remote PC emulating an ANSI, VT102, Televideo, or Wyse terminal.

Use the **O** (ANSI compatibility) option to select between ANSI.SYS 2.x emulation and ANSI.SYS 3.x emulation. When you are using PROCOMM PLUS to emulate an ANSI terminal, PROCOMM PLUS also emulates the screen driver ANSI.SYS, a file distributed with the operating system (DOS) and commonly loaded through the CONFIG.SYS file (see Chapter 10, "Terminal Emulation"). PROCOMM PLUS does not use DOS's ANSI.SYS at all. Many host systems (primarily bulletin board systems) can display characters on your screen in color through use of ANSI codes. Some host systems, however, expect you to be using the version of ANSI.SYS that is distributed with DOS 2.x (2.0, 2.1, or 2.11, and so on), and other systems assume that you are using ANSI.SYS from DOS 3.x.

Customizing Color

Most people find the default screen colors used by PROCOMM PLUS pleasing to the eye; but no two people have exactly the same taste. You may decide that you don't care for one or more of the color schemes used by PROCOMM PLUS for a screen, window, or menu. For elements other than the Terminal mode screen, use the Setup Utility Color Options screen to tailor the screen colors more to your liking. For instructions for customizing Terminal mode screen colors, refer to "Adjusting Terminal Color."

In order to display the Color Options screen, select COLOR OPTIONS from the Setup Utility Main Menu (not to be confused with the COLOR OPTIONS choice on the Terminal Options menu, which displays the Terminal Color Options screen). The Setup Utility displays the screen shown in figure 8.13.

Fig. 8.13.

The first Color Options screen, with instructions.

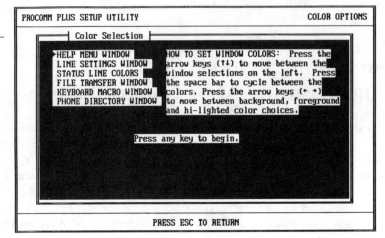

```
PROCOMM PLUS SETUP UTILITY                              COLOR OPTIONS
    ┤ Color Selection ├
  ►HELP MENU WINDOW      HOW TO SET WINDOW COLORS:  Press the
   LINE SETTINGS WINDOW  arrow keys (↑↓) to move between the
   STATUS LINE COLORS    window selections on the left.  Press
   FILE TRANSFER WINDOW  the space bar to cycle between the
   KEYBOARD MACRO WINDOW colors. Press the arrow keys (← →)
   PHONE DIRECTORY WINDOW to move between background, foreground
                         and hi-lighted color choices.


              Press any key to begin.




                    PRESS ESC TO RETURN
```

TIP   A frequent reason for modifying the default PROCOMM PLUS color scheme is the use of a black-and-white monitor with a color graphics adapter. When PROCOMM PLUS is set for use on a color screen, significant portions of menus and screens are hard or impossible to read on black-and-white monitors. A similar situation also occurs on laptop computers with LCD screens. This problem has two easy remedies. First, you can use the PCINSTAL program to set the color scheme properly for a black-and-white screen. (This technique also works for LCD laptop screens.) Second, if you installed PROCOMM PLUS assuming that you were going to use it on a color system, use the following command to start the program:

PCPLUS /B

Either procedure results in a color scheme that is easy to read on black-and-white screens and on LCD laptop screens. Refer to Appendix A, "Installing and Starting PROCOMM PLUS," for more details about using PCINSTAL and starting PROCOMM PLUS.

The screen shown in figure 8.13 is actually the first of two Color Options screens. This first screen includes a paragraph of instructions for using the screens. As soon as you press any key, the Setup Utility displays the second Color Options screen, shown in figure 8.14.

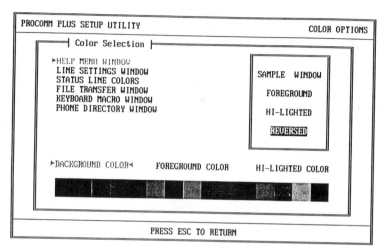

Fig. 8.14.

The second Color Options screen, without instructions.

All the "action" in the Color Options screen takes place in the box entitled Color Selection. You will find four areas of interest inside this box. The top left quadrant lists the names of six windows and screens used by PROCOMM PLUS. (Table 8.3 correlates these names with the screen names used in this book.) A triangular-shaped pointer to the left of the name marks the selected window or screen. The name of the selected window or screen is also highlighted (displayed in a different color from the other characters on the screen). In figure 8.14, the HELP MENU WINDOW is selected. Use the up arrow or the down arrow to select a different window or screen.

The top right quadrant of the Color Options screen shows a sample window. Within this window, the Setup Utility displays an example of the current color scheme for the selected window or screen. The sample window shows the selected window's background color as well as the color of normal characters (FOREGROUND), high-intensity characters (HI-LIGHTED), and inverse video (REVERSED).

Across the bottom portion of the Color Options screen, you see a color bar divided into 16 sections. Above this bar are 3 labels:

- BACKGROUND COLOR denotes the color of the screen's background, as opposed to the characters.

Table 8.3
Window and Screen Names Listed in the Color Options Screen

Name Listed in Color Options Screen	Corresponding PROCOMM PLUS Screen
HELP MENU WINDOW	Command Menu
LINE SETTINGS WINDOW	Line/Port Setup window
STATUS LINE COLORS	Terminal mode screen status line
FILE TRANSFER WINDOW	File-transfer Progress window
KEYBOARD MACRO WINDOW	Keyboard Macros screen
PHONE DIRECTORY WINDOW	Dialing Directory screen

- FOREGROUND COLOR is the color of normal characters.

- HI-LIGHTED COLOR is the color of highlighted or emphasized characters.

Each label represents a different characteristic of the color scheme. The Setup Utility enables you to set the color of each color characteristic independently. The characteristic you are currently setting is displayed in highlighted text and is marked by a pair of triangular-shaped pointers. The Color Options screen shown in figure 8.14 is setting background color for the Help Menu Window (the Command Menu).

Only six keys are used in the Color Options screen. These keys and their functions are listed in table 8.4. Use the up arrow or down arrow to select the window or screen to be redecorated. Use the left or right arrow to select the color characteristic to be changed. And finally, press the space bar to cycle through the available colors. The Setup Utility enables you to choose among blue, green, cyan, red, magenta, brown, and white for a background color. You can set the foreground color and highlight color to any of the 16 colors displayed in the color bar.

CAUTION: Be careful not to set foreground and background to the same color, or you will render the selected window or screen "invisible."

When you have made all the desired color adjustments, press Esc to accept the new colors and to return to the Setup Utility Main Menu.

Table 8.4
Color Options Screen Keys

Key	Function
↑ or ↓	Select the window or screen
← or →	Select the color characteristic
Space bar	Select a different color
Esc	Accept the new colors and return to the Main Menu

Setting ASCII Transfer Options

ASCII—American Standard Code for Information Interchange—forms the lowest common denominator among the countless programs that run on a PC and among the many different types of computers. "ASCII," in Chapter 5, "Transferring Files," describes how to use PROCOMM PLUS's ASCII file-transfer protocol to send to a remote computer, files made up entirely of ASCII characters.

The ASCII file-transfer protocol performs no error check. Through the Setup Utility ASCII Transfer Options screen, however, PROCOMM PLUS does give you control of several aspects of an ASCII upload or download. From the Setup Utility Main Menu, select ASCII TRANSFER OPTIONS. The Setup Utility displays the screen shown in figure 8.15. The sections that follow discuss the options listed in this screen.

```
 PROCOMM PLUS SETUP UTILITY                    ASCII TRANSFER OPTIONS
  A- Echo locally ................ NO     K- Strip 8th bit .............. NO
  B- Expand blank lines .......... YES
  C- Expand tabs ................. YES
  D- Character pacing (millisec).. 15
  E- Line pacing (1/10 sec)....... 10
  F- Pace character .............. 0
  G- CR translation (upload) ..... NONE
  H- LF translation (upload) ..... STRIP
  I- CR translation (download) ... NONE
  J- LF translation (download) ... NONE
 Alt-Z: Help    Press the letter of the option to change:    Esc: Exit
```

Fig. 8.15

The Setup Utility ASCII Transfer Options screen.

When you are finished making changes to the values in this screen, press Esc (Exit) once to return to the Setup Utility Main Menu.

Using Local Echo

By default, PROCOMM PLUS does not display on your screen the ASCII characters sent by an ASCII upload procedure. Instead, during the transmission, PROCOMM PLUS displays a line count at the left end of the status line. This count indicates the number of lines of characters that PROCOMM PLUS has successfully transmitted.

If you prefer to monitor visually the ASCII text as it is sent, turn on *local echo*. From the ASCII Transfer Options screen, select **A** (Echo locally). When the cursor moves into the entry space, press any key to toggle the entry from NO to YES. Press Enter to accept this change. The next time you execute an ASCII upload, PROCOMM PLUS displays on your screen each line of characters as it is transmitted to the remote computer, just as if you were typing the characters at your keyboard.

Expanding Blank Lines and Tabs

Many message systems on bulletin boards interpret a completely empty line as the end of your message; but you may occasionally want to include one or more blank lines in the middle of a message. When you are typing the message on-line, you can include a blank line in a message by pressing the space bar. This action inserts a space character (ASCII decimal 32); so even though the line appears to be empty, it actually contains an ASCII character, and the message system doesn't interpret the line as the end of your message.

By default, during an ASCII upload, PROCOMM PLUS adds a space character to any empty line (that is, a line that contains only a carriage return or carriage return and line feed). This added space character enables you to use the ASCII upload procedure to send an ASCII file containing blank lines as a message to a message system (or electronic mail system) that interprets an empty line as the end of a message. In the unlikely event that you need to disable this feature, select **B** (Expand blank lines) from the ASCII Transfer Options screen, and press any key to toggle the value of the entry to NO.

When you press the Tab key in PCEDIT (discussed in Chapter 6), the cursor moves eight spaces to the right, but the editor generates the ASCII horizontal tab character (ASCII decimal 9). A bulletin board message system or electronic mail system may not, however, interpret the tab character in the same way. The system may insert five space characters for a tab, for example, or ignore tabs altogether. To ensure that your message maintains the appearance you intend during an ASCII upload, PROCOMM PLUS by default converts each tab character to eight space characters. You can turn off this feature, however, so that PROCOMM PLUS transmits the tab character. Select **C** (Expand tabs) from the ASCII Transfer Options screen, and press any key to toggle the entry value from YES to NO.

Pacing ASCII File Transfers

When you transmit an ASCII file using the ASCII file-transfer protocol, PROCOMM PLUS uses a feature known as *character pacing*. This feature separates transmission of succeeding characters by a short interval during which nothing is sent. This interval helps the remote system distinguish one character from the next. A message system, for example, may "expect" you to type messages on-line so may not be prepared to process an uninterrupted flow of ASCII characters. The character pacing feature mitigates this problem by delaying the transmission of each character by a set interval of time.

By default, PROCOMM PLUS inserts a 15-millisecond delay between characters. To change this number, select **D** (Character pacing (millisec)) from the ASCII Transfer Options screen. When the cursor moves into the entry space, use the keys listed in table 8.2 to edit the entry value. Changing the value to 0 disables the feature entirely and allows the fastest possible upload to systems that can accept characters at this rate.

In addition to character pacing, PROCOMM PLUS also uses *line pacing* during an upload using ASCII file-transfer protocol. Line pacing separates transmission of succeeding lines of characters by a interval during which nothing is sent. Many remote messaging systems process one line of characters at a time, looking first for an editing command and then adding the characters to the uploaded file. A short delay between lines gives the remote system more time to process each line.

By default, PROCOMM PLUS inserts an interval of 10 tenths of a second, or 1 second, between transmitted lines. Unfortunately, the line pacing feature can significantly slow the file-transfer process. If you routinely send long ASCII files, you may want to shorten the line pacing interval or to eliminate line pacing completely. Select **E** (Line pacing (1/10 sec)) and then edit the entry value. Changing the value to 0 disables the feature.

In most situations, you use the ASCII file-transfer protocol to send an ASCII file to a message system or electronic mail system that expects you to type the message on-line. Occasionally, though, you may use this protocol to transmit a file to a computer that is using its own ASCII file-transfer protocol to download the file. Some protocols use a flow-control device known as a *pace character*. As soon as the remote computer receives a carriage return, indicating that the computer has received a complete line of characters, the remote computer replies by sending to your computer a predetermined character known as the pace character. For this feature to work properly, your computer must not send the next line of characters until your computer receives the pace character.

By default, PROCOMM PLUS does not wait for a pace character. The program does permit you to activate this feature if you need it. To enable the pace character feature, select **F** (Pace character) from the ASCII Transfer Options screen. Determine the pace character that is going to be sent by the remote computer and type the ASCII decimal code for this character (see the ASCII table in Appendix D of the PROCOMM PLUS documentation). Press Enter to accept the change.

Set the pace character back to 0 to disable the feature.

Translating CR and LF

Four settings on the ASCII Transfer Options screen control how PROCOMM PLUS translates each carriage return (CR) and line feed (LF) in files uploaded to a remote computer and in files downloaded from a remote computer. These options are

G (CR translation (upload))
H (LF translation (upload))
I (CR translation (download))
J (LF translation (download))

Using these options, you can cause PROCOMM PLUS to add a line feed to each carriage return or to remove all carriage returns. Similarly, you can cause PROCOMM PLUS to add a carriage return to each line feed or to remove all line feeds.

As explained in "Setting CR Translation," earlier in this chapter, most host computers, such as on-line services and bulletin boards (including PROCOMM PLUS Host mode), add a line feed (LF) to each carriage return (CR).

Because the most common use of the ASCII file-transfer protocol is to send and receive files to and from message systems and electronic mail systems, the default CR and LF translations for uploading and downloading ASCII files are set for that purpose. You should not have to change the default settings of the four CR/LF translation options in order to send ASCII files to a message system or to an electronic mail system.

By default, when you are uploading a file using the ASCII file-transfer protocol, PROCOMM PLUS does not add a line feed to each carriage return. On the contrary, because the remote computer is expected to add a line feed to each carriage return, PROCOMM PLUS strips all line feeds from the transmitted file.

Occasionally, however, you may need to send an ASCII file to a computer that does not add a line feed to each carriage return. In that case, select **H** (LF translation (upload)) from the ASCII Transfer Options screen. Press any key twice to change the entry value from STRIP to ADD CR and then to NONE. Press Enter to accept the change. The next time you use the ASCII upload file-transfer protocol, PROCOMM PLUS sends the file without stripping line feeds.

By default, when you download a file using the ASCII file-transfer protocol, PROCOMM PLUS passes the incoming file through "untouched," neither adding nor removing carriage returns or line feeds. If, however, you notice that the files you receive don't include any line feeds (everything overwrites the same line on the screen), make the following adjustment. Select **I** (CR translation (download)) from the ASCII Transfer Options screen. Press any key twice to change the entry value from NONE to STRIP and then to ADD LF. The next time you download a file, PROCOMM PLUS adds a line feed for every carriage return.

Stripping the 8th Bit

The IBM character set is different in two important respects from the ASCII character set recognized by other computers—referred to in this chapter as the generic ASCII character set. The generic ASCII character set includes only the first 128 characters of the IBM character set (decimal values 0 through 127); and the generic ASCII character set can be represented using just 7 of the 8 bits available in each byte of a PC's memory (RAM) or storage (disk).

In practical use, however, these two distinctions between the generic ASCII character set and the IBM character set seldom create a problem. Most people do not use the extended IBM characters (decimal 128 through 255) in normal correspondence. PROCOMM PLUS, however, allows you to do ASCII transfers using the full IBM character set as long as option **K** (Strip 8th bit) is set to NO (its default value). If you have to upload a file that uses the full IBM character set (for instance, a WordStar file or a file containing IBM box-drawing characters) to a system that supports only the generic ASCII character set, you may set this option to YES. You can also set this option to YES if you want to download a WordStar file and "translate it down" to generic ASCII.

When you use the ASCII file-transfer protocol to send an ASCII file to a computer that can handle only 7 data bits, activate the PROCOMM PLUS feature that strips the 8th bit. Select **K** (Strip 8th bit) from the ASCII Transfer Options screen. Press any key to toggle the entry value from NO to YES. Press Enter to accept the change.

Setting File and Path Options

The Setup Utility File/Path Options screen is used to specify file or path names for a number of different purposes. In this one screen, you can specify the following defaults:

- File name for the log file (discussed in "Capturing the Session to Disk," in Chapter 4)

- File name for screen snapshots (discussed in "Taking a Snapshot," in Chapter 4)

- DOS path for files downloaded using a file-transfer protocol

- Name of the text editor activated by the Alt-A (Editor) command

- Names for three separate programs that can be called from within PROCOMM PLUS

To display the File/Path Options screen, select FILE/PATH OPTIONS from the Setup Utility Main Menu. The Setup Utility displays the screen shown in figure 8.16.

When you first install PROCOMM PLUS, three entries on the File/Path Options screen are already filled in, as shown in figure 8.16. The default file name for the log file is PCPLUS.LOG. The default file name for screen snapshots is PCPLUS.SCR. And the editor invoked by the Alt-A (Editor) command is PCEDIT. To change any of these values, first press the corresponding letter to move the cursor into the entry area. Then use the cursor-movement and editing keys listed in table 8.2 to modify the entry. Press Enter to accept the change.

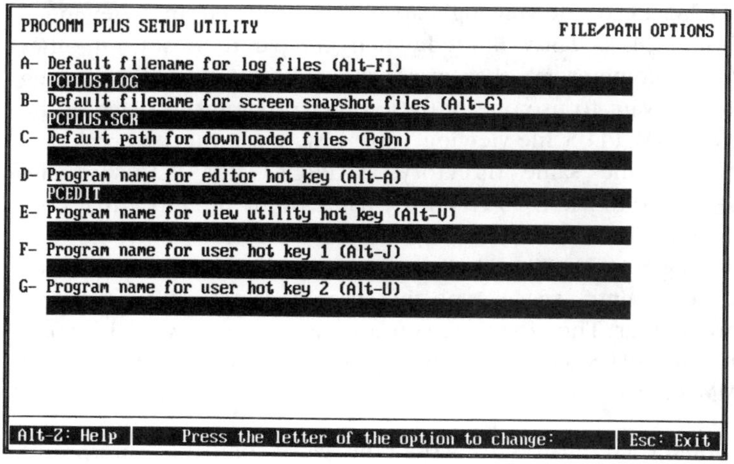

```
PROCOMM PLUS SETUP UTILITY                        FILE/PATH OPTIONS

A- Default filename for log files (Alt-F1)
   PCPLUS.LOG
B- Default filename for screen snapshot files (Alt-G)
   PCPLUS.SCR
C- Default path for downloaded files (PgDn)

D- Program name for editor hot key (Alt-A)
   PCEDIT
E- Program name for view utility hot key (Alt-V)

F- Program name for user hot key 1 (Alt-J)

G- Program name for user hot key 2 (Alt-U)




  Alt-Z: Help   │   Press the letter of the option to change:   │ Esc: Exit
```

Fig. 8.16.

The File/Path Options screen.

The third option listed on the File/Path Options screen, **C** (Default path for downloaded files (PgDn)), is not filled in with a default value when you first install PROCOMM PLUS. Instead, PROCOMM PLUS normally downloads files into the current working directory (see "Changing the Current Working Directory," in Chapter 5). To specify a default download directory, select **C** (Default path for downloaded files (PgDn)) from the File/Path Options screen. Type the full DOS path for the directory into which you want files downloaded.

TIP When you designate a download directory, make sure that the directory exists. Otherwise, PROCOMM PLUS will abort every attempted download without any explanation. The most common way to fall into this trap is to designate a download directory that exists and later remove the directory from your disk. The next time you try to download a file, PROCOMM PLUS aborts the download. Because the program worked fine before you removed the directory, you may scratch your head for a long time trying to determine what went wrong. If you remove the download directory, make sure that you change the download directory entry on the File/Path Options screen to a directory which exists or that you delete the entry entirely.

Keystroke commands that can access another program from within PROCOMM PLUS are referred to as *hot-key* commands. The command Alt-A (Editor) is a hot-key command. The File/Path Options screen enables you to activate up to three more hot-key commands.

The PROCOMM PLUS file-viewing capability, discussed in "Viewing a File," in Chapter 4, is fairly limited when compared to several other readily available commercial, shareware, and public domain file-viewing programs. If you want to use one of these other programs in place of the built-in PROCOMM PLUS file-viewing capability, copy the external program's files into the same directory as PROCOMM PLUS (usually C:\PCPLUS), or add to the DOS PATH the name of the directory that contains the program. Then select **E** (Program name for view utility hot key (Alt-V)) from the Setup Utility File/Path Options screen. When the cursor is displayed in the entry space, type the start-up command for the program, and press Enter. The next time you use the Alt-V (View a File) command, PROCOMM PLUS will use the new program to view the file instead of PROCOMM PLUS's built-in file-viewing capability.

> **TIP** For hot-key commands to work, the DOS command interpreter program, COMMAND.COM, must be in the root directory of your boot disk or in a directory specified by the SET COMSPEC command in your AUTOEXEC.BAT file. Refer to Appendix A, "Installing and Starting PROCOMM PLUS," and to your DOS manual for information about the AUTOEXEC.BAT file and the SET COMSPEC command.

The PROCOMM PLUS Setup Utility File/Path Options screen also enables you to attach two more programs to hot-key commands. You can attach one program to the keystroke command Alt-J and the other to the command Alt-U.

To attach a program to the command Alt-J, select **F** (Program name for user hot key 1 (Alt-J)) from the File/Path Options screen. Type the start-up command (including full DOS path and any necessary start-up parameters) in the entry area, and press Enter. You can then start this external program from within PROCOMM PLUS by pressing Alt-J. Use the same procedure to attach a different program to the Alt-U command through the File/Path Options screen selection **G** (Program name for user hot key 2 (Alt-U)).

TIP | The Alt-J and Alt-U hot-key commands work best when you use a batch file to start the external program. For example, suppose that you want to be able to activate Lotus 1-2-3 quickly from within PROCOMM PLUS. When you need to take a quick look at a spreadsheet while you are on-line, you don't want to disconnect and quit PROCOMM PLUS. To solve this problem, first use an ASCII text editor, such as PCEDIT, to create a DOS file named 123.BAT, similar to the following:

```
ECHO OFF
CLS
CD \LOTUS
123
CD \PCPLUS
```

Save this file in the directory that contains the PROCOMM PLUS program files. If you have installed 1-2-3 in a directory other than C:\LOTUS, substitute that directory name for LOTUS in the third line of this ASCII file.

To attach 1-2-3 to the command Alt-J, select **F** (Program name for user hot key 1 (Alt-J)) from the File/Path Options screen. Type *123* in the entry area, and press Enter. Now when you are in the Terminal mode screen, you can press Alt-J, and PROCOMM PLUS immediately suspends itself and runs 1-2-3. As soon as you leave 1-2-3, you return to the Terminal mode screen. If you are on-line, PROCOMM PLUS will not disconnect the line when you activate 1-2-3 through this hot-key command. The host system to which you are connected, however, may log you off if you take too long to return to the on-line session.

Changing KERMIT Options

As explained in Chapter 5, "Transferring Files," KERMIT is both a program and a file-transfer protocol. This section of the chapter describes how to customize PROCOMM PLUS's implementation of the file-transfer protocol KERMIT.

Since its introduction in 1981, KERMIT has had many enhancements and has been implemented on countless different computer systems. Consequently, not all versions of KERMIT have exactly the same features. To maintain complete compatibility among all KERMIT implementations, at the beginning of the transmission, the protocol performs a unique "hand shake," which can be described as *feature negotiation*.

Even with this hand shake, two implementations of KERMIT possibly may not communicate properly. If you experience difficulty sending or receiving files with another KERMIT implementation, you may need to use the Setup Utility Kermit Options screen. From the Setup Utility Main Menu, select **KERMIT OPTIONS**. The Setup Utility displays the Kermit Options screen, shown in figure 8.17. Contact the operator or system administrator of the other computer and determine whether you need to change the values listed in this screen.

Fig. 8.17.

The Kermit Options screen.

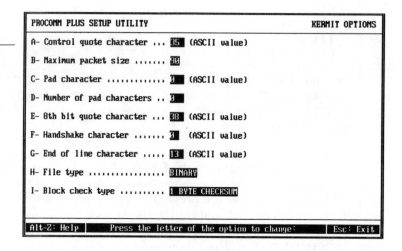

```
PROCOMM PLUS SETUP UTILITY                                KERMIT OPTIONS

A- Control quote character ... 35  (ASCII value)

B- Maximum packet size ....... 90

C- Pad character ............. 0   (ASCII value)

D- Number of pad characters .. 0

E- 8th bit quote character ... 38  (ASCII value)

F- Handshake character ....... 0   (ASCII value)

G- End of line character ..... 13  (ASCII value)

H- File type ................. BINARY

I- Block check type .......... 1 BYTE CHECKSUM


 Alt-Z: Help │    Press the letter of the option to change:    │ Esc: Exit
```

Once you have made any necessary adjustments to the Kermit options, press Esc to return to the Setup Utility Main Menu.

Changing Protocol Options

The Setup Utility Protocol Options screen enables you to add up to three external file-transfer protocol programs for uploading and downloading files. This screen also includes a setting for activating the so-called *relaxed* version of the XMODEM file-transfer protocol and a setting that causes PROCOMM PLUS to delete files that result from aborted download procedures.

To display the Protocol Options screen, select **PROTOCOL OPTIONS** from the Setup Utility Main Menu. The Setup Utility displays the screen shown in figure 8.18.

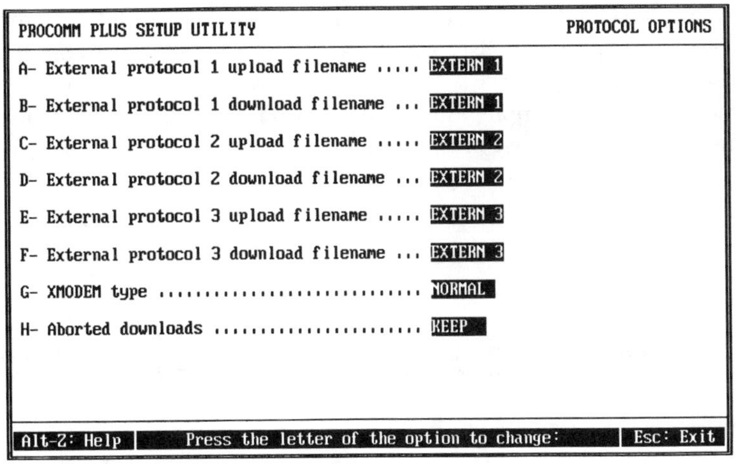

```
┌─────────────────────────────────────────────────────────────────┐
│ PROCOMM PLUS SETUP UTILITY                        PROTOCOL OPTIONS │
│ A- External protocol 1 upload filename ..... EXTERN 1             │
│ B- External protocol 1 download filename ... EXTERN 1             │
│ C- External protocol 2 upload filename ..... EXTERN 2             │
│ D- External protocol 2 download filename ... EXTERN 2             │
│ E- External protocol 3 upload filename ..... EXTERN 3             │
│ F- External protocol 3 download filename ... EXTERN 3             │
│ G- XMODEM type ............................. NORMAL               │
│ H- Aborted downloads ....................... KEEP                │
│                                                                   │
│                                                                   │
│ Alt-Z: Help │   Press the letter of the option to change:  │ Esc: Exit │
└─────────────────────────────────────────────────────────────────┘
```

Fig. 8.18.

The Protocol Options screen.

Adding an External Protocol

Even though PROCOMM PLUS supplies 13 different built-in file-transfer protocols, it also provides a means for you to invoke up to 3 other protocols from within PROCOMM PLUS. These protocols are referred to as *external protocols*. For example, you may want to add the ZMODEM protocol from Chuck Forsberg and the CompuServe B+ protocol from CompuServe. Neither of these protocols is included in PROCOMM PLUS, but both have enhanced features that warrant your attention.

Use the first six options listed on the Protocol Options screen to specify external programs that perform file uploading or file downloading. Use option **A**, **C**, or **E** to add an upload protocol; and use option **B**, **D**, or **F** to add a download protocol.

When you select an external protocol option, the cursor moves into the corresponding entry space. Use the cursor-movement and editing keys listed in table 8.2 to replace the current entry with the start-up command for the protocol program you want to use. For example, to start ZMODEM as an upload protocol, you choose **A** (External protocol 1 upload file-name) from the Protocol Options screen and then type *zmodemu* as the entry. The actual command you specify depends on the external protocol program you are adding and on whether you are using a DOS batch file to activate the program. Refer to Appendix C, "Using External Protocols," for more specific instructions, using ZMODEM as an example.

After you have specified an external upload protocol in the Protocol Options screen, using PgUp (Upload Files) from the Terminal mode screen causes PROCOMM PLUS to list the new protocol in the Upload menu. In the ZMODEM example, PROCOMM PLUS lists ZMODEMU as upload protocol number 14, as shown in figure 8.19. You can then type *14* and press Enter to activate this file-transfer protocol.

```
┤ Upload Protocols ├

 1) XMODEM       5) TELINK      9) WXMODEM      13) YMODEM-G BATCH
 2) KERMIT       6) MODEM7     10) IMODEM       14) ZMODEMU
 3) YMODEM       7) SEALINK    11) YMODEM-G     15) EXTERN 2
 4) ASCII        8) COMPUSERVE B 12) YMODEM BATCH 16) EXTERN 3

 Your Selection:     (or press ENTER for XMODEM)
```

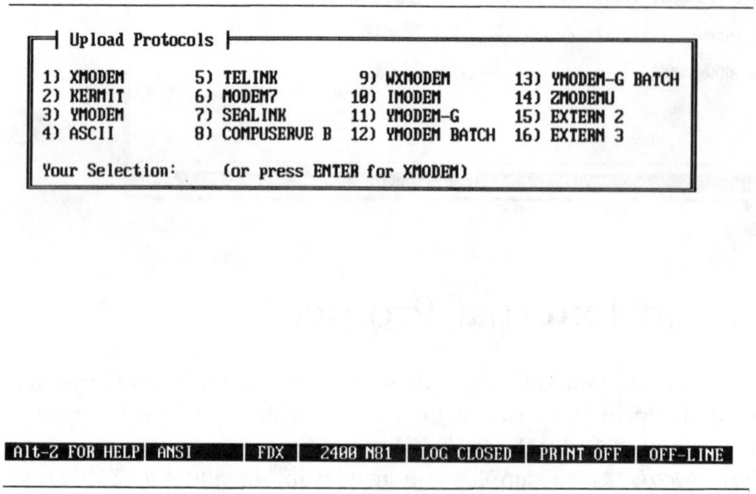

Fig. 8.19.

The Upload menu with ZMODEMU added as protocol number 14.

```
Alt-Z FOR HELP  ANSI      FDX    2400 N81   LOG CLOSED   PRINT OFF   OFF-LINE
```

Before PROCOMM PLUS runs the external program, the program displays a small window containing the prompt Enter parameters. At this point you can type additional start-up parameters for the external program, and press Enter. PROCOMM PLUS then temporarily suspends itself and runs the external program. When you later quit from the external program, normally after a successful file transfer, PROCOMM PLUS returns you to the Terminal mode screen. For a specific example of how to use an external protocol in this manner, refer to Appendix C.

Relaxing XMODEM

The most widely available PC-based file-transfer protocol is known as XMODEM. The protocols YMODEM, YMODEM BATCH, MODEM7, WXMODEM, TELINK, and SEALINK are derived from XMODEM. The "XMODEM" section of Chapter 5 discusses how to use this protocol. XMODEM is designed to provide a reliable and yet easily implemented method of transferring files between small computers. Its design does not, therefore, account for the substantial transmission delays that can be introduced when a connection is made over long-distance telephone lines or over a packet-switching public data network.

When you begin an XMODEM download (or start to receive a file with one of the other XMODEM-derived file-transfer protocols), PROCOMM PLUS repeatedly sends out the character *C* to the sending computer as a handshake to indicate that PROCOMM PLUS's XMODEM supports the CRC error-checking method. If PROCOMM PLUS doesn't receive an appropriate response to these characters, the program falls back to the checksum error-checking method. The delay caused by long-distance telephone lines and public data networks can cause PROCOMM PLUS to fall back prematurely. The Relaxed XMODEM mode, however, sends out the *C*s at a slower pace and waits a longer period of time for a response.

To select the Relaxed XMODEM mode, choose **G** (XMODEM type) from the Protocol Options screen. Then press any key to toggle the entry value from NORMAL to RELAXED. Press Enter to accept the change. The next time you use XMODEM (or one of the XMODEM derivatives), it permits a longer delay for receiving a reply to the handshake.

Keeping Aborted Downloads

When you begin a download using a PROCOMM PLUS file-transfer protocol, PROCOMM PLUS opens a file on your disk to receive the downloaded data. As the transfer progresses, PROCOMM PLUS periodically adds data to this download file. The program closes the download file at the completion of the transfer. By default, if a file transfer is aborted before the entire original file is received, PROCOMM PLUS closes the download file, even though it contains only a portion of the original file. PROCOMM PLUS keeps the download file on-disk; this file contains all data that was received before the transfer aborted.

In many cases, a partial file is of little use. A spreadsheet file, for example, is of no benefit unless you have the whole thing. Your spreadsheet program will not be able to load a partial spreadsheet; and it is unlikely that such an abbreviated spreadsheet would be meaningful even if it could be loaded. PROCOMM PLUS therefore provides an option that causes PROCOMM PLUS simply to abandon incomplete download files.

Select **H** (Aborted downloads) from the Protocol Options screen. Then press any key to toggle the entry value from KEEP to DELETE. Press Enter to accept the change.

Chapter Summary

This chapter has taught you how to customize the many program settings that control the way PROCOMM PLUS operates. The text has discussed how to use the Line/Port Setup window to set the COM port and line settings and how to use the Setup Utility to modify modem, terminal, general, color, and file-transfer options. As you no doubt have discovered by now, PROCOMM PLUS works pretty well right off the rack. Now that you know how to tailor its many features, you should be able make adjustments until PROCOMM PLUS fits your needs like a designer suit.

Using Host Mode

One of the most popular uses for PROCOMM PLUS and other personal computer-based communications programs is to connect to electronic bulletin board systems. Traditionally, these bulletin board systems are run by computer hobbyists for computer hobbyists. A growing number of businesses, however, are discovering that PC-based bulletin boards can provide a convenient and inexpensive company-wide electronic mail service. A centrally located PC running bulletin board software can serve as an unattended company electronic post office. By means of a telephone line, the bulletin board receives electronic mail and files from across town, across the country, or around the world. This chapter describes the Host mode, the rudimentary bulletin board system built into PROCOMM PLUS.

The chapter first describes the steps you should take before opening your electronic post office. These preliminary steps include specifying a welcome message, establishing a list of registered users, and deciding whether the system is to be open or closed to new unregistered users. The chapter then describes several ways to invoke Host mode and describes the operation of Host mode, first from a user's perspective and then from your perspective as the Host mode system operator. The text explains how a user logs on to the system and describes the Host mode features available to the user, including uploading and downloading of files and electronic mail. The chapter then describes the functions the system operator can perform in Host mode. Finally, the text discusses administrative matters related to Host mode, including management of Host mode files and administration of the Host mode electronic mail facility. When you have completed this chapter, you will be ready to use PROCOMM PLUS to set up your own bulletin board system.

Preparing Host Mode

As with virtually every other feature of PROCOMM PLUS, you can use PROCOMM PLUS's built-in bulletin board system with no special preparation. Because the system is configured when you take it out of the box, you can immediately invoke Host mode and permit users to call in, log on, and use this system as an electronic bulletin board. In order to maintain reasonable control over your computer's resources, however, you probably will want to take certain actions before permitting a user to log on to Host mode. This section of Chapter 9 describes how to set several Host mode parameters through the PROCOMM PLUS Setup Utility and also explains how to edit the disk file PCPLUS.USR in order to establish a list of authorized users.

The PROCOMM PLUS Setup Utility's Host Mode Options screen is used to modify several Host mode parameters. To display the Host Mode Options screen, from the Terminal mode screen press Alt-S (Setup Facility) and select HOST_MODE_OPTIONS from the Setup Utility Main menu. The Setup Utility displays the screen shown in figure 9.1. The following few sections describe when and how to modify the six options listed on this screen.

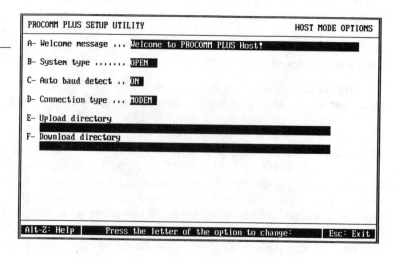

Fig. 9.1.

The Host Mode Options screen.

```
 PROCOMM PLUS SETUP UTILITY                              HOST MODE OPTIONS

 A- Welcome message ... Welcome to PROCOMM PLUS Host!

 B- System type ....... OPEN

 C- Auto baud detect .. ON

 D- Connection type ... MODEM

 E- Upload directory

 F- Download directory




 Alt-Z: Help      Press the letter of the option to change:     Esc: Exit
```

When you have finished modifying the Host mode options, press Esc (Exit) to return to the Setup Utility Main menu. Finally, select SAVE_SETUP_OPTIONS to save the new Host mode settings, and press Esc (Exit) again to return to the Terminal mode screen.

Modifying the Host Mode Welcome Message

With the first option on the Host Mode Options screen, you can customize the Host mode welcome message. Each time a user logs on to Host mode, PROCOMM PLUS displays this message on the user's screen. The default message is Welcome to PROCOMM PLUS Host! but you can certainly think of a more appropriate welcome message for your bulletin board system. To modify this message, select **A** (Welcome message) from the Host Mode Options screen. Then use the cursor-movement and editing keys listed in table 9.1 to enter the new welcome message. Press Enter to accept the change.

Table 9.1
Setup Utility Cursor-Movement and Editing Keys

Key	Function
←	Move the cursor one space to the left
→	Move the cursor one space to the right
Home	Move the cursor to the left end of the entry
End	Move the cursor one space to the right of the last character in the entry
Insert	Toggle Insert/Overtype mode
Delete	Delete the character at cursor
Backspace	Delete the character to the left of cursor
Tab	Delete the entire entry
Ctrl-End	Delete the characters from cursor to right end

For example, suppose that you are the national sales manager for Terry's T-Shirts, Inc. You are tired of playing telephone tag, so you want to set up PROCOMM PLUS Host mode to act as an electronic post office for your four regional sales offices. At the close of each business day, you want each regional sales manager to send you a status report, using the Host mode electronic mail facility. Every Friday afternoon each regional office also has to send you a report in the form of a spreadsheet. You will consolidate the four regional reports into one spreadsheet, which you, in turn, will submit first thing Monday morning to Terry, the CEO.

An appropriate welcome message for your bulletin board might be as follows:

You are connected to Terry's T-Shirts, Inc., BBS

In order to make this message the Host mode welcome message, from the Terminal mode screen, press Alt-S (Setup Facility) to display the Setup Utility Main menu. Then select **HOST_MODE_OPTIONS**. When the Setup Utility displays the Host Mode Options screen, select **A** (Welcome message). Press Tab to delete the current message; then type the new message. The entry space accommodates a message up to 50 characters long. Press Enter to accept the new message.

TIP | PROCOMM PLUS also lets you create a log-on message of unlimited length, in addition to the welcome message. Refer to "Creating a Log-on Message File," later in this chapter.

Setting the System Type

B (System type) is the second option listed on the Host Mode Options screen. The default value is OPEN, the least restrictive bulletin board system type. When your Host mode system is open, anyone who connects to your computer (while it is running PROCOMM PLUS in Host mode) can log on and use the bulletin board. For many systems, however, like the Terry's T-Shirts' bulletin board, an open system is not appropriate. A closed system gives you more control over who can access your bulletin board.

In order to switch to a closed system, select **B** (System type) from the Host Mode Options screen and then press any key. The Setup Utility toggles the setting from OPEN to CLOSED. Press Enter to accept the change. Now only users whose names you have included in the PCPLUS.USR file can log on to your bulletin board. (PCPLUS.USR is an ASCII file that contains a list of authorized users. Adding users to this file is covered in the section "Building the User File.")

Using Auto Baud Detect

The third option on the Host Mode Options screen, **C** (Auto baud detect), determines whether PROCOMM PLUS automatically matches the PROCOMM PLUS transmission rate to the calling computer's transmission rate. By default, this option is set to ON, and you seldom (if ever) have a reason to change this setting. When a user connects to your computer in Host mode, PROCOMM PLUS detects the actual connect speed and adjusts the PROCOMM PLUS baud setting up or down accordingly.

TIP | The Host mode auto baud detect feature works independently from the auto baud detect option on the Modem Options screen (see "Activating Auto Baud Detect," in Chapter 8). In other words, the **H** (Auto baud detect) option on the Modem Options screen can be set to OFF, and the Host mode auto baud detect feature still operates as long as the **C** (Auto baud detect) option on the Host Mode Options screen is set to ON. Neither auto baud detect option works properly, however, if the entries on the Modem Result Messages screen are not spelled correctly (see "Setting Result Messages," in Chapter 8).

For example, you might set PROCOMM PLUS's transmission speed to 2400 bps and then activate Host mode. (Refer to Appendix A, "Installing and Starting PROCOMM PLUS," and to "Assigning a COM Port and Line Settings," in Chapter 8, for discussions on setting the transmission speed.) One of your sales representatives is calling in, however, using a laptop computer that contains only a 1200-bps modem. When your modem and the salesperson's modem connect, they "negotiate" a transmission rate that both modems can handle—1200 bps. With the auto baud detect option set to ON, PROCOMM PLUS detects the connect speed of 1200 bps and adjusts the PROCOMM PLUS baud rate accordingly. Host mode displays the welcome message on the user's screen, and the user can now log on to Host mode. Without this feature, the modems would connect at 1200 bps, but the sales representative would see only "happy faces" and other strange characters on the screen instead of the Host mode welcome message. PROCOMM PLUS would still be operating under the assumption that the connection was made at 2400 bps.

Setting the Connection Type

The fourth option on the Host Mode Options screen specifies whether the user will be connected by MODEM or DIRECT. Most often you will use the PROCOMM PLUS Host mode to permit remote users to connect to your computer through a modem over telephone lines. In most situations, therefore, the **D** (Connection type) setting on the Host Mode Options screen should be MODEM. This setting is the default.

Occasionally, however, you may want to transfer data between two computers that are physically close enough to be connected by a short cable (50 feet or less). Such a connection is referred to as a *direct connection*. This connection requires a special type of cable. If you connect the two computers by a standard RS-232 serial cable, they cannot communicate; instead, you must connect the serial ports of the computers by a special cable known as a *null modem* (which means literally "no modem"). This connection is sometimes referred to as "hard wiring" the computers together.

A null modem looks like a normal serial cable but is configured to connect two DTE (Data Terminal Equipment) RS-232 devices. A PC's serial port is normally configured as a DTE device. (Refer to Appendix B, "Installing a Modem," for a description of the RS-232 standard and a diagram that shows how to construct a null modem cable.)

For example, suppose that your desktop PC has a 5 1/4-inch floppy disk drive, but your laptop PC has a 3 1/2-inch disk drive. How can you trade files between the two machines? You could buy a 3 1/2-inch drive for the desktop PC or perhaps an external 5 1/4-inch drive for the laptop PC. Either of these solutions, however, would probably cost at least $100—more if you pay full retail, and even more if you don't want to install the new drive yourself. As an alternative, you can buy a null modem cable for as little as $7 by mail order and no more than $30 full retail. Then you can use PROCOMM PLUS to send files in either direction over the null modem cable between the two computers.

After you have connected two PCs with a null modem cable, load PROCOMM PLUS on both machines. With the **D** (Connection type) option set to DIRECT on one PC, activate Host mode (see "Invoking and Exiting Host Mode," in this chapter). PROCOMM PLUS on the Host mode machine immediately sends the welcome message to the other machine. You then control the session from the user computer, the one not running in Host mode.

TIP More than likely, you will need to use the Alt-P (Line/Port Setup) command to set the COM port and transmission speed (see "Assigning a COM Port and Line Settings," in Chapter 8). When using a null modem cable to connect to another PC, you must specify the COM port to which the cable is connected. Unless you have disconnected your modem and are using the same COM port, you will have to change the COM port setting. And because you are not using a modem, you are not constrained by the modem's maximum speed. You can transfer files between computers connected by a null modem, using the fastest transmission speed that both computers can accommodate.

The easiest way to determine the maximum transmission speed between a pair of computers is by trial and error. Set both computers to PROCOMM PLUS's fastest speed, 115,900 bps, and attempt a file transfer. If the file transfer aborts, try a slower speed. When you complete a successful transfer, you have discovered the maximum mutual speed. Use this speed when you transfer files between these two computers over a null modem cable.

Specifying Upload and Download Directories

The fifth option on the Host Mode Options screen enables you to assign a DOS directory into which users can send files. Because users will be uploading these files to the bulletin board, this directory is referred to as the *upload directory*. Similarly, the last option on the Host Mode Options screen enables you to assign a directory from which users can download files, referred to as the *download directory*.

By default, PROCOMM PLUS places files uploaded through Host mode into the current working directory, and users are permitted to download files from the current working directory. The result is that any file uploaded by one user can be downloaded by any subsequent user. Indeed, all files contained in the current working directory are available to all users for downloading. The capability to assign separate upload and download directories solves this obvious system security problem. All user-uploaded files go exclusively into the designated upload directory; and users can download files only from the specified download directory. (Note: This limitation does not apply to privileged users—privilege level 1. See "Understanding User Privileges," in this chapter.)

To specify an upload directory, select **E** (Upload directory) from the Host Mode Options screen. In the entry space provided, type the directory name, and press Enter. Similarly, to designate a download directory, choose **F** (Download directory) from the Host Mode Options screen. Type the directory name, and press Enter.

Assume, for example, that the current working directory is C:\PCPLUS. You create a DOS directory named C:\PCPLUS\HOST-UP as the Host mode upload directory and the directory C:\PCPLUS\HOST-DWN for use as the Host mode download directory.

To specify C:\PCPLUS\HOST-UP as the upload directory, select **E** (Upload directory) from the Host Mode Options screen. In the entry space, type *host-up*, and press Enter. To designate C:\PCPLUS\HOST-DWN as the download directory, choose **F** (Download directory) from the Host Mode Options screen. Type *host-dwn*, and press Enter. Whenever a user sends (uploads) a file to the bulletin board, PROCOMM PLUS places the file in the directory C:\PCPLUS\HOST-UP. Users are, however, permitted to download files only from the directory C:\PCPLUS\HOST-DWN.

TIP When you designate an upload or download directory, make sure that the directory exists. Otherwise, PROCOMM PLUS Host mode aborts every attempted upload or download. The program gives no explanation but displays only the message TRANSFER ABORTED. The most common way to fall into this trap is to designate an existing directory as a Host mode upload (or download) directory and then later remove the directory from your disk.

Building the User File

As explained in "Setting the System Type," you can set up Host mode as an open system or a closed system. When the system is open, the default condition, anyone is permitted to log on to the Host mode bulletin board. As soon as the user specifies a first name, a last name, and a password, any new user is given immediate access to the system. On the other hand, a closed system is just the opposite. No one can log on remotely to a closed Host mode system until the system operator has added the user's name and password to the list of authorized users. This list is kept in the file called PCPLUS.USR.

In either an open or a closed system, a user file must be maintained. In an open system, PROCOMM PLUS helps you maintain this file by automatically adding each new name when the new user logs on for the first time. In a closed system, the system operator must maintain the user file. Even in an open system, however, you should keep control over the user file because it includes every user's name and password.

The Host mode user file is an ASCII file kept in the directory that contains the other PROCOMM PLUS program files (usually C:\PCPLUS). When you first install PROCOMM PLUS, the file consists of three sample user file entries. To view this file, press Alt-V (View a File), type *pcplus.usr*, and press Enter. PROCOMM PLUS displays the file shown in figure 9.2.

You can use PCEDIT, the text editor that is distributed with PROCOMM PLUS, to edit the user file; or you can use any other text editor or word processor program that is capable of editing an ASCII file (refer to Chapter 6, "Using the PROCOMM PLUS Editor"). Before you use Host mode, you should use a text editor or word processor to delete the three sample entries and to add the users who will need access to your system.

Each line, or *entry*, in the user file represents one authorized user. Each entry consists of four fields separated by semicolons in the format

    LASTNAME;FIRSTNAME;PASSWORD;N

```
LAST:FIRST:PASSWORD:0   * Sample entries for the user file.  Be sure *
DOE:JOHN:SALLY:0        * to erase these entries before going online *
USER:JOE:HEIDI:0        * in host mode                               *
```

Fig. 9.2.

*PCPLUS.USR
containing three
sample entries.*

```
Home: Top of file      PgUp: Previous page      PgDn: Next page      ESC: Exit
```

TIP The PROCOMM PLUS installation program places the PCPLUS.USR file in the same directory as the PROCOMM PLUS program files, usually C:\PCPLUS (see Appendix A, "Installing and Starting PROCOMM PLUS"). When a user signs on to Host mode, PROCOMM PLUS looks first for the user file on the current working directory (which may not be the same directory that contains the PROCOMM PLUS program files—see "Changing the Current Working Directory," in Chapter 5). If the program can find no PCPLUS.USR file on the working directory, PROCOMM PLUS then looks for PCPLUS.USR in the directory that contains the program files, the directory specified in the SET PCPLUS= command in your AUTOEXEC.BAT file (the use of this SET command is explained in Appendix A).

LASTNAME is the user's last name, *FIRSTNAME* is the user's first name, *PASSWORD* is the user's password, and *N* is a single digit (either 1 or 0) that represents the user's privilege level. The user's privilege level is assigned by the system operator and determines what Host mode features are available to that user.

The LASTNAME and FIRSTNAME fields can each be from 1 to 30 characters long, but their combined length can be no more than 30 characters. Both names must be stored in the user file in all uppercase letters. The user PASSWORD field can contain from 1 to 8 characters. The password too is stored in all caps.

You can have one entry with an empty last name field. This entry is typically used for the system operator, whose name is usually the abbreviated title *SYSOP*. To indicate an empty last name field, start the entry with a semicolon. For example, you might add the following entry to your user file:

    ;SYSOP;MYPW;1

This entry enables you to log on as SYSOP, using the password MYPW. Your privilege level is 1 whenever you log on as SYSOP (refer to "Understanding User Privileges," later in this chapter, for more information about privilege level).

> **TIP** | If you include more than one entry with an empty lsst name field, PROCOMM PLUS recognizes only the first. The only other time PROCOMM PLUS permits logging on with an empty last name field is when a user's first name is exactly 30 characters long, an unlikely occurrence.

As shown in figure 9.2, you can add text to the right of each entry. PROCOMM PLUS ignores any text to the right of the privilege number. You can use this area of the file to store comments about each entry. Indeed, when PROCOMM PLUS adds a new user to the user file, the program places the comment * NEW USER * to the right of the new entry.

> **TIP** | Refer also to the tip in "Testing the System," later in this chapter, for another method of adding users to the user file.

Invoking and Exiting Host Mode

You invoke Host mode from the Terminal mode screen, normally before you are on-line to a remote user. When you press Alt-Q (Host Mode), PROCOMM PLUS displays the message Initializing in the status line. PROCOMM PLUS sends to your modem a software command that places the modem in Answer mode (refer to "Adjusting the Number of Rings before Auto Answer," in Chapter 8). After the modem is in Answer mode, ready to pick up the next incoming call, PROCOMM PLUS changes the status line message to Waiting ...., as shown in figure 9.3.

Fig. 9.3.

Waiting for a call in Host mode.

```
 Waiting ....  HOST MODE (Alt-Z for HELP)
```

TIP You can also invoke Host mode by using the ASPECT script command HOST. Refer to Chapter 11, "An Overview of the ASPECT Script Language," for more information, including how to cause a script to run at PROCOMM PLUS start-up and how to use the TEF utility to run a script at a predetermined time.

When a user calls and connects to your computer, PROCOMM PLUS changes the status line message to Logging on. (When the connection type is direct, by null modem cable, PROCOMM PLUS skips directly to this step.) PROCOMM PLUS also displays the connect speed (transmission speed at which the two modems are communicating). For example, if a user connects to your computer at 2400 bps, the status line displays the message 2400.

During log on and throughout the Host mode session, you can see both the messages Host mode sends to the user and the user's responses. As soon as the user successfully logs on to the bulletin board, your status line message changes again. Now the message is User On-line, as shown in figure 9.4. The last section of the status line also displays the user's name.

```
You are connected to Terry's T-Shirts, Inc. BBS


First name: Sam;Spade
SAM SPADE
Is this correct (Y/N)? Y

Password: ******
```

Fig. 9.4.

The Host mode screen, indicating that a user is on-line.

```
F)iles  U)pload  D)ownload
H)elp  T)ime  C)hat  G)oodbye
R)ead mail  L)eave mail

Your choice?
 User On-line  HOST MODE (Alt-Z for HELP)    2400  SAM SPADE
```

After the user logs off, PROCOMM PLUS "hangs up" the telephone line and displays the status line message DISCONNECTING. Then, as PROCOMM PLUS again places the modem in Answer mode, the program displays the message Recycling.... When the modem is ready to accept another call, PROCOMM PLUS displays the status line message Waiting.....

PROCOMM PLUS also lets you start Host mode after you are already connected to another PC. When you press Alt-Q (Host Mode), PROCOMM PLUS asks Hangup line(Y/N)? and begins to sound a beep every 3 seconds for 30 seconds. If you respond No, PROCOMM PLUS switches to the Host mode without disconnecting the remote user. Host mode prompts the user for a first name. The user then must log on in order to use Host mode. If you respond Yes (or fail to respond within 30 seconds), PROCOMM PLUS disconnects the remote user, places the modem in Answer mode, and displays the status line message Waiting.....

When you finish using Host mode, press Esc (Exit). PROCOMM PLUS displays the message Exiting HOST in the status line, and then returns to the Terminal mode screen.

TIP   If a user is on-line when you attempt to exit from Host mode, PROCOMM PLUS asks Hangup line(Y/N)? and begins to sound a beep every 3 seconds for 30 seconds. If you respond No, PROCOMM PLUS returns to the Terminal mode screen without disconnecting the remote user. If you respond Yes (or fail to respond at all within 30 seconds), PROCOMM PLUS hangs up the telephone line, unceremoniously disconnecting the remote user.

Logging On to Host Mode

PROCOMM PLUS Host mode is intended to be used as a small-scale unattended bulletin board system. When PROCOMM PLUS is in Host mode, most commands are issued by a user at a remote PC. The steps required to call and connect to a PC running PROCOMM PLUS Host mode are similar to the steps required to call and connect to any PC-based bulletin board. Refer to Chapter 3, "Building Your Dialing Directory," and to Chapter 4, "A Session with PROCOMM PLUS," for information on using PROCOMM PLUS to dial and connect to an electronic bulletin board. In order to connect to your bulletin board, a user's computer can be running almost any communications software. In other words, a remote user does not have to be running PROCOMM PLUS in order to connect to and use PROCOMM PLUS running in Host mode on your PC.

Once a remote user connects to your computer, Host mode sends the welcome message and prompts for the user's first name, as shown in figure 9.5. The user then types a name. The name can contain from 1 to 30 characters, including letters, numbers, and other displayable keyboard characters. Host mode does not distinguish between upper- and lowercase letters.

PROCOMM PLUS on-line to Terry's T-Shirts, Inc. at 2400 baud

Fig. 9.5.

Logging on to Host mode.

You are connected to Terry's T-Shirts Inc. BBS

First name:

| Alt-Z FOR HELP | ANSI | FDX | 2400 N81 | LOG CLOSED | PRINT OFF | ON-LINE |

After typing the first name, the user presses Enter. Host mode then prompts the user for Last name. Again, the user types a name of from 1 to 30 characters (letters, numbers, and other displayable keyboard characters). When the user presses Enter at the last name, Host mode displays the first and last names in all caps and prompts Is this correct (Y/N)? (see fig. 9.6).

TIP As a shortcut, the user is permitted to type both the first and last names at the First name prompt, separating the names by a space or a semicolon. Host mode then skips the Last name prompt and immediately asks whether the full name is correct.

If the displayed user name is not correct, the user responds No, and Host mode again prompts the user to enter First name and then Last name.

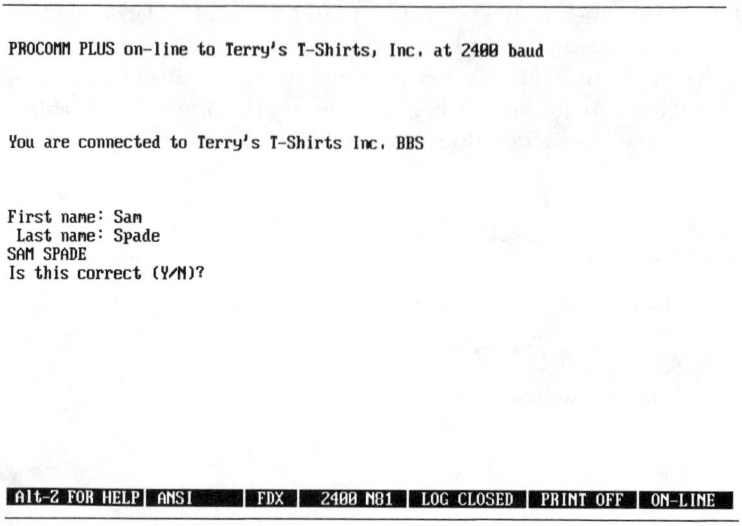

Fig. 9.6.

Confirming first and last names.

```
PROCOMM PLUS on-line to Terry's T-Shirts, Inc. at 2400 baud


You are connected to Terry's T-Shirts Inc. BBS


First name: Sam
 Last name: Spade
SAM SPADE
Is this correct (Y/N)?
```

`Alt-Z FOR HELP | ANSI   | FDX | 2400 N81 | LOG CLOSED | PRINT OFF | ON-LINE`

If the caller confirms that the user name is typed correctly, PROCOMM PLUS checks this name against the user file. If PROCOMM PLUS cannot find the caller's name in the user file, what occurs next depends on whether the system is closed or open:

- *Closed system.* Only users whose names you have already added to the user file, *authorized users*, may log on to the system. PROCOMM PLUS disconnects unauthorized users, anyone whose name is not found in the user file.

- *Open system.* PROCOMM PLUS permits any user to log on. If a caller enters a name not already in the user file, PROCOMM PLUS prompts the user to enter a password. The user types any string of characters consisting of from 1 to 8 characters, and presses Enter. This string of characters becomes the user's personal password for use each time he or she logs on to the system, unless the Host mode operator changes that password in the PCPLUS.USR file. As the user types the password, Host mode echoes only asterisks (*) to the user's screen. This security precaution helps the user keep his or her password private.

  PROCOMM PLUS Host mode then prompts the user to Please verify. To verify means to enter the password a second time to ensure that Host mode received the password correctly the first time. After the user enters the password the second time and presses Enter, PROCOMM PLUS adds the caller as a new user in the user file and displays the main Host mode menu, shown near the bottom of the screen in figure 9.7.

```
You are connected to Terry's T-Shirts Inc, BBS


First name: Sam
 Last name: Spade
SAM SPADE
Is this correct (Y/N)? Y

Enter a password: ******
Please verify: ******




F)iles U)pload D)ownload
H)elp T)ime C)hat G)oodbye
R)ead mail L)eave mail

Your choice?
 Alt-Z FOR HELP  ANSI    FDX   2400 N81  LOG CLOSED  PRINT OFF  ON-LINE
```

Fig. 9.7.

The log-on procedure for a new user on an open Host mode system.

Whether a caller is logging on to a closed system or an open system, PROCOMM PLUS prompts a previously authorized caller to Enter a password. The caller is given three chances to enter the correct password. PROCOMM PLUS disconnects any caller who cannot in three tries enter the password stored in the user file with the caller's name. An authorized user does not have to enter the password a second time for verification (compare fig. 9.8 with fig. 9.7).

Understanding User Privileges

Once the caller enters the correct password, PROCOMM PLUS displays one of two Host mode menus, depending on the user's privilege level. Any new user logging on to an open system for the first time is assigned privilege level 0. These users, called *normal users*, see the following menu:

```
F)iles  U)pload  D)ownload
H)elp  T)ime  C)hat  G)oodbye
R)ead mail  L)eave mail

Your choice?
```

Fig. 9.8.

*The log-on
procedure for an
authorized user
on either an open
or a closed Host
mode system.*

```
You are connected to Terry's T-Shirts, Inc. BBS


First name: Sam;Spade
SAM SPADE
Is this correct (Y/N)? Y

Password: ******




F)iles  U)pload  D)ownload
H)elp  T)ime  C)hat  G)oodbye
R)ead mail  L)eave mail

Your choice?
Alt-Z FOR HELP| ANSI      |  FDX  | 2400 N81 | LOG CLOSED | PRINT OFF | ON-LINE
```

When the user has privilege level *1*, referred to as a *privileged user* or a *superuser*, PROCOMM PLUS adds two more options to the menu seen by a normal user:

    S)hell A)bort

You designate a privileged user by using a text editor to add an entry or to modify the privilege-level number in an entry in the PCPLUS.USR file. You often grant the system operator privileged user status. Figure 9.9 shows the log-on procedure for the privileged user SYSOP, whose user file entry was discussed in "Building the User File," earlier in this chapter.

Fig. 9.9.

*The log-on
procedure for a
privileged user
named SYSOP.*

```
You are connected to Terry's T-Shirts Inc. BBS


First name: sysop
SYSOP
Is this correct (Y/N)? Y

Password: ****




F)iles  U)pload  D)ownload
H)elp  T)ime  C)hat  G)oodbye
R)ead mail  L)eave mail
S)hell  A)bort

Your choice?
Alt-Z FOR HELP| ANSI      |  FDX  | 2400 N81 | LOG CLOSED | PRINT OFF | ON-LINE
```

The next section discusses the function of each menu option available on either version of the main Host mode menu.

TIP | You can also display a log-on message to every user by creating a file named PCPLUS.NWS and placing this file in the directory that contains the PROCOMM PLUS program files. PROCOMM PLUS displays this message after a caller logs on, just before the program displays the main Host mode menu. Refer to "Maintaining Host Mode Files," later in this chapter, for more information.

Using the Main Host Mode Menu

As explained in the preceding section, the main Host mode menu for a normal user contains nine options. The main Host mode menu for a privileged user (superuser) contains two additional options. The following paragraphs explain, from the user's perspective, how to use each option.

Viewing the Download Directory

A primary reason for using Host mode is to enable remote users to upload and download files easily to and from your computer without your being at your computer. The first command on the main Host mode menu displays a list of all the files available for downloading.

When a remote user selects **F**iles from the main Host mode menu, PROCOMM PLUS Host mode displays the following prompt on the user's screen:

```
Enter FILE SPEC: (Carriage Return = *.*)
 >
```

At this prompt, the caller presses Enter to see a list of all the files in the Host mode download directory. If you did not specify a download directory on the Host Mode Options screen, Host mode lists files from your current working directory and returns to the main Host mode menu (refer to "Specifying Upload and Download Directories," in this chapter). The user can also use the DOS wild-card characters * and ? to display only a subset of the files in this directory. When Host mode has listed all matching file names, it returns to the main Host mode menu.

For example, suppose that the sales manager for your company's Southern region has logged on to Host mode and wants to download a copy of the consolidated sales report for August, 1989, but she cannot remember the name of the file. She first selects **Files** from the main Host mode menu. Then she types *.wk1* to the right of the › and presses Enter. Host mode lists all the files in the download directory that have the file name extension .WK1. In this case, two .WK1 files, SALES789.WK1 and SALES889.WK1, are in the download directory (see fig. 9.10). The SALES889.WK1 file is apparently the August '89 spreadsheet the regional sales manager is looking for.

Fig. 9.10.

Listing all files in the download directory with file name extension .WK1.

```
F)iles  U)pload  D)ownload
H)elp  T)ime  C)hat  G)oodbye
R)ead mail  L)eave mail

Your choice? F
Enter FILE SPEC: (Carriage Return = *.*)
> *.WK1

(Press Ctrl-C to abort display)
SALES789 WK1    3504   7-31-89  22:22
SALES889 WK1    3040   8-31-89  20:25




F)iles  U)pload  D)ownload
H)elp  T)ime  C)hat  G)oodbye
R)ead mail  L)eave mail

Your choice?
```

`Alt-Z FOR HELP | ANSI      | FDX  | 2400 N81 | LOG CLOSED | PRINT OFF | ON-LINE`

Sometimes more files may be in the download directory than will fit vertically on the screen. To give the caller a chance to review all the file names before they scroll off the screen, Host mode displays a maximum of 23 file names and then pauses. The message -MORE- displayed after the 23rd file name indicates that more file names follow. The user presses any key to continue the display.

The user can cancel the **Files** command before it finishes listing all requested file names. As soon as the user presses Ctrl-C, Host mode stops listing file names and returns to the main Host mode menu.

TIP | When the caller is a privileged user—privilege level 1—the **Files** command lists the host system's current working directory rather than the download directory. Also, when Host mode asks for a FILE SPEC, a privileged user is permitted to specify any directory on the host system by typing the complete DOS path.

Downloading a File

Downloading a file from Host mode is similar to downloading a file from most other bulletin board systems. The user first selects **D**ownload from the main Host mode menu. Host mode then prompts the user to choose from among nine file-transfer protocols:

```
A)scii  K)ermit  S)ealink  X)modem  Y)modem Batch
I)modem  T)elink  W)xmodem  G)Ymodem-G Batch
Your choice?
```

The user chooses a file-transfer protocol that is also supported by the user's communication program (refer to "File-Transfer Protocols," in Chapter 5, for guidelines for deciding which protocol to use).

Once the caller selects a file-transfer protocol, Host mode prompts for a File name. If the user has chosen **K**ermit, **S**ealink, **Y**modem Batch, **T**elink, or **G**Ymodem-G Batch, the user can specify multiple files by using the DOS wild-card characters. With any of the nine protocols, the user can specify a single file name, and then press Enter.

TIP | A privileged user (superuser) can specify a file from any directory on the host system by including the complete DOS path. Normal users can download files only from the current working directory or from the download directory if one is specified.

Host mode prompts the user to Begin your PROTOCOL transfer procedure.... The word *PROTOCOL* is replaced by the name of the file-transfer protocol the caller selected. The caller then executes the proper commands at the remote computer to begin the download procedure, using the same file-transfer protocol that the user instructed the Host system to use. When the transfer is complete, Host mode displays the message TRANSFER COMPLETE and returns to the main Host mode menu.

For example, the Southern region sales manager is ready to download the August, 1989, sales report, the file SALES889.WK1. From the main Host mode menu, she selects **D**ownload. Host mode displays the list of nine available file-transfer protocols. Her communications software supports the YMODEM BATCH file-transfer protocol, so she selects **Y**modem Batch.

Host mode prompts for a file name. In response, the sales manager types the file name *sales889.wk1*, and presses Enter. Host mode displays the following prompt:

```
Begin your YMODEM BATCH transfer procedure...
```

The sales manager then issues the appropriate command to her communications software to begin downloading the file, using the YMODEM BATCH file-transfer protocol. (Refer to the Chapter 5, "Transferring Files," for a complete discussion of how to use PROCOMM PLUS to download files using a file-transfer protocol.)

When the file transfer is finished, Host mode sends the message TRANSFER COMPLETE and returns to the main Host mode menu. The complete downloading procedure, as seen on the sales manager's screen, is shown in figure 9.11.

Fig. 9.11.

Downloading a file from Host mode.

```
F)iles  U)pload  D)ownload
H)elp  T)ime  C)hat  G)oodbye
R)ead mail  L)eave mail

Your choice? D

A)scii  K)ermit  S)ealink  X)modem  Y)modem Batch
I)modem  T)elink  W)xmodem  G)Ymodem-G Batch
Your choice? Y


File name? SALES889.WK1

Begin your  YMODEM BATCH  transfer procedure...
TRANSFER COMPLETE


F)iles  U)pload  D)ownload
H)elp  T)ime  C)hat  G)oodbye
R)ead mail  L)eave mail

Your choice?
Alt-Z FOR HELP | ANSI    | FDX  | 2400 N81 | LOG CLOSED | PRINT OFF | ON-LINE
```

Uploading a File

Uploading a file to Host mode is similar to uploading a file to most other bulletin board systems. The user first selects Upload from the main Host mode menu. Host mode then prompts the user to choose from among nine file-transfer protocols:

```
A)scii  K)ermit  S)ealink  X)modem  Y)modem Batch
I)modem  T)elink  W)xmodem  G)Ymodem-G Batch
Your choice?
```

The user chooses a file-transfer protocol that is also supported by the user's communication program (refer to "File-Transfer Protocols," in Chapter 5, for guidelines for deciding which protocol to use).

After the caller selects a file-transfer protocol, Host mode prompts for a file name. If the user has chosen **K**ermit, **S**ealink, **Y**modem Batch, **T**elink, or **GY**modem-G Batch, the user can specify multiple files using the DOS wild-card characters. With any of the nine protocols, the user can specify a single file name and then press Enter.

TIP A privileged user (superuser) can upload a file to any directory on the host system by including the complete DOS path. Normal users can upload files only to the current directory or to the upload directory if one is specified. Protocols capable of sending multiple files transfer files to the current directory, or to the upload directory if one is specified, no matter what the user specifies at the Filename prompt and regardless of the user's privilege level.

TIP The file name used on the Host system does not have to be the same as the file name on the user's disk. Sometimes the file name may already be used in the host system's upload directory. When the user selects **A**scii, **X**modem, **I**modem, or **W**xmodem and tries to upload a file using an existing file name, Host mode displays the message < = = File(s) already exist and asks again for a file name. The user must type a different file name before Host mode will permit the file to be uploaded. Then, whenever the user executes the Upload command on his or her own computer, the user must specify the name of the file as the name exists on the user's disk.

A user can avoid this potentially confusing procedure by always using KERMIT, SEALINK, YMODEM BATCH, TELINK, or GYMODEM-G BATCH. When a user uploads a file to Host mode using one of these five protocols, all of which are capable of sending multiple files, PROCOMM PLUS automatically renames the uploaded file if another with the same name already exists. PROCOMM PLUS replaces the first character in the name of the file to be uploaded with a dollar sign ($).

Host mode next asks the user to type a Description. The user can enter a description of up to 40 characters or leave this line blank. This description is added to the host system's upload directory file (see "Reviewing the Upload Directory File," later in this chapter).

After the user presses Enter, Host mode prompts the user to Begin your PROTOCOL transfer procedure.... The word *PROTOCOL* is replaced by the name of the file-transfer protocol the user selected. The caller then executes the proper commands to begin the upload procedure on his or her own computer, using the same file-transfer protocol as the caller instructed the host system to use. When the transfer is compete, Host mode displays the message TRANSFER COMPLETE and returns to the main Host mode menu.

For example, the Southern regional sales manager is ready to upload the weekly sales report for September 22, 1989, contained on her disk in the file SO989.WK1.

From the main Host mode menu, the sales manager selects Upload. Host mode displays the list of nine available file-transfer protocols. Her communications software supports the YMODEM BATCH file-transfer protocol, so she selects **Y**modem Batch.

Host mode prompts for a file name. In response, the sales manager types the file name, *so989.wk1*, and presses Enter. Host mode then prompts the caller to enter the description. She types the following description and presses Enter:

Southern region sales report 9/22/89

Host mode displays the following prompt:

Begin your YMODEM BATCH transfer procedure...

The sales manager issues the appropriate command to her communications software to begin uploading the file, using the YMODEM BATCH file-transfer protocol. (Refer to Chapter 5, "Transferring Files," for a complete discussion of how to use PROCOMM PLUS to upload files by using a file-transfer protocol.)

When the file transfer is finished, Host mode sends the message TRANSFER COMPLETE and returns to the main Host mode menu. The complete upload procedure, as seen on the sales manager's screen, is shown in figure 9.12.

Getting Help

It is possible, even likely, that someone will log on to your Host mode system without really knowing how to use it. The main Host mode menu therefore includes the **Help** option.

When the user selects **Help** from the main Host mode menu, Host mode displays the first of six screens of information. This information explains, from the user's perspective, how the Host mode system operates. The first help screen is displayed in figure 9.13.

```
F)iles  U)pload  D)ownload
H)elp  T)ime  C)hat  G)oodbye
R)ead mail  L)eave mail

Your choice? U

A)scii  K)ermit  S)ealink  X)modem  Y)modem Batch
I)modem  T)elink  W)xmodem  G)Ymodem-G Batch
Your choice? Y

File name? S092289.WK1
             |.....................|.....................|
Description: Southern region sales report 9/22/89.

Begin your  YMODEM BATCH  transfer procedure...
TRANSFER COMPLETE



F)iles  U)pload  D)ownload
H)elp  T)ime  C)hat  G)oodbye
R)ead mail  L)eave mail

Your choice?
```
`Alt-Z FOR HELP| ANSI     |  FDX  |  2400 N81  | LOG CLOSED | PRINT OFF | ON-LINE`

Fig. 9.12.

Uploading a file to Host mode.

```
(Press Ctrl-C to abort display)

PROCOMM PLUS HOST MODE HELP


Every user is presented the following menu after successfully logging
on and viewing the optional news file.

    ┌──────────────────────────────────┐
    │ F)iles  U)pload  D)ownload        │
    │ H)elp  T)ime  C)hat  G)oodbye     │
    │ R)ead mail  L)eave mail           │
    └──────────────────────────────────┘

If the user has superuser status, then the following line is added to
the prompt:

    ┌──────────────────────────────────┐
    │ S)hell  A)bort                    │
    └──────────────────────────────────┘

An available function is invoked by pressing the first letter of the name.
No carriage return is needed so to download, for example, just press <D>.


-MORE-
```
`Alt-Z FOR HELP| ANSI     |  FDX  |  2400 N81  | LOG CLOSED | PRINT OFF | ON-LINE`

Fig. 9.13.

The first of six help screens on using the Host mode system, from the remote user's perspective.

TIP | You can customize the PROCOMM PLUS help screens. (Refer to "Customizing the Help File," later in this chapter.) Consider advising users of your system to capture to disk and/or print a copy of all six help screens displayed by the **Help** command. This copy will provide a handy manual for reference whenever someone logs on to your Host mode system.

Checking the Time

Depending on the purpose of your bulletin board, a user may be calling long distance and therefore be interested in how long he or she has been connected. If you have many users competing for time on the system, you may have to impose a time limit on each connection. Whatever the reason, PROCOMM PLUS Host mode lets a caller determine the time at which the connection was made and the current time of day.

When the caller selects **Time** from the main Host mode menu, Host mode displays the message:

```
Online at: HH:MM:SSxM
It is now: HH:MM:SSxM
```

HH is hours, *MM* is minutes, *SS* is seconds, and *x* is either A or P. Figure 9.14 shows the screen of a user who connected at 05:05:04 p.m. and is still on-line at 05:09:46 p.m. The user has therefore been connected for 4 minutes, 42 seconds.

Fig. 9.14.

Checking the time.

```
F)iles U)pload D)ownload
H)elp T)ime C)hat G)oodbye
R)ead mail L)eave mail

Your choice? T

Online at: 05:05:04PM
It is now: 05:09:46PM


F)iles U)pload D)ownload
H)elp T)ime C)hat G)oodbye
R)ead mail L)eave mail

Your choice?




Alt-Z FOR HELP| ANSI    |  FDX |  2400 N81 | LOG CLOSED | PRINT OFF | ON-LINE
```

TIP | The Host mode history file, PCPLUS.HST, keeps a running log of all Host mode transactions, including the times each user logged on and logged off. Refer to "Reviewing the History File," later in this chapter.

Chatting with the Host Mode Operator

Although PROCOMM PLUS Host mode is intended as a simple, unattended bulletin board system, probably someone is within the vicinity of your computer during significant portions of the day. PROCOMM PLUS therefore provides the Chat feature, through which the user can page the host system's operator.

When the remote user wants to "converse" interactively with the host system's operator, the user selects Chat from the main Host mode menu. Host mode displays the message Paging Host operator... on the user's screen and sounds a beep on the host system once each second for 10 seconds. If the operator does not come on-line within 10 seconds, Host mode sends the message ...Host operator has been paged and returns to the main Host mode menu.

Any time the host system operator presses F1 (Chat) (whether or not in response to the Chat command page by a caller), Host mode sends the message Host operator online to the caller's screen. The operator (you or someone else at your computer) and the caller can then type and receive information interactively (refer to "Using Chat Mode," in Chapter 4). If the caller is also using PROCOMM PLUS to communicate, the caller may also switch to Chat mode. Once the user and operator finish their on-line conversation, the operator terminates Chat mode. Host mode sends the message Exiting chat mode to the user's screen, and returns to the main Host mode menu. (Refer also to "Interacting with a User On-Line," later in this chapter, and to "Using Chat Mode," in Chapter 4.)

Saying Goodbye

When the user is ready to disconnect from Host mode, the user selects Goodbye. Host mode sends the following message to the remote user's screen:

```
Online at: HH:MM:SSxM
It is now: HH:MM:SSxM

Goodbye USER NAME
```

HH is hours, *MM* is minutes, *SS* is seconds, *x* is either *A* or *P*, and *USER NAME* is the user's log-on name.

Host mode causes its modem to "hang up" the telephone and prepares the modem for answering the next call. The status line message on the host system's screen again says Waiting .....

If the user hangs up without "saying Goodbye" or if the telephone connection is lost for some other reason, your modem sends to Host mode a message that the modem no longer detects a carrier signal. In response, Host mode recycles, displays the Waiting .... message in the status line, and sends to the modem a command that prepares it to answer the next call.

Sending Electronic Mail

One of the most impressive capabilities of PROCOMM PLUS Host mode is the electronic mail, or E-mail, feature. This feature enables a user to send both public and private messages. A public message can be read by any user; but a private message is accessible only by the user to whom it is addressed and the user who created it.

When a user wants to send a public or private message (mail), the user selects Leave Mail from the main Host mode menu. Host mode first asks the user for the addressee, by displaying the prompt To. The user types the first and last name of the individual intended to receive the mail and presses Enter.

TIP If you have a user entry for SYSOP, as described in "Building the User File," earlier in this chapter, the user can send mail to the system operator using the single name SYSOP.

After the user specifies the addressee, Host mode asks for the subject. To the right of the prompt Re, the user has the option of indicating what the message is about. The user can type a description of the subject up to 28 characters in length (all displayable characters are acceptable). The user then presses Enter.

Host mode displays the following message:

```
Private Mail(Y/N)?
```

The user responds Y if the message should be read only by the addressee or N if the message is to be available for any user on the system.

The Host mode E-mail system then displays the user's entries so far. The system displays a prompt similar to the following:

```
To: ADDRESSEE NAME
From: USER NAME
  Re: SUBJECT
Private
```

ADDRESSEE NAME is the name of the intended recipient of the message, *USER NAME* is the name of the current user, and *SUBJECT* is the subject of the message, if any, specified by the user. If the message is a public message, the prompt does not include the word *Private*.

Host mode then asks Is this correct (Y/N)? The user responds N if any line is incorrect. Otherwise, the user responds Y to indicate that the information is correct. Host mode then displays 1: at the beginning of a new line. This line is the first message entry line.

For example, suppose that your assistant, Sam Spade, is sending a message to Victor Laslo, manager of the Eastern sales region, concerning the upcoming October sales meeting. Sam selects **Leave Mail** from the main Host mode menu, types *Victor Laslo* at the To prompt, and presses Enter. Sam then types *October Sales Meeting* at the Re prompt and again presses Enter. Finally, Sam indicates that the message is intended only for Victor by responding Y to the Private Mail(Y/N)? prompt. Host mode displays the information Sam has entered so far and asks whether the information is correct. Sam responds Y, so Host mode displays the first message entry line. This entire sequence is shown in figure 9.15.

```
F)iles  U)pload  D)ownload
H)elp  T)ime  C)hat  G)oodbye
R)ead mail  L)eave mail

Your choice? L
  To: Victor Laslo
  Re: October Sales Meeting
Private Mail(Y/N)? Y

  To: VICTOR LASLO
From: SAM SPADE
  Re: OCTOBER SALES MEETING
Private

Is this correct (Y/N)?
```

Fig. 9.15.

Leaving a private message.

```
Alt-Z FOR HELP│ ANSI    │  FDX  │ 2400 N81 │ LOG CLOSED │ PRINT OFF │ ON-LINE
```

Once Host mode displays the first message entry line, Sam can type the message line-by-line or can use the ASCII file-transfer protocol (or equivalent, if the user is not using PROCOMM PLUS) to send a previously typed ASCII message to the host system.

If Sam types the message while on-line, he can edit only the current line. He must press Enter at the end of each line, and after he presses Enter, the line can no longer be changed. Host mode interprets a blank line as the end of the message.

TIP To include a blank line within the message, the user presses the space bar, causing the blank space character (ASCII decimal 32) to be sent to Host mode, before the user presses Enter.

When Sam finishes the message and presses Enter at a blank line, Host mode displays the Leave Mail Host mode menu, which contains the following four options:

- **S)ave.** The sender chooses this option to save the message and send it to the addressee. The message is actually saved to a file named PCPLUS.MSG on the host system's disk (see "Administering the Electronic Mail Facility," in this chapter). Host mode displays the message Saving message... and then returns to the main Host mode menu.

- **A)bort.** If the sender selects Abort, Host mode asks Abort Message (Y/N)? In order to return to the main Host menu without saving the message, the user responds Y. The user might select this option by mistake, however, so the user can respond N and return to the Leave Mail menu.

- **D)isplay.** Sometimes a message is too long to display all on one screen. By the time the user types the last line, the first portion of the message has scrolled off the screen. If the user chooses **D**isplay, Host mode displays the entire message, 23 lines at a time, starting at the beginning of the message. At the end of each screen, Host mode displays the message -MORE-, and the user presses any key to see the next screen. After displaying the entire message, Host mode returns to the Leave Mail menu.

- **C)ontinue.** This option enables the user to continue entering text at the end of the message. Host mode places the cursor at the beginning of a blank line. The user should choose this option in order to finish typing the message, when he or she has accidentally pressed Enter at a completely blank line or has pressed Enter purposely and chosen **D**isplay to display the first part of the message.

For example, after Sam types the message to Victor Laslo, he presses Enter at a blank line; Host mode displays the Leave Mail menu, as shown in figure 9.16.

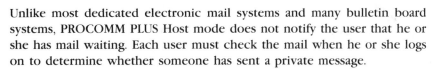

```
   To: VICTOR LASLO
 From: SAM SPADE
   Re: OCTOBER SALES MEETING
Private

Is this correct (Y/N)? Y
  1: Vic
  2:
  3: The October sales meeting has been rescheduled for
  4: Oct. 18 at 11:00 in my office.  I hope this doesn't
  5: conflict with your appointments.  It is very
  6: important that we all put our heads together this
  7: month to discuss all the Christmas promotions we
  8: have planned.
  9:
 10: I received your weekly report yesterday.  Terry was
 11: impressed.
 12:
 13: Keep up the good work!  See you on the 18th.
 14:
 15: Sam
 16:

S)ave  A)bort  D)isplay  C)ontinue. ?
 Alt-Z FOR HELP | ANSI |      | FDX | 2400 N81 | LOG CLOSED | PRINT OFF | ON-LINE
```

Fig. 9.16.

A private message followed by the Leave Mail Host mode menu.

Ultimately, the message is either sent or aborted, and Host mode returns to the main Host mode menu.

Reading Electronic Mail

Unlike most dedicated electronic mail systems and many bulletin board systems, PROCOMM PLUS Host mode does not notify the user that he or she has mail waiting. Each user must check the mail when he or she logs on to determine whether someone has sent a private message.

To check and read mail, the caller selects **R**ead from the main Host mode menu. Host mode first indicates the number of messages currently stored on the system. All private and public messages are included in this number; but all messages are not always *accessible* to every user. Host mode messages that are accessible by a particular user include all public messages, any private message addressed to that user, and any private message sent by that user.

After Host mode displays the number of messages, Host mode displays the Read Mail Host mode menu, which contains the following options:

```
F)orward read
S)earch mail
I)ndividual read
Q)uit
```

To return to the main Host mode menu, the user selects **Quit**. The other options are discussed in the following sections.

Reading Messages in Sequence

The user selects **Forward Read** in order to read multiple messages in sequence. This option is used frequently on a system where the majority of messages are public and users normally read all messages. When a user selects **Forward Read**, Host mode displays the following prompt:

```
Starting message (<CR> for first):
```

This prompt means that the user can type the number of the first message he or she wants to read, and press Enter. Host mode will then display all accessible messages in sequence, starting with the specified message. Alternatively, the user can simply press Enter without typing a message number in order to see all accessible messages, beginning with message number 1.

For example, suppose that 12 public messages are stored on the system and the user wants to read the last 4 messages, starting with the 9th message; the user types the number 9 and presses Enter. In order to see all 12 messages, the user just presses Enter.

If Host mode finds no accessible messages in the range of message numbers specified, the program displays the following message:

```
No mail found!
```

TIP Messages are stored in chronological order. Host mode indicates the date and time the message was sent just below the Re line in the message (see fig. 9.17).

Some messages are too long to display all on one screen. When a message is more than 23 lines long, Host mode displays 23 lines at a time and places the message -MORE- in the 24th line. The user then presses any key to see the next 23 lines of the message. This procedure continues until the entire message has been displayed.

At the end of each message, Host mode gives the user three options:

- **R)**eply. The user selects this option to send back to the message's author a message on the same subject.

- **Q)**uit. The user chooses this selection to quit reading messages and return to the main Host mode menu.

- (<CR> for another). The user presses Enter to read the next accessible message, without replying to the current message.

At the end of each message written by the current user, as well as each message addressed to the current user, Host mode adds one more option (see fig. 9.17):

- **D**)elete. Either the user who wrote the message or the user to whom it is addressed can issue this command to delete the message. A deleted message actually remains in the PCPLUS.MSG file until removed by the system operator, but the message can no longer be displayed by the Read Mail options.

```
Msg: 7
  To: VICTOR LASLO
From: SAM SPADE
  Re: OCTOBER SALES MEETING
10/02/89 18:45:41
Private

Vic

The October sales meeting has been rescheduled for
Oct. 18 at 11:00 in my office. I hope this doesn't
conflict with your appointments. It is very
important that we all put our heads together this
month to discuss all the Christmas promotions we
have planned.

I received your weekly report yesterday. Terry was
impressed.

Keep up the good work! See you on the 18th.

Sam

R)eply Q)uit D)elete (<CR> for another): ?
Alt-Z FOR HELP  ANSI      FDX    2400 N81   LOG CLOSED   PRINT OFF   ON-LINE
```

Fig. 9.17.

Reading private mail.

TIP | When the user selects **Delete** to remove a message, Host mode immediately displays the deleted message again. Notice, however, that the program adds the message `Deleted` below the date and time. The next time the user executes the **R**ead File option, the deleted message is not displayed.

When Host mode has displayed all accessible messages, it returns to the main Host mode menu.

Searching for Specific Messages

The **S**earch option enables a user to find specific messages to read. When a user chooses this option from the Read Mail menu, Host mode prompts `Which field` and lists three options: **T**o, **F**rom, and **S**ubject. Each option corresponds to one of the first three entries, or *fields*, in a Host mode message: To, From, and Re, respectively.

A user chooses **T**o when looking for accessible messages addressed to a particular user. This option corresponds to the To line in each message. After the user selects this option, Host mode prompts for a Search string. The user then types the desired user name and presses Enter. Host mode searches the message file for all mail addressed to the specified user and displays the following prompt:

```
Starting message (<CR> for first):
```

The user presses Enter to display the first message.

For example, suppose that Victor Laslo, manager of the Eastern sales region, logs on to Host mode to check his electronic mail. He selects **R**ead Mail from the main Host mode menu and then **S**earch from the Read Mail menu. When asked Which field, Victor chooses **T**o and types *Victor Laslo* as the search string. He presses Enter, and Host mode searches for his mail. He presses Enter again, and Host mode displays his first message, as shown in figure 9.17.

When Host mode finds no matching message at all, the program displays the message No mail found!

Using a procedure similar to that used with the **T**o option, a user can read all messages sent by a particular user or on a given subject using the **F**rom and **S**ubject options, respectively.

Reading a Single Message

When the user wants to read a single message for which the user already knows the message number, the user selects **I**ndividual Read from the Read Mail menu. Host mode displays the following prompt:

```
Message number:
```

The user types the message number, and presses Enter. Host mode displays the specified message if it exists and is accessible by that user. When the specified message does not exist in the message file or is not accessible by the current user, Host mode simply displays the Message number prompt again.

Accessing DOS from a Remote Computer

One of the two additional options available on a privileged user's main Host mode menu is the Shell option. This option enables a privileged user operating from a remote computer to access DOS (the operating system) on the Host system. Host mode is temporarily suspended, and the DOS operating system prompt is displayed on the privileged user's screen.

For example, the system operator is usually granted privileged user status (superuser status). Figure 9.18 shows what happens when the privileged user SYSOP selects Shell.

```
SYSOP
Is this correct (Y/N)? Y

Password: ****
```

Fig. 9.18.

Selecting the Shell option.

```
F)iles  U)pload  D)ownload
H)elp  T)ime  C)hat  G)oodbye
R)ead mail  L)eave mail
S)hell  A)bort

Your choice? S

Microsoft(R) MS-DOS(R)  Version 3.30
          (C)Copyright Microsoft Corp 1981-1987


Enter 'EXIT' to return to PROCOMM PLUS
C:\PCPLUS>
```
| Alt-Z FOR HELP | ANSI | | FDX | 2400 N81 | LOG CLOSED | PRINT OFF | ON-LINE |

This command has the same effect as pressing Alt-F4 (DOS Gateway) has when you are using the PROCOMM PLUS Terminal mode screen, but the user is accessing DOS on the host computer rather than on his or her own computer.

TIP Because PROCOMM PLUS Host mode uses DOS services to enable a remote user to access DOS on the host system, the user will not be able to use Shell if the version of DOS you are using does not support the COM port to which your modem is connected (refer to "Altering Port Assignments," in Chapter 8, for further discussion on COM ports).

At first, you may think that the Shell command enables a privileged user to run other applications remotely. PROCOMM PLUS is not, however, designed as a remote-control program (other than for message and file transfer). The Shell option simply makes use of a fairly limited DOS facility. Attempt to run other programs only if you know that they use DOS video services instead of writing directly to the screen. You can, however, maintain files and directories by using the usual DOS commands.

Because of the potential for damage to the system—deleted system files, formatted disks, and so on—this option is best reserved for use only by the system operator and then only to perform such mundane DOS functions as copying and renaming files. This option can, however, be particularly useful in advanced applications using PROCOMM PLUS script files (refer also to Chapter 11, "An Overview of the ASPECT Script Language").

Aborting Host Mode from a Remote Computer

The second option available exclusively to privileged users is the **Abort** option. This option enables a privileged user to abort Host mode, returning the host system to PROCOMM PLUS's Terminal mode.

When a privileged user selects **Abort**, Host mode displays the following prompt:

```
Abort Host mode (Y/N)
```

The user responds Y to cause Host mode to abort and return to Terminal mode, the same effect as if the host system's operator had pressed Esc (Exit). Host mode displays the message

```
Aborting...
```

Then the program asks, Hangup line (Y/N)? The privileged user then responds N to maintain the connection. The remote computer is still on-line to the host system, but the host system is no longer in Host mode. Consequently, the remote user can no longer execute commands on the host computer, except through a script or through remote script commands (see the discussion of the **M** (Remote commands) in "Adjusting General Options," in Chapter 8). This feature is also a convenient way to prevent subsequent callers from connecting with Host mode.

Understanding Operator Rights and Responsibilities

PROCOMM PLUS's Host mode is intended to provide access to remote users; but someone who has direct access to the host computer must be given the responsibility to perform certain administrative tasks in order to keep the system running efficiently. This individual is often referred to as the *system operator*, or *SYSOP*.

As far as PROCOMM PLUS is concerned, anyone who has access to the host system's keyboard is the system operator. In other words, PROCOMM PLUS assumes that any command executed at the host computer's keyboard is performed by the system operator.

The system operator's duties are fairly simple. Obviously, someone must be responsible for turning on Host mode when users are expected to call (refer to "Invoking and Exiting Host Mode," earlier in this chapter.) Also, in order to encourage efficient use, someone must be available (within reasonable limits) while new users are on-line to help them get comfortable with Host mode's features. Finally, someone must periodically check the various important system files for needed maintenance.

Table 9.2 lists the commands available to the system operator from the host computer's keyboard. These commands are covered in the paragraphs that follow.

Table 9.2
Host Mode System Operator Commands

Command	Meaning
F1	Enter Chat mode to converse with user
F2	Log on to Host mode from the host computer
Ctrl-X	Log off current user and recycle for next caller
Esc	Quit Host mode and return to Terminal mode

TIP | Even though you may be the system operator, you too may occasionally need some help on a particular Host mode operation. While PROCOMM PLUS is in Host mode, the Alt-Z (Help) command displays three screens of information about using Host mode from the system operator's perspective.

Interacting with a User On-Line

During the time a user is on-line to Host mode, the host computer's screen displays an exact copy of what the user sees on the remote computer's screen. When you are in the vicinity of the host computer, you can monitor the activity of a caller by watching the host system's screen.

Be aware, however, that while a user is on-line to Host mode, the host computer's keyboard and the remote keyboard are in "lock step." If you touch one of the displayable keys (letters, numbers, or symbols), the corresponding character displays on both the host system's screen and the remote user's screen.

Occasionally, a user may have a question for the system operator while the user is on-line. The user can select Chat from the main Host mode menu. Host mode sounds a beep on the host system once each second for 10 seconds. If the operator does not come on-line within 10 seconds, Host mode returns to the main Host mode menu.

TIP | The beep, or alarm, will not sound if the **C** (Alarm sound) option on the Setup Utility General Options screen is set to OFF, but the user's screen will still say `Paging operator....` Use this feature to turn off the alarm if you use Host mode at home and leave the system on late at night. When you do this, however, add a line to the log-on message to the effect that no operator is available so that a frustrated user won't keep paging you to no avail.

Any time the operator presses F1 (Chat) (whether or not in response to the Chat command page by a user), the host system's screen switches to Chat mode. The operator (you or someone else at your computer) can then type and receive information interactively (refer to "Using Chat Mode," in Chapter 4, and to "Chatting with the Host Mode Operator" in this chapter). After the user and operator finish their on-line conversation, the operator presses Esc (Exit) to terminate Chat mode.

Testing the System

From time to time, you may want to test Host mode, but you probably don't want to go to the trouble of calling in from a remote computer. PROCOMM PLUS enables you to log on directly from the host computer's keyboard just for this purpose.

In order for you to log on to Host mode from the host system's keyboard, no user can be on-line. When you press F2, Host mode asks whether you want to Continue to answer calls? If you expect a call you don't want to miss while you are checking the system, respond Yes to this prompt. Even though you will be using Host mode on your computer, PROCOMM PLUS will answer any incoming call. As soon as you hear the telephone ring and your modem answer the call, you can log off the system so that Host mode can let the user log on. Until you log off, the user will be staring at a blank screen.

If you respond No to the Continue to answer calls? prompt, PROCOMM PLUS will not answer incoming calls for the time you are using the system. Your telephone will still ring, however, if a user calls during this time.

Next, Host mode asks whether you want to Logon as SYSOP? Answer Yes, and Host mode logs you on as a privileged user. If you respond No to this prompt, Host mode displays the system welcome message, and you must then go through the log-on procedure as if you were calling from a remote computer.

TIP | The testing log-on procedure also provides a convenient way to add a new user's name to the user file (PCPLUS.USR). Instead of logging on as SYSOP, log on using the name of a user. If the name doesn't already exist in the PCPLUS.USR file, PROCOMM PLUS adds the name. This method of adding a new user works even if the system is closed. The next time you view the contents of PCPLUS.USR, it will contain the new user's name, password, and privilege level, with the note * NEW USER *.

When you have finished testing the system, perhaps reading messages or leaving mail, you select **G**oodbye from the main Host mode menu to log off. Host mode then recycles and prepares to answer the next call. Indeed, if you instructed PROCOMM PLUS to continue taking calls while you were on the system, the program may have already answered the next call.

Maintaining Host Mode Files

Several files stored on the host computer are important to your Host mode bulletin board system. Two optional files—the *log-on message file* and the *help file*—may need to be changed only occasionally. But PROCOMM PLUS continually updates the *history file*, the *upload directory file*, and two *mail files*, so you need to know how to review and purge these files routinely to maintain control of your system.

Creating a Log-on Message File

In addition to the Host mode welcome message, which can be no longer than 50 characters, you can create a *log-on message* of unlimited length. Use a text editor, such as PCEDIT, to create an ASCII file named PCPLUS.NWS. Store the file in the directory that contains the PROCOMM PLUS program files (the directory specified by the SET PCPLUS= command in the AUTOEXEC.BAT file—see Appendix A). Host mode displays this file, if it exists, immediately after a user logs on to the system. If the file is longer than 23 lines, Host mode displays 23 lines at a time, adding the notation -MORE- as the 24th line. The user presses any key to see the next screen of information. After PROCOMM PLUS has displayed the entire message, Host mode displays the main Host mode menu.

As suggested by the file name extension .NWS, this file is convenient for distributing "news"—information that will be of interest to the majority of your users. You can even cause the message to display in color by including ANSI escape sequences (refer to your DOS manual for more information about ANSI). Users must use ANSI or VT102 terminal emulation in order to see any color enhancements to Host mode screens.

Customizing the Help File

As explained in "Getting Help," earlier in this chapter, the **H**elp option on the main Host mode menu provides several screens of information helpful to users. These help messages are good information about how to use Host mode, but they are not specific to your bulletin board. Consequently, PROCOMM PLUS enables you to modify, enhance, or replace this file in order to supply your users with more pertinent information.

To edit the help file, use PCEDIT or another text editor to modify the file PCPLUS.HHP. This file is initially distributed on the PROCOMM PLUS Supplemental Diskette. After you use the Installation Utility to place PROCOMM PLUS on your computer (see Appendix A), copy PCPLUS.HHP from the Supplemental Diskette into the directory that contains the PROCOMM PLUS program files (usually C:\PCPLUS). Then you can modify this file, adding or deleting information as you see fit in order to tailor the message to the needs of your users. If you prefer, you can replace the file with an entirely different help message, but be sure to use the same file name.

TIP A convenient practice is to write a short user's manual for your system, using the distributed PCPLUS.HHP file as a model. Then use this manual as the help message file. Encourage users to capture the file to disk and print it as a handy reference.

Reviewing the History File

Every time a user logs on to your Host mode system, PROCOMM PLUS adds a few more lines to a file named PCPLUS.HST, the Host mode *history file*. In PCPLUS.HST, Host mode records the date, time, and user name (when applicable) for each of the following events:

- Host mode goes on-line

- User logs on

- User leaves message(s) (message number(s) also recorded)

- User downloads file(s) (file name(s) also recorded)

- User executes **C**hat command

- User executes **S**hell command

- User executes **A**bort command

- User logs off

- User enters three incorrect passwords in succession

Figure 9.19 shows the entries created in PCPLUS.HST when Sam Spade leaves a message and downloads several spreadsheet files from Host mode.

```
*********************************************
*********************************************
* Host online at  4:55:19PM on 10/03/89  *
*********************************************
*********************************************
10/03/89  10:52:39PM
          SAM SPADE online at   2400 baud
10:57:14  Left Message:   4
11:02:49  Downloaded:       DOWNLOAD\*.WK1
11:05:34  Logged off
*********************************************
*********************************************
*** Offline at 09:15:41AM on 10/04/89  ***
*********************************************
```

Fig. 9.19.

The Host mode history file, PCPLUS.HST.

You can view the history file by using the Alt-V (View a File) command. If you operate a busy Host mode system, however, this history file can grow quickly. So use PCEDIT or another ASCII text editor routinely to delete the old entries from this file.

Reviewing the Upload Directory File

A third important Host mode file is the *upload directory file*. Each time a user uploads a file to the upload directory, Host mode records in the file PCPLUS.ULD the date, time, file name, user's name, and file description. To view this file, press Alt-V (View a File), type *pcplus.uld*, and press Enter. PROCOMM PLUS displays a file similar to the one shown in figure 9.20.

Fig. 9.20.

The Host mode upload directory file, PCPLUS.ULD.

```
08/29/89 21:15:25 we889.wk1    FROM WALT HOUSTON    >>> Western region sal
08/29/89 22:18:37 so889.wk1    FROM ELSA RAINS      >>> Southern region sa
08/31/89 16:05:28 no889.wk1    FROM PETE GREEN      >>> Northern region sa
08/31/89 20:02:16 ea889.wk1    FROM VICTOR LASLO    >>> Eastern region sal
09/01/89 01:05:57 modems.arc   FROM WALT BRUCE      >>> Sample modem setup
09/29/89 21:30:32 so989.wk1    FROM ELSA RAINS      >>> Southern region sa
09/29/89 22:19:27 no989.wk1    FROM PETE GREEN      >>> Northern region sa
10/02/89 08:50:07 ea989.wk1    FROM VICTOR LASLO    >>> Eastern region sal
10/02/89 11:16:57 we989.wk1    FROM WALT HOUSTON    >>> Western region sal
```

```
Home: Top of file      PgUp: Previous page      PgDn: Next page      ESC: Exit
```

As you remove files from the upload directory, you can use PCEDIT or another text editor to remove the corresponding lines from this file.

TIP | Some file-transfer protocols available to the Host mode user (including KERMIT, SEALINK, YMODEM BATCH, TELINK, or GYMODEM-G BATCH) transmit multiple files and therefore send the file names from the remote computer to the host system. If a file by the same name already exists on the host, these protocols change the name. The end result is that the file name listed in the PCPLUS.ULD file may not be the same as the actual file name on the disk. For example, the user may use a wild card in the file specification, such as *.wk1. The file-transfer protocol then sends to the user's working directory all files that have a file name extension .WK1. Host mode records in PCPLUS.ULD only the specification *.wk1, not the names of the files actually sent. In other words, you have to examine the upload directory itself to determine the actual names of the files sent.

TIP | You may want to keep a disk file or a printout of a complete list of all files uploaded to the Host mode. This list conveniently documents the source of each file. The list also provides a textual description of each file, a luxury not afforded by a normal DOS directory listing.

Administering the Electronic Mail Facility

Unlike the other Host mode related files, the files that contain messages left on the Host mode electronic mail system cannot be maintained by using an ASCII text editor. Instead, DATASTORM distributes a separate utility program, PCMAIL.EXE, that enables you to perform a number of maintenance functions on the Host mode mail files.

PCMAIL is a separate program distributed on the PROCOMM PLUS Supplemental Diskette. After you use the Installation Utility (see Appendix A) to place PROCOMM PLUS on your computer, copy PCMAIL.EXE from the Supplemental Diskette into the directory that contains the PROCOMM PLUS program files (usually C:\PCPLUS).

PCMAIL is intended to operate on two files, PCPLUS.MSG (the message file) and PCPLUS.HDR (the *header file*). Together PCPLUS.MSG and PCPLUS.HDR constitute the Host mode *message base*. The message file, PCPLUS.MSG, contains the text of the messages; and the header file, PCPLUS.HDR, contains the sender's name, the addressee's name, the subject, the date and time, and a notation indicating whether the message is private or deleted.

PROCOMM PLUS creates these two files on the working directory the first time a Host mode user leaves mail. PROCOMM PLUS then updates the files as users add and delete messages using the Host mode electronic mail facility. PCMAIL enables you, as the system operator, to review, add, and remove permanently messages from this message base.

Before you can start PCMAIL, you must be at the DOS prompt. Exit to DOS either by quitting from PROCOMM PLUS using Alt-X (Exit) or by using Alt-F4 (DOS Gateway). The directory that contains PCPLUS.EXE and PCMAIL.EXE should be the current DOS directory. At the DOS prompt, type *pcmail*, and press Enter. You see the PCMAIL Main Options menu, shown in figure 9.21.

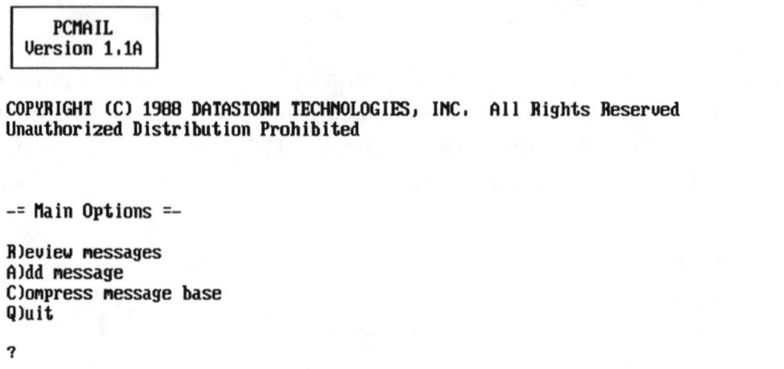

Fig. 9.21.

The PCMAIL Main Options menu.

```
┌─────────────┐
│    PCMAIL   │
│ Version 1.1A│
└─────────────┘

COPYRIGHT (C) 1988 DATASTORM TECHNOLOGIES, INC.  All Rights Reserved
Unauthorized Distribution Prohibited


-= Main Options =-

R)eview messages
A)dd message
C)ompress message base
Q)uit

?
```

TIP The most convenient way to start PCMAIL is to assign the program to one of the two hot-key commands available in PROCOMM PLUS. For example, you can assign PCMAIL to the command key Alt-J. Then when you want to perform maintenance on the message base, you simply press Alt-J. PROCOMM PLUS temporarily suspends itself and loads PCMAIL. Used in this manner, PCMAIL operates as if it were a part of the PROCOMM PLUS program itself. Refer to "Setting File and Path Options," in Chapter 8, for more information on how to assign a program to a hot-key command.

PCMAIL looks and feels similar to the PROCOMM PLUS Host mode electronic mail facility. The PCMAIL Main Options menu options work in similar ways.

You select **Review Messages** from the PCMAIL Main Options menu to read all or some of the messages contained in the message base. When you select this option, PCMAIL tells you the total number of messages currently in the message base. PCMAIL then displays the Review Options menu:

```
F)orward read
S)earch mail
I)ndividual read
Q)uit
```

Each of these options operates in a manner similar to the corresponding option available on the Read Mail Host mode menu, explained in "Reading Electronic Mail," earlier in this chapter.

Use **Forward Read** to browse through the message base sequentially and **Search** to find one or more particular messages to review. First, PCMAIL displays a message header (addressee, sender, subject, date and time, private, deleted). PCMAIL then provides six options:

- **View**. Use this option to view the contents of the message.

- **Reply**. Select this choice in order to leave a reply addressed To the message sender. Your reply message is From SYSOP.

- **Delete**. Choose this selection to toggle on or off the Deleted status in the header.

- **Private**. Use this option to toggle on or off the Private status in the header.

- **Quit**. This choice returns to the PCMAIL Main Options menu.

- **(‹CR› for another)**. Press Enter to review another message.

When you already know the number of the message you want to read, select **Individual Read** from the Review Options menu. PCMAIL displays the message header, the content of the message, and then the list of options.

For example, suppose that you decide to review message 3, left on the system by your assistant, Sam Spade. You start PCMAIL, select **Review Messages**, and choose **Individual Read**. PCMAIL prompts you for the Message number, so you type 3, and press Enter. PCMAIL displays the screen shown in figure 9.22.

Fig. 9.22.

Using PCMAIL to review a message.

```
? I
Message number: 3


 Msg: 3
  To: SYSOP
From: SAM SPADE
  Re: DELETED MAIL
10/04/89 01:57:14


Please try to recover a message addressed to me
from Victor Laslo dated 9/21/89. The subject was
"Freight Charges." I thought I saved a copy but
can't seem to find it. Hope you can come up with
a copy for me.

Thanks,

Sam

R)eply  D)elete  P)rivate  Q)uit  (<CR> for another): ?
```

You can also use PCMAIL to send a message. Select **Add** from the Main Options menu. PCMAIL asks for the addressee, subject, and an indication of whether the message is private, and then lets you compose the message. Refer to "Sending Electronic Mail," earlier in this chapter, for more on this subject. A message sent using PCMAIL is listed as From SYSOP.

The third option on the PCMAIL Main Options menu is **Compress Message Base**. Use this command to remove permanently all messages marked Deleted. When you choose this option, PCMAIL displays the question Compress message base(Y/N)? Respond Y to cause PCMAIL to reconstruct both the message file (PCPLUS.MSG) and the header file (PCPLUS.HDR) to remove permanently the messages marked Deleted.

As PCMAIL compresses the message base, the program displays the message Compressing, please wait.... PCMAIL then announces:

    Compression complete!
    Started with n message(s), m were deleted leaving x message(s)

where n, m, and x are positive integers. PCMAIL then quits, returning you to DOS (or to PROCOMM PLUS if you used a hot-key command to start PCMAIL).

When PCMAIL compresses the message base, PCMAIL renumbers messages that come after a message that is removed. For example, suppose that message number 2 is marked Deleted, but messages 1 and 3 are not deleted. When you execute the Compress Message Base option, the original message number 2 is permanently removed, and the original message number 3 becomes message number 2.

TIP You may want to establish a policy of deleting from the message base all messages past a given age. For example, you might decide that all messages posted for more than 30 days will be permanently removed. You should encourage users to delete the mail they have read; but deleted messages are removed from the message base only when you execute the Compress Message Base command. Inform your users that messages more than 30 days old will be removed, whether or not marked Deleted by a user.

Choose Quit from the PCMAIL Main Options menu to return to DOS.

Chapter Summary

This chapter describes the rudimentary bulletin board system built into PROCOMM PLUS, known as the Host mode. You first learned how to prepare Host mode for use by others, including how to specify a welcome message, establish a list of registered users, and designate the system as open or closed to new unregistered users. The chapter then described several ways to invoke Host mode. Next, the text explored Host mode, first from a user's perspective and then from your perspective as the host system operator. The text explained how a user logs on to the system, and then described the Host mode features available to the user, including uploading and downloading of files and electronic mail. The chapter then described the functions the system operator can perform in Host mode and discussed a number of administrative functions that must be performed by the system operator, including management of Host mode files and administration of the Host mode electronic mail facility. Now that you have completed this chapter, you are ready to set up your own bulletin board system using PROCOMM PLUS.

10

Terminal Emulation

Terminals are special "slave" computers whose sole purpose is to connect to a larger computer for entry and retrieval of data. When PROCOMM PLUS *emulates* a particular type of terminal, PROCOMM PLUS "impersonates" that terminal in order to communicate with a host computer that expects only certain types of terminals to be connected. This chapter describes the 16 terminal emulations available in PROCOMM PLUS: DEC VT52 and VT102; ANSI; IBM 3101 and 3270/950; TeleVideo 910, 920, 925, 950 and 955; Lear Siegler ADM 3/5; Heath Zenith 19; ADDS Viewpoint; Wyse 50 and 100; and TTY. The chapter also explains how you can customize a terminal emulation by modifying the keyboard mapping, the translation table, or both.

Understanding Terminal Emulation

Terminal emulation performs a function analogous to that of a United Nations translator. If you have ever seen or read about the United Nations General Assembly, you are probably aware that an army of translators are always at work so that representatives of all nationalities can understand what is being said, regardless of the language spoken. Each representative can listen over earphones to a simultaneous translation of the proceedings into his or her native language.

When PROCOMM PLUS is emulating a particular type of terminal, PROCOMM PLUS too is performing a simultaneous translation between different "languages," but PROCOMM PLUS translates in two directions at once. Each time you press a key on your keyboard, PROCOMM PLUS converts the keystroke into the code that would be generated by a real terminal; this code is the "language" the host minicomputer or mainframe computer expects to receive. This conversion of outgoing keystrokes is referred to as *keyboard mapping*. At the same time, the host computer is sending to your computer codes intended to control the screen and printer of a real terminal; these codes are in a "language" the terminal would understand. PROCOMM PLUS also translates these incoming codes into codes your PC understands.

Just as the U.N. needs translators for more than one pair of spoken languages, PROCOMM PLUS also needs to emulate more than one type of terminal. Terminals from different manufacturers often don't speak exactly the same "language," and not all host computers are designed to work with the same type of terminal. Unless PROCOMM PLUS emulates a terminal that speaks and understands the "language" spoken and understood by the host computer to which your computer is connected, effective communication cannot take place. PROCOMM PLUS therefore gives you 16 terminal emulations to choose from, in an effort to provide at least one emulation that each host computer can understand.

Each emulation maps your PC's keyboard in a different way and expects a different set of screen (and sometimes printer) control signals from the host computer. Standard typewriter keys, A–Z, and 0–9, are universally understood by other computers, regardless of the type of terminal PROCOMM PLUS is emulating. This understanding is possible because all terminals emulated by PROCOMM PLUS send the generic ASCII character codes for these keys. The remaining, so-called *special,* keys on the keyboard, however, are the crux of the issue. These keys, used alone or with another key (Shift or Ctrl) can be programmed (mapped) to send a different code to the remote computer for each different type of terminal emulation. Table 10.1 lists the PC keystrokes that are mappable by PROCOMM PLUS.

Each key listed in table 10.1 can be programmed, or *mapped*, to send a special code (refer also to "Changing the Keyboard Mapping," later in this chapter). Through this mapping, PROCOMM PLUS makes your keyboard act like the keyboard of a real terminal. Each time you press a mapped key, PROCOMM PLUS sends the code that would be sent by pressing a corresponding key on a real terminal's keyboard. Each of PROCOMM PLUS's 16 terminal emulations uses a particular keyboard mapping (listed later in the chapter). All the mappings are stored together on the disk in the file PCPLUS.KBD in the same directory as the PROCOMM PLUS program files.

Always keep in mind that terminal emulation is never 100 percent effective. Just as all PCs are not alike, each type of terminal has its own special features and capabilities. Although the power and flexibility of your PC enables programmers to make it act like many different types of terminals,

Table 10.1
PROCOMM PLUS Mappable PC Keystrokes

PC, PC AT, and Enhanced Type Keyboards	*Enhanced Keyboard Only*
Tab	
Shift-tab (Backtab)	
Ins	Gray Insert
Del	Gray Delete
Backspace	
Home	Gray Home
End	Gray End
Enter	
↑	Gray ↑
↓	Gray ↓
←	Gray ←
→	Gray →
	Gray Page Up
	Gray Page Down
Ctrl-Home	
Ctrl-End	
Ctrl-PgUp	
Ctrl-PgDn	
Ctrl-Backspace	

Keypad characters:
　　*
　　—
　　+
　　.
　　/
　　Enter
　　0 through 9

Function keys:
　　F1 through F10
　　F11 and F12
　　Shift-F1 through Shift-F10
　　Shift-F11 and Shift-F12
　　Ctrl-F1 through Ctrl-F10
　　Ctrl-F11 and Ctrl-F12

your PC cannot always perform every special function of every type of terminal. For example, some terminals have special screen capabilities, such as 132 columns, that cannot be accomplished on your PC because of hardware limitations, or that are just not implemented by PROCOMM PLUS.

When you install PROCOMM PLUS using the Installation Utility (see Appendix A, "Installing and Starting PROCOMM PLUS"), you choose the type of terminal you want the program to emulate. You can change this default setting by using the Setup Utility (refer to "Setting the Default Terminal Emulation," in Chapter 8), and you can override a default setting by choosing a different terminal emulation in a dialing directory entry (see "Adding an Entry," in Chapter 3).

Choosing a Terminal Emulation

The sections that follow describe briefly each terminal emulation available in PROCOMM PLUS and list the standard and special features. Use these descriptions to help you decide which emulation is appropriate in each case. Each section also includes a table showing the keyboard mapping used by PROCOMM PLUS for that terminal emulation. It is important to know which key on your PC's keyboard is mapped to each key on a real terminal's keyboard.

Ultimately, the terminal emulation you select for any communications session is determined by the computer with which you want to communicate. Make sure that the terminal emulation you select is supported by the host computer. Many host computers or the protocol converters that act as translators at the host computer are capable of supporting several different terminal types, so you may also have to notify the host computer which emulation you are using.

DEC VT52 and VT102

Digital Equipment Corporation (DEC) has for many years been a primary manufacturer of minicomputers. DEC computers are widely used in business, industry, government, and education; and DEC computers support several different DEC terminals. PROCOMM PLUS can emulate two of the most popular DEC terminals: the VT52 and the VT102.

TIP | If at all possible, obtain the following materials:

- A manual or some other form of instructions for the type of terminal you have instructed PROCOMM PLUS to emulate. A diagram of the terminal's keyboard is especially helpful.

- Instructions from the operator of the host computer. These instructions should explain how to use the terminal effectively with the host computer.

Compare the terminal's documentation with the corresponding keyboard mapping table listed in the appropriate table in this chapter to determine which keys on your keyboard correspond to the keys on the real terminal's keyboard. The keyboard mappings, including the hexadecimal codes transmitted to the host, are also listed in Appendix F of the PROCOMM PLUS documentation. Use the instructions from the host computer operator to determine the effect each key has on the application you want to use on the host.

The PROCOMM PLUS VT52 emulation supports the following VT52 features:

- Full duplex

- Half duplex

- Keypad application mode

- Full-screen cursor control

- Erase functions

- Printer control functions, including both dedicated (to printer only) and transparent (to screen and printer) printing

- Full display attributes

The VT52 emulation lets you use the numbers on your numeric keypad and gives you access to the VT52 keypad application mode. The emulation accomplishes this capability by mapping the PC's ten function keys and the Shift-function-key combinations to the VT52 numeric keypad.

This mapping is shown in figure 10.1. The label on the top of each key in the figure represents the PC keystroke. The label on the front of each key in the figure represents the equivalent keystroke on the DEC VT52 terminal's keypad application mode keys. This mapping makes the most sense when the function keys on your keyboard are oriented vertically (as on the original IBM PC and IBM Personal Computer AT keyboards).

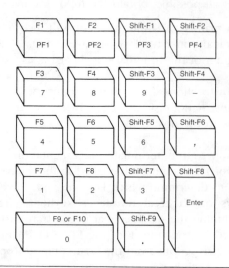

Table 10.2
DEC VT52 and VT102 Keyboard Mapping
(see also figures 10.1 and 10.2)

PROCOMM PLUS Key	VT52 Function
Enter	Enter
Tab	Horizontal tab
Delete	Character delete
Home	Home cursor
↑	Cursor up
↓	Cursor down
←	Cursor left
→	Cursor right
Ctrl-PgDn	Clear screen
Ctrl-PgUp	Delete line
Ctrl-J	Line feed

You can use the PROCOMM PLUS VT102 emulation when the host computer expects a DEC VT100 terminal or a VT102 terminal. These two terminals are similar, VT102 having a few more features. Both terminals are more powerful than the VT52 terminal. The functions supported by the PROCOMM PLUS VT102 emulation are the following:

- Full duplex

- Half duplex

- Set/reset modes

- Scroll region

- Special graphics character set

- U.S. and U.K. character sets

- Keypad application mode

- Full-screen cursor control

- Erase functions

- Insert/delete lines

- Programmable tabs

- Printer control functions, including both dedicated and transparent printing

- Full display attributes including ANSI color graphics

The VT102 132-column display feature is not supported by the PROCOMM PLUS emulation.

When you have set the **J** (Enquiry (ENQ)) option on the Setup Utility Terminal Options screen to ON, PROCOMM PLUS responds to an incoming ENQ character in a special way (refer to "Assigning a Response to ENQ," in Chapter 8). The **J** (Enquiry (ENQ)) option on the Terminal Options screen specifies the way that PROCOMM PLUS responds to the ENQ (ASCII decimal 5) character (generated by pressing Ctrl-E at either end of the connection). By default, this feature is set to OFF, and PROCOMM PLUS doesn't respond at all to the ENQ character. If you set this option to ON, PROCOMM PLUS executes the keyboard macro Alt-0 when the program receives the ENQ character (refer to Chapter 7, "Automating PROCOMM PLUS with Macros and Script Files," for information on how to create a keyboard macro). This response is often called an *answerback message*.

PROCOMM PLUS's VT102 keyboard mapping of the keypad application mode is not the same as the mapping for VT52. The VT102 emulation mapping uses function keys F1 through F4 and your PC's numeric keypad, as illustrated in figure 10.2. Because function keys map to function keys and numeric keys map to numeric keys, this implementation is clearer than the VT52 keyboard mapping and therefore easier to use.

Fig. 10.2.

Mapping of PC numeric keypad to VT102 keypad application mode keys.

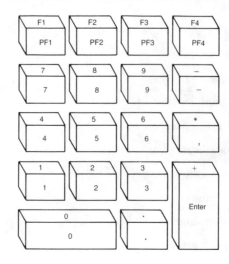

In addition to the keypad application mode keys, PROCOMM PLUS maps a number of other keys to VT102 functions. This mapping is summarized in table 10.2 (the same as for VT52).

ANSI

When you use PROCOMM PLUS primarily to access electronic bulletin boards, you should set the default terminal emulation to ANSI. *ANSI* stands for American National Standards Institute, but in this context, ANSI refers to the terminal emulation standard recommended by the American National Standards Institute. The ANSI standard's widespread use by bulletin boards stems from its capability to create colorful screens and to provide full-screen control over the cursor.

ANSI's special display attributes are accomplished through the host computer's use of a set of codes known collectively as *ANSI escape sequences.* The codes are called escape sequences because each begins with the ASCII ESC character (ASCII decimal 27). For example, the ANSI escape sequence to set the screen to white characters (foreground) on a blue background is

ESC[37;44m

DOS (IBM PC DOS or Microsoft MS-DOS) is distributed with a driver file named ANSI.SYS. This file normally defines the ANSI escape sequences that can be recognized and acted on by your computer. (Refer to your DOS manual for a complete explanation of ANSI.SYS and ANSI escape sequences.) PROCOMM PLUS, however, doesn't use the ANSI driver provided by DOS; PROCOMM PLUS uses its own emulation of ANSI.SYS. PROCOMM PLUS responds to ANSI escape sequences sent by a host computer in the same manner as DOS's ANSI.SYS responds, however. And because two versions of ANSI.SYS exist—one supplied with DOS 2.*x* and the other distributed with DOS 3.*x*—PROCOMM PLUS provides two versions of its ANSI emulation.

Use the **O** (ANSI compatibility) option on the Setup Utility General Options screen to select ANSI.SYS 2.*x* emulation or ANSI.SYS 3.*x* emulation (refer to "Adjusting General Options," in Chapter 8). Some host systems expect you to use the version of ANSI.SYS that is distributed with DOS 2.*x* (2.0, 2.1, or 2.11, and so on), but other systems assume that you are using ANSI.SYS from DOS 3.*x*.

With regard to keyboard mapping, PROCOMM PLUS's ANSI terminal emulation is the same as in VT52 emulation, described in the preceding section of this chapter. Screen handling is more like the VT102 emulation, except that normal, boldface, and reverse video are handled (internally) in a different manner.

IBM 3101

A popular asynchronous IBM terminal is the IBM 3101 series. These terminals are normally used with IBM's Time Sharing Option (TSO). PROCOMM PLUS's emulation of the IBM 3101 terminals supports the most commonly used features of Model 1x and Model 2x IBM 3101 series terminals, including

- Full duplex

- Half duplex

- Full IBM character set

- Scroll on/off

- Program function keys

- Erase functions

Block mode transfer of data to and from the host computer is, however, not supported by the PROCOMM PLUS 3101 emulation. Data is sent and received one character at a time.

Table 10.3 lists the PROCOMM PLUS keyboard mapping used during IBM 3101 emulation.

Table 10.3
IBM 3101 Keyboard Mapping

PROCOMM PLUS Key	3101 Function
Enter	Enter
Tab	Horizontal tab
Delete	Character delete
Home	Home cursor
↑	Cursor up
↓	Cursor down
←	Cursor left
→	Cursor right
Ctrl-PgDn	Clear screen
End	Erase to end of line
Ctrl-End	Erase to end of screen
F1 through F8	Program Function 1 (PF1) through Program Function 8 (PF8)

TIP | If you are using the IBM 3101 emulation to connect to an IBM mainframe by dialing an asynchronous-to-synchronous protocol converter, consider using the IBM 3270/950 emulation instead. When the protocol converter sends signals from your computer to the mainframe, the converter is most likely emulating a 3270 terminal. The PROCOMM PLUS 3270/950 terminal emulation provides keyboard mapping that enables you to use all 24 program function keys available on a 3270 keyboard. If you use the 3101 emulation, you can use only the first eight program function keys.

IBM 3270/950

The IBM 3270/950 terminal emulation actually doesn't emulate one terminal but two. The keyboard mapping emulates an IBM 3270, and the screen-handling features emulate a TeleVideo 950 terminal. To understand the logic behind this "hybrid" emulation, you need a little background information.

Real IBM 3270 terminals connect by synchronous modem to IBM mainframe computers and afford many advanced features. Complete emulation of a 3270 terminal using a PC, however, requires you to purchase a synchronous modem and a "terminal on a card" integrated circuit board for installation in your PC and may require you to use a special more costly telephone line, known as a *leased line.*

Sometimes the benefits of true 3270 emulation justify the cost. If you plan to use your PC exclusively or primarily as a terminal to an IBM mainframe, consider going all the way. Data transfer is several times faster, and you have access to all the 3270 features. But if you only occasionally need to connect to a host computer that supports a 3270 terminal and you more often connect to systems that support asynchronous communications, PROCOMM PLUS's emulation may suit your needs well.

In order to connect your PC to an IBM mainframe through an asynchronous modem, you must go through a protocol converter. The *protocol converter* changes your computer's asynchronous signal into the synchronous signal required by the mainframe computer.

A protocol converter also performs terminal emulation. The protocol converter enables many different types of terminals to dial in, and then the protocol converter performs a two-way translation between the mainframe and each terminal. For example, an IBM 7171 protocol converter can accept connections from DEC VT102 terminals, TeleVideo 950 terminals, and IBM 3101 terminals (among many others) and can make all these terminals look to the mainframe computer like IBM 3270 terminals. At the same time, the protocol converter translates signals sent by the mainframe into signals the different terminals can interpret and act on.

The PROCOMM PLUS IBM 3270/950 emulation is intended for use through a protocol converter. The keyboard mapping enables you to send through an IBM 7171 protocol converter the codes necessary to emulate all the keys on a 3270 keyboard.

IBM 3270 keyboards have 24 function keys. Figure 10.3 shows how the PROCOMM PLUS 3270/950 emulation maps the function keys on your computer's keyboard to these 24 function keys. The label on the top of each key in the figure represents the PC keystroke. The label on the front of each key in the figure represents the equivalent function key on the 3270 terminal's keyboard. When two labels are on the top of a key, the second is optional and is available only on an Enhanced keyboard.

The remaining 3270 functions are mapped to your computer's keys as listed in table 10.4.

Fig. 10.3.

Mapping of PC function keys to IBM 3270 function keys.

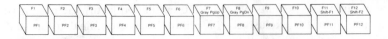

Table 10.4
IBM 3270/950 Keyboard Mapping

PC, PC AT, and Enhanced Keyboard	Enhanced Keyboard Only	3270 Function
Enter		Enter
Keypad +	Keypad Enter	Clear
Ctrl-Home		Clear
Ctrl-F5		PA1
Ctrl-F6		PA2
Ctrl-F7		PA3
Delete	Gray Delete	Character delete
Ctrl-Backspace		Character delete
Ins	Gray Insert	Toggle Insert mode
End	Gray Delete	Erase to end of field
Ctrl-End		Erase to end of field
Ctrl-F8		Erase input
↑	Gray ↑	Cursor up
↓	Gray ↓	Cursor down
←	Gray ←	Cursor left
→	Gray →	Cursor right
Home	Gray Home	Home cursor

Table 10.4—*continued*

PC, PC AT, and Enhanced Keyboard	Enhanced Keyboard Only	3270 Function
Ctrl-F9		Column tab
Ctrl-F10	Column backtab	
Tab		Field tab
Shift-Tab		Field backtab
Ctrl-PgUp		Indent
Ctrl-PgDn		Undent
Ctrl-J		New line
Ctrl-Enter		New line
Ctrl-X		Type-ahead purge
Ctrl-S		Pacing start
Ctrl-Q		Pacing stop
Ctrl-T		Keyboard unlock
Ctrl-R		Char-error reset
Ctrl-G		Master reset
Esc-*		Redisplay

For addressing and controlling your PC's screen, this emulation uses the TeleVideo 950 protocol. You therefore must inform the protocol converter that you are using a TeleVideo 950 terminal. PROCOMM PLUS then takes advantage of the protocol converter's capability to translate screen-handling commands from the mainframe into the proper codes for a TeleVideo 950 terminal. PROCOMM PLUS, in turn, performs a second translation and sends the proper screen- and printer-control commands to your PC's hardware.

TeleVideo 910, 920, 925, 950, and 955

TeleVideo Systems manufactures a number of popular terminals. PROCOMM PLUS includes emulations for several TeleVideo 900 series terminals, including 910, 920, 925, 950, and 955. (These protocols are listed in the Setup Utility Terminal Options screen and in the Dialing Directory screen as TVI 910, TVI 920, and so on.) For each of these terminal types, PROCOMM PLUS supports the following features:

- Full duplex
- Half duplex
- Program function keys

- Full-screen cursor control

- Erase functions

- Protected fields

- Full display attributes

- Special graphics set

- User-loadable status line

- Printer functions, including both dedicated and transparent printing

The keyboard mapping for all the TeleVideo terminals is listed in table 10.5.

Table 10.5
TeleVideo 900 Series Keyboard Mapping

PROCOMM PLUS Key	*900 Series Function*
Enter	Enter
Tab	Horizontal tab
Shift-Tab	Reverse tab
Insert	Insert character
Ctrl-Home	Insert line
Delete	Delete character
Ctrl-PgUp	Delete line
Home	Home cursor
↑	Cursor up
↓	Cursor down
←	Cursor left
→	Cursor right
Ctrl-PgDn	Clear screen
End	Erase to end of line
Ctrl-End	Erase to end of page (screen)
F1 through F10	Function 1 (F1) through Function 10 (F10)
Shift-F1	Function 11 (F11)
Shift-F3	Shift erase to end of line
Shift-F4	Shift erase to end of page
Shift-F5	Shift line insert
Shift-F6	Shift line delete
Shift-F7	Shift character insert
Shift-F8	Shift character delete
Alt-F10	FUNCT

In order to use the Alt-F10 (FUNCT) feature, first press Alt-F10; then press the key you want to use with FUNCT. PROCOMM PLUS sends the proper code to the host computer.

The primary difference among the TeleVideo emulations is the code PROCOMM PLUS sends to the host computer when you press the down arrow. The TeleVideo 910 and 920 emulations send the LF (line feed) character (ASCII decimal 10—same as Ctrl-J); but the 925, 950, and 955 emulations send the SYN character (ASCII decimal 22—same as Ctrl-V).

Lear Siegler ADM 3/5

Another popular series of terminals is manufactured by Lear Siegler. The PROCOMM PLUS ADM 3/5 emulation (listed in the Terminal Options screen and Dialing Directory screen as ADM 5) is used in place of Lear Siegler ADM-3/5 series terminals. The ADM 3/5 emulation supports the following functions:

- Full duplex

- Half duplex

- Full character set

- Erase functions

- Full-screen cursor control

The key mapping for the ADM 3/5 emulation is listed in table 10.6.

Table 10.6
Lear Siegler ADM-3/5 Series Keyboard Mapping

PROCOMM PLUS Key	ADM-3/5 Function
Enter	Enter
Tab	Horizontal tab
Delete	Delete character
Home	Home cursor
↑	Cursor up
↓	Cursor down
←	Cursor left
→	Cursor right
Ctrl-PgDn	Clear screen
End	Erase to end of line

Heath/Zenith 19

PROCOMM PLUS also provides an emulation of the Heath/Zenith 19 terminal (listed as HEATH 19 on the Dialing Directory screen and on the Terminal Options screen). PROCOMM PLUS's emulation supports the following functions:

- Full duplex

- Half duplex

- Full character set

- Program function keys

- Erase functions

- Full-screen cursor control

- Display attributes

The key mapping for the emulation of Heath/Zenith 19 is listed in table 10.7.

Table 10.7
Heath/Zenith 19 Keyboard Mapping

PROCOMM PLUS Key	Heath/Zenith 19 Function
Enter	Enter
Tab	Horizontal tab
Delete	Character delete
Home	Home cursor
↑	Cursor up
↓	Cursor down
←	Cursor left
→	Cursor right
Ctrl-PgDn	Clear screen
End	Erase to end of line
Ctrl-Home	Insert line
Ctrl-PgUp	Delete line
F1 through F10	Program Function 1 (PF1) through Program Function 10 (PF10)

ADDS Viewpoint

Another terminal supported by PROCOMM PLUS is the ADDS Viewpoint (listed in the Dialing Directory screen and the Setup Utility Terminal Options screen as ADDS VP). PROCOMM PLUS supports the following functions:

- Full duplex
- Half duplex
- Function keys
- Erase functions
- Insert and delete functions
- Full-screen cursor control
- Display attributes

Table 10.8 lists the keyboard mapping for the ADDS Viewpoint terminal emulation.

Table 10.8
ADDS Viewpoint Keyboard Mapping

PROCOMM PLUS Key	ADDS Viewpoint Function
Enter	Enter
Tab	Horizontal tab
Home	Home cursor
↑	Cursor up
↓	Cursor down
←	Cursor left
→	Cursor right
Ctrl-PgDn	Clear screen
End	Erase to end of line
Ctrl-End	Erase to end of screen
Ins	Insert character
Ctrl-Home	Insert line
Delete	Delete character
Ctrl-PgUp	Delete line
F1 through F8	Function 1 (F1) through Function 8 (F8)
Shift-F1 through Shift-F8	Shift-Function 1 through Shift-Function 8

Wyse 50 and 100

PROCOMM PLUS includes emulation of Wyse 50 terminals and Wyse 100 terminals. PROCOMM PLUS supports the following functions of each type of Wyse terminal:

- Full duplex

- Half duplex

- Program function keys

- Erase functions

- Protected fields

- Full-screen cursor control

- Full display attributes

- Printer functions, including both dedicated and transparent printing (Wyse 50 only)

Table 10.9 lists the keyboard mapping for Wyse 50 terminal emulation. Table 10.10 lists keyboard mapping for Wyse 100 terminal emulation.

Table 10.9
Wyse 50 Keyboard Mapping

PC, PC AT, and Enhanced Keyboard	Enhanced Keyboard Only	Wyse 50 Function
Enter		Enter
Tab		Horizontal tab
Shift-Tab		Reverse tab
Insert		Insert character
Ctrl-Home		Insert line
Delete		Delete character
Home		Home cursor
↑		Cursor up
↓		Cursor down
←		Cursor left
→		Cursor right
Ctrl-PgDN		Clear screen
End		Erase to end of line
Ctrl-End		Erase to end of screen
Ctrl-PgUp		Delete line

Table 10.9—*Continued*

PC, PC AT, and Enhanced Keyboard	Enhanced Keyboard Only	Wyse 50 Function
F1 through F10		Function 1 (F1) through Function 10 (F10)
Shift-F1		Function 11 (F11)
	F11	Function 11 (F11)
Shift-F2		Function 12 (F12)
	F12	Function 12 (F12)
Shift-F3 through Shift-F6		Function 13 (F13) through Function 16 (F16)
Ctrl-F1		Shift line erase
Ctrl-F2		Shift page erase
Ctrl-F3		Shift line insert
Ctrl-F4		Shift line delete
Ctrl-F5		Shift character insert
Ctrl-F6		Shift character delete
Alt-F10		FUNCT

In order to use the Alt-F10 (FUNCT) feature, first press Alt-F10; then press the key you want to use with FUNCT. PROCOMM PLUS sends the proper code to the host computer.

Table 10.10
Wyse 100 Keyboard Mapping

PROCOMM PLUS Key	Wyse 100 Function
Enter	Enter
Tab	Horizontal tab
Shift-Tab	Reverse tab
Insert	Insert character
Ctrl-Home	Insert line
Delete	Delete character
Ctrl-PgUp	Delete line
Home	Home cursor
↑	Cursor up
↓	Cursor down

Table 10.10—*Continued*

PROCOMM PLUS Key	Wyse 100 Function
←	Cursor left
→	Cursor right
Ctrl-PgDn	Clear screen
End	Erase to end of line
Ctrl-End	Erase to end of screen
F1 through F8	Function 1 (F1) through Function 8 (8)
Shift-F1 through Shift-F8	Shift Function 1 to Shift Function 8

TTY

A TTY, or Teletype, terminal is a video display terminal version of the original one-line-at-a-time typewriter-style terminal. If you have never used one of these old Teletype-style terminals, you may have seen one. They look much like a typewriter, having a keyboard on the front and a paper feed and printer on the back. As you type a command to the host computer, the command is printed on the paper. Any response from the host is also printed on the paper. Once a line scrolls up past the printing element, the line cannot be recovered.

The familiar DOS command line uses a TTY approach. Even though your computer's screen can display at least 25 lines of text, the DOS command line uses only one line at a time. Once you press Enter, the cursor moves to the next command line, and you cannot go back up to correct a mistake.

When you select PROCOMM PLUS's TTY terminal emulation, you get basic one-line-at-a-time control of your screen, with no special screen attributes, such as line drawing or color. The only special keys mapped by PROCOMM PLUS are the Tab, Backspace, and Enter keys and the keys available on your numeric keypad with Num Lock activated (0–9 * − + . / and Enter). Use TTY emulation only if absolutely required by the host you are calling.

Changing the Keyboard Mapping

Sometimes a terminal emulation's keyboard mapping doesn't include a function you need. This omission may occur because the host computer's support of a particular type of terminal is slightly different from what PROCOMM PLUS expects, because you are dialing in through a protocol converter that is performing a second-level terminal emulation, or simply because you want to add a function that is not part of the terminal emulation. Whatever the reason, PROCOMM PLUS enables you to customize any terminal emulation keyboard mapping.

All the terminal emulation mappings are stored on the disk in the file PCPLUS.KBD, in the same directory as the PROCOMM PLUS program files.

CAUTION: Before you make any changes to PCPLUS.KBD, make a disk copy of the file. Then you can easily return the file to its default settings. Of course, you should never change the PCPLUS.KBD file—or any other file—on the distributed PROCOMM PLUS diskettes. You should always work from an installed copy of the original diskettes.

To alter a keyboard mapping, from the Terminal mode screen, press Alt-F8 (Key Mapping). PROCOMM PLUS displays the Keyboard Mapping screen. For example, the Keyboard Mapping screen in figure 10.4 shows the mapping for the ANSI terminal emulation.

```
   PROCOMM PLUS      F1 ..... ^[OP    S-F1 .... ^[OR    C-F1
   KEYBOARD MAPPING   F2 ..... ^[OQ    S-F2 .... ^[OS    C-F2
     Version 1.1B     F3 ..... ^[Ow    S-F3 .... ^[Oy    C-F3
                      F4 ..... ^[Ox    S-F4 .... ^[Om    C-F4
                      F5 ..... ^[Ot    S-F5 .... ^[Ou    C-F5
 KEYPAD * *           F6 ..... ^[Ou    S-F6 .... ^[Ol    C-F6
 KEYPAD - -           F7 ..... ^[Oq    S-F7 .... ^[Os    C-F7
 KEYPAD + +           F8 ..... ^[Or    S-F8 .... ^[OM    C-F8
 KEYPAD . .           F9 ..... ^[Op    S-F9 .... ^[On    C-F9
 KEYPAD / /           F10 .... ^[Op    S-F10 ... ^[OM    C-F10
 KEY ENTER ^M         F11 ....          S-F11 ...          C-F11
                      F12 ....          S-F12 ...          C-F12

 TAB ....... ^I       KEYPAD 0 0       GREY CUP  ^[[A    CURUP ^[[A
 BACKTAB ...          KEYPAD 1 1       GREY CDN  ^[[B    CURDN ^[[B
 INSERT ....          KEYPAD 2 2       GREY CLF  ^[[D    CURLF ^[[D
 DELETE .... »DEL«    KEYPAD 3 3       GREY CRT  ^[[C    CURRT ^[[C
 BACKSPACE . ^H       KEYPAD 4 4       GREY INS
 C-HOME .... ^[[L     KEYPAD 5 5       GREY DEL  »DEL«   HOME  ^[[H
 C-END .....          KEYPAD 6 6       GREY HOME ^[[H    END   ^[[K
 C-PGUP .... ^[[M     KEYPAD 7 7       GREY END  ^[[K
 C-PGDN .... ^[[H^[[2J KEYPAD 8 8      GREY PGUP          ENTER ^M
 C-BACKSPACE »DEL«    KEYPAD 9 9       GREY PGDN
 Emulation: ANSI      PgUp/PgDn/Space: Next emulation   Esc: Exit   Alt-Z: Help
```

Fig. 10.4.

The Keyboard Mapping screen showing the mapping for the ANSI terminal emulation.

TIP The Keyboard Mapping screen is displayed by an external program, PCKEYMAP.EXE, which is distributed on the PROCOMM PLUS Supplemental Diskette. After installing PROCOMM PLUS using the Installation Utility, copy PCKEYMAP.EXE from the Supplemental Diskette into the directory that contains the PROCOMM PLUS programs (refer to Appendix A, "Installing and Starting PROCOMM PLUS"). If you execute the Alt-F8 (Key Mapping) command without first copying PCKEYMAP.EXE into the appropriate directory, PROCOMM PLUS displays the message EXTERNAL PROGRAM ERROR: File or path not found.

When you press Alt-F8 (Key Mapping), the Keyboard Mapping screen shows the mapping for the current terminal emulation. The screen lists the 79 keystrokes that are programmable by PROCOMM PLUS. (These keystrokes are listed in table 10.1.) To the right of each keystroke, the Keyboard Mapping screen lists the PROCOMM PLUS macro code for the actual keystrokes that are sent to the remote computer when you press that keystroke (refer to "Creating Macros," in Chapter 7). For example, figure 10.4 shows the macro code ^[OP listed to the right of the function key F1. The macro code ^[translates into the ASCII character ESC; therefore, when you are using the ANSI terminal emulation and press F1, PROCOMM PLUS sends to the remote computer the ESC character followed by the letters OP.

Refer to table 7.2 in Chapter 7, "Automating PROCOMM PLUS with Macros and Script Files," for a list of the most frequently used macro control codes. Refer also to the ASCII table in "Translating Control Codes," in Appendix D of the PROCOMM PLUS documentation, for other macro control codes.

TIP The keyboard mappings given in Appendix F of the PROCOMM PLUS documentation list the hexadecimal codes sent by each keystroke. For example, the following three codes are listed to the right of F1 keystroke in the keyboard mapping for ANSI terminal emulation:

1B 4F 50

These hexadecimal codes are, respectively, the codes for the ESC character, the letter *O*, and the letter *P*—the same three characters represented by the macro codes listed in the Keyboard Mapping screen for the ANSI terminal emulation.

On the Keyboard Mapping screen, you can cycle through the keyboard mappings for all 16 available terminal emulations by pressing PgUp, PgDn, or the space bar. The keyboard mappings are arranged in the following order:

VT52
VT102
ANSI
Heath 19
IBM 3101
ADDS VP
ADM 5
TVI 910
TVI 920
TVI 925
TVI 950
TVI 955
Wyse 50
Wyse 100
IBM 3270/950
TTY

Each time you press the space bar or the PgDn key, the Keyboard Mapping screen displays the keyboard mapping for the next terminal on this list. For example, if you start at the ANSI mapping and press the PgDn key or the space bar, the Keyboard Mapping screen displays the mapping for the Heath 19 terminal emulation. Each time you press the PgUp key, the Keyboard Mapping screen displays the mapping for the preceding terminal emulation in this list.

When the Keyboard Mapping screen displays the mapping you want to modify, press the key for which you want to change or add mapping. The Keyboard Mapping screen displays in inverse video the entry space to the right of the subject keystroke. Use the cursor-movement and editing keys listed in table 10.11 to edit the entry. Each entry can contain from 1 to 10 characters. Press Enter after you finish the new entry.

When you have finished making changes or additions to the keyboard mappings, press Esc (Exit). The Keyboard Mapping screen displays a small blinking window containing the question, Save changes (Y/N)? Respond Yes in order to save the changes and additions you have made. The changes will have no effect if you respond No to this question.

After you indicate whether you want the keyboard mapping changes saved, PROCOMM PLUS returns to the Terminal mode screen.

Table 10.11
Keyboard Mapping Entry, Cursor-Movement, and Editing Keys

Key	Function
←	Move the cursor one space to the left
→	Move the cursor one space to the right
Home	Move the cursor to the left end of the entry
End	Move the cursor one space to the right of the last character in the entry
Insert	Toggle Insert/Overtype modes
Delete	Delete the character at cursor
Backspace	Delete the character to the left of cursor
Tab	Delete the entire entry
Ctrl-End	Delete the characters from cursor to right end

TIP Most keyboard mappings do not use all the programmable keys. You are free to assign macros to any keystrokes that are not used by a terminal emulation. Indeed, you can create a set of macros for use exclusively with a terminal emulation. Use the Keyboard Mapping screen to display the standard keyboard mapping for the terminal emulation. Assign macro codes to keystrokes that are not already used. Then when you select this emulation, PROCOMM PLUS also loads the macros. These terminal emulation specific macros are added to the Alt-key macros discussed in Chapter 7.

For example, you may find that when you are using PROCOMM PLUS to communicate with electronic bulletin board systems, you routinely type the word *OPEN*. This word, followed by a number, is usually the command required to use the DOORS feature, which is available on many popular BBSs. You normally have the ANSI terminal emulation active when you connect to a BBS, and you notice that the Ctrl-F1 keystroke is not used by the emulation's standard keyboard mapping. You decide to add to the keyboard mapping a macro that types the word *OPEN*.

To add this macro, press Alt-F8 (Key Mapping) to display the Keyboard Mapping screen, and press the space bar until the ANSI keyboard mapping is displayed (again see fig. 10.4). Then press

Ctrl-F1. The Keyboard Mapping screen moves the cursor into the entry space to the right of C-F1. Type *open* in the entry space, and press Enter. Finally, press Esc (Exit), and choose Yes to save the change. The next time you connect to a bulletin board using the ANSI terminal emulation, you can just press Ctrl-F1 in order to cause PROCOMM PLUS to type *OPEN*.

Changing the Translation Table

In addition to controlling outgoing characters with keyboard mapping, you can also use PROCOMM PLUS to control the display of incoming characters on your screen. Normally, the current terminal emulation interprets data coming in from a host computer and displays the appropriate characters. Occasionally, however, you may want to display characters that are different from the characters actually being transmitted. When you want PROCOMM PLUS to perform such a translation, press Alt-W (Translate Table) to display the first Translation Table screen, shown in figure 10.5.

╡ TRANSLATION TABLE ╞								
0: 0	16: 16	32: 32	48: 48	64: 64	80: 80	96: 96	112:112
1: 1	17: 17	33: 33	49: 49	65: 65	81: 81	97: 97	113:113
2: 2	18: 18	34: 34	50: 50	66: 66	82: 82	98: 98	114:114
3: 3	19: 19	35: 35	51: 51	67: 67	83: 83	99: 99	115:115
4: 4	20: 20	36: 36	52: 52	68: 68	84: 84	100:100	116:116
5: 5	21: 21	37: 37	53: 53	69: 69	85: 85	101:101	117:117
6: 6	22: 22	38: 38	54: 54	70: 70	86: 86	102:102	118:118
7: 7	23: 23	39: 39	55: 55	71: 71	87: 87	103:103	119:119
8: 8	24: 24	40: 40	56: 56	72: 72	88: 88	104:104	120:120
9: 9	25: 25	41: 41	57: 57	73: 73	89: 89	105:105	121:121
10: 10	26: 26	42: 42	58: 58	74: 74	90: 90	106:106	122:122
11: 11	27: 27	43: 43	59: 59	75: 75	91: 91	107:107	123:123
12: 12	28: 28	44: 44	60: 60	76: 76	92: 92	108:108	124:124
13: 13	29: 29	45: 45	61: 61	77: 77	93: 93	109:109	125:125
14: 14	30: 30	46: 46	62: 62	78: 78	94: 94	110:110	126:126
15: 15	31: 31	47: 47	63: 63	79: 79	95: 95	111:111	127:127

F1► Save   F2► Toggle Screens   F3► Table On   F4► Table Off   ESC► Exit

Translation Table INACTIVE

NUMBER TO CHANGE ⟶   NEW VALUE ⟶

Fig. 10.5.

The first Translation Table screen.

The translation table consists of a total of 256 entries in two screens, one entry for each character in the IBM extended character set. Each entry contains two numbers separated by a colon (:). The first number represents the decimal ASCII code for a character that is received from the remote computer (refer to the ASCII Table in Appendix D of the PROCOMM PLUS documentation). The second number represents the decimal ASCII code for the character that PROCOMM PLUS displays when the first character is received and the translation table is active.

The translation table is normally (by default) inactive. You have two ways to activate this table:

- *Temporarily*. To activate the table for use during a single session of PROCOMM PLUS, from the Terminal mode screen, press Alt-W to display the translation table, and press F3 (Table On). PROCOMM PLUS changes the message near the bottom of the screen from Translation Table INACTIVE to Translation Table ACTIVE. Press Esc to return to the Terminal mode screen. The next time you start PROCOMM PLUS, the Translation Table setting will return to the default condition.

- *Permanently*. You use the Setup Utility to activate the translation table and change the default setting so that PROCOMM PLUS activates the translation table when you start the program. (Refer to "Adjusting General Options," in Chapter 8, for information on changing this setting.) To make this change permanent, be sure to save the setup options. If you do not save the setup options, the setting is effective only for the current PROCOMM PLUS session.

After you activate the translation table, PROCOMM PLUS compares each incoming character with its entry in the translation table. When the number on the right of the colon is the same as the number on the left, PROCOMM PLUS passes the character through unaltered. But when the number on the right is different from the number on the left, PROCOMM PLUS displays the character represented by the number on the right.

When the translation table is active, you have two ways to turn it off:

- *Temporarily*. To deactivate the table during a single session of PROCOMM PLUS, from the Terminal mode screen, press Alt-W to display the translation table, and press F4 (Table Off). PROCOMM PLUS changes the message near the bottom of the screen from Translation Table ACTIVE to Translation Table INACTIVE. Press Esc to return to the Terminal mode screen. The next time you start PROCOMM PLUS, the translation table setting will return to the default condition.

- *Permanently*. You use the Setup Utility to deactivate the translation table and change the default setting so that PROCOMM PLUS does not use the translation table when you start the program. (Refer to "Adjusting General Options," in Chapter 8, for information on changing this setting.) To make this change permanent, be sure to save the setup options. If you do not save the setup options, the setting is effective only for the current PROCOMM PLUS session.

As you see in figure 10.5, initially, the numbers to the right of each colon are exactly the same as the numbers to the left. In order to cause PROCOMM PLUS to translate a particular incoming character into a different character on the screen, first display the Translation Table screen that contains the entry for the character you want to translate. If you don't see the character you want to modify on the first screen, press F2 (Toggle Screens) to switch to the second screen. Next, type the ASCII decimal code for the character you want PROCOMM PLUS to translate, and press Enter. The number appears to the right of the prompt NUMBER TO CHANGE. Finally, type the entry for NEW VALUE for this character, and press Enter.

TIP | To prevent a character from displaying at all, type *0* as the new value.

After you have made all your desired changes, press F1 (Save) to save the new translation table. The next time you go on-line with the translation table active, PROCOMM PLUS uses the new translations.

Chapter Summary

This chapter has described the 16 terminal emulations available in PROCOMM PLUS: DEC VT52 and VT102; ANSI; IBM 3101 and 3270/950; TeleVideo 910, 920, 925, 950 and 955; Lear Siegler ADM 3/5; Heath Zenith 19; ADDS Viewpoint; Wyse 50 and 100; and TTY. The chapter also has described how you can customize a terminal emulation by modifying the keyboard mapping and the translation table.

Turn now to Chapter 11, the last chapter in the book. Chapter 11 presents an overview of the ASPECT script language, discusses the commands available, and offers some suggestions on how to develop scripts with ASPECT.

11

An Overview of the ASPECT Script Language

This last chapter in the book provides an overview of the programmable capabilities available through the PROCOMM PLUS script language, ASPECT. You have already seen, in Chapter 7, "Automating PROCOMM PLUS with Macros and Script Files," how to record scripts that automate the log-on sequence to connect to another computer. This chapter introduces you to all the major features of the ASPECT programming language.

The ASPECT script language is a high-level communications programming language suitable for use in developing full-featured communications applications. Although extensive coverage of this programmable side of PROCOMM PLUS is beyond the scope of this book, this chapter is intended to help you develop a good feel for the overall capabilities of ASPECT. Use this chapter to learn the basic ways you can create, run, and debug (correct errors in) ASPECT programs and to discover the wide array of ASPECT commands available. This information will give you a good start at learning to program with ASPECT.

What Is ASPECT?

ASPECT is a high-level communications programming language by means of which you create structured programs that control PROCOMM PLUS. ASPECT is the language used by the PROCOMM PLUS Record mode, which is discussed in Chapter 7. By modifying scripts created by Record mode and by writing your own scripts, you can develop many useful communications applications.

The ASPECT script language is similar to the QuickBASIC programming language. If you are already experienced at programming in QuickBASIC, the structure and commands of ASPECT should be easy for you to understand. If you are a novice at programming, however, you should take time to learn some fundamental programming concepts before you try to develop a complex script with ASPECT. Either of the following books is a good resource for programming fundamentals:

- *Using QuickBASIC 4* by Phil Feldman and Tom Rugg

- *Using Turbo Pascal* by Michael Yester

After you have gained expertise, you may want to advance your knowledge by reading *QuickBASIC Advanced Techniques*, by Peter Aitken. All three books are published by Que Corporation.

Understanding ASPECT Scripts

ASPECT programs are referred to as *scripts*. Chapter 7 first introduces you to scripts and demonstrates how to record as a script the log-on sequence to connect to another computer. Chapter 7 also discusses several sample script files, which are distributed on the PROCOMM PLUS disks. In addition, Chapter 7 explains how to edit a script and how to run a script from start-up, from the Dialing Directory screen, and from the Terminal mode screen. This chapter picks up where Chapter 7 leaves off, introducing you to all the commands and capabilities available in ASPECT.

A PROCOMM PLUS script file consists of a series of lines of ASCII characters, each line containing one command from the ASPECT script language. Chapter 7 introduces you to three of these commands: WAITFOR, PAUSE, and TRANSMIT. PROCOMM PLUS executes commands one by one from top to bottom, unless the program encounters a command that causes execution to branch to some other portion of the script. Any text or other characters that appear to the right of a semicolon (;) are ignored when PROCOMM PLUS executes the script. This text is used to incorporate comments, often called *internal documentation*, into the script for your future reference. The DOS file name of each ASPECT script must end in the extension .ASP.

ASPECT scripts are used to control and automate the operation of PROCOMM PLUS. They simplify even further the use of this already easy-to-use communications program. For example, you can create several scripts, executed through the dialing directory, each of which logs on to a different computer system. Or you can create a single script that condenses the use of PROCOMM PLUS down to choosing options from a menu system of your own design. You can even create scripts that prompt the user for textual input, which, in turn, is used by PROCOMM PLUS in the script.

Comparing ASPECT to Other Languages

The ASPECT programming language is similar in many ways to traditional programming languages like QuickBASIC. Three important similarities are

- *Placement of commands.* ASPECT is a free-form programming language. Because you can start a command anywhere on a line, you can indent commands to enhance code readability. You cannot, however, place more than one ASPECT command on one line or split one command on several lines.

- *Modularization.* ASPECT enables you to *modularize* your programs. That is, you can break the program into more manageable pieces by using one script to execute one or more other scripts. In ASPECT, this technique is referred to as *chaining* script files.

- *Structure.* You can use ASPECT program control commands to create *structured* programs, scripts that are internally modularized.

In spite of these similarities, ASPECT is not exactly like every other programming language. To learn to use ASPECT successfully and efficiently, you need to become familiar with its features and capabilities. But after all is said and done, the only way to learn to program is to roll up your sleeves and write programs. This chapter will help you get started.

Creating an ASPECT Script

PROCOMM PLUS provides two ways to create an ASPECT script: you can use PROCOMM PLUS's Record mode, or you can use a text editor to modify an existing script or to write a script from scratch.

Record mode records your keystrokes while you are on-line to a remote computer. Record mode also records the remote system's prompts. This mode is ideal for recording the log-on sequence for connecting to another computer. (Refer to Chapter 7, "Automating PROCOMM PLUS with Macros and Script Files," for a complete discussion of using the Record mode.)

You can write a script from scratch or edit a script by using PCEDIT or any other text editor or word processor that can produce an ASCII file. The file name extension of each script file must be .ASP. Lines can be up to 132 characters long. Because you can start a command anywhere on a line, you can enhance code readability by indenting some commands. You cannot, however, have more than one ASPECT command on one line or break up one command on to several lines.

You can type ASPECT commands in upper- or lowercase letters, but the commands must be completely spelled out. The available ASPECT commands are simply introduced in this chapter. All the commands are listed and completely explained in the PROCOMM PLUS documentation.

Internally documenting, or explaining the operation of, the script is easy because PROCOMM PLUS ignores anything typed to the right of a semicolon (;) unless the semicolon is part of text enclosed in double quotation marks (for example, "Sam;Spade;Falcon"). Blank lines do not affect the operation of a script, so you can use them as often as you like to make the script easier to read.

You may want to take a look at the scripts supplied on the PROCOMM PLUS Supplemental Diskette and at the examples presented in this chapter and in the PROCOMM PLUS documentation in order to get a better idea of how ASPECT is used. (See table 7.3, for a list of the sample ASPECT script files supplied with PROCOMM PLUS.)

Running a Script

PROCOMM PLUS provides numerous options for executing, or running, the script. You can run an ASPECT script from the DOS prompt at start-up of PROCOMM PLUS, from the Dialing Directory screen, from the Terminal mode screen, from within another script, or from another computer. You can even put a script on a "timer" so that the script will start and execute on its own at a predetermined time.

Once PROCOMM PLUS begins executing the script, the program displays the name of the script in the left section of the Terminal mode screen status line.

From the DOS Prompt

PROCOMM PLUS provides three ways to run a script from DOS. All three methods cause a script to run immediately when PROCOMM PLUS is started.

First, you can cause PROCOMM PLUS to run a script immediately on start-up by creating a script with the file name PROFILE.ASP. Change to the DOS directory that contains this script. Then type *pcplus*, and press Enter. PROCOMM PLUS loads and stops at the PROCOMM PLUS logo screen. After you press any key, PROCOMM PLUS displays the Terminal mode screen and then runs the script. A script named PROFILE.ASP runs *every* time you start PROCOMM PLUS.

Second, you can run any ASPECT script from DOS by adding the script name to the PROCOMM PLUS start-up command. Change to the DOS directory that contains the script you want to run. Type the following:

pcplus /f*scriptname*

In this command, replace *scriptname* with the file name of the script. You do not have to type the .ASP file name extension. Do not leave a space after /*f*. When you press Enter, PROCOMM PLUS starts and immediately plays the script, without stopping at the logo screen. If you also create a script named PROFILE.ASP in the same DOS directory, PROCOMM PLUS executes PROFILE.ASP first and then runs the script you specify in the start-up command.

For example, to run from DOS a script stored in the file MENU.ASP, type the following and press Enter:

pcplus /fmenu

The third way to start PROCOMM PLUS from DOS employs a special utility program, TEF.EXE, included on the PROCOMM PLUS Supplemental Diskette. TEF stands for *Timed Execution Facility*. This program is like the automatic timers built into many modern kitchen appliances, such as coffee pots and microwave ovens. TEF lets you set the time for the script to execute. This feature is most often used to perform lengthy uploads and downloads at off-peak hours when host computer usage is usually lighter and connect charges are often less expensive.

To use this facility, change to the DOS directory that contains the script file you want to execute. Type *tef*, and press Enter (Note: If you are using a monochrome monitor with a color graphics adapter or are using a laptop computer with a monochrome LCD screen, type *tef /b*, and press Enter). PROCOMM PLUS displays the TEF screen shown in figure 11.1. Type the name of the script (with or without the .ASP file name extension), and press Enter. TEF then prompts, Enter the hour to start execution (0-23). Type the hour you want the script to execute, using a 24-hour clock (military time—midnight is 0). When you press Enter, TEF displays the prompt, Enter the minute to start execution (0-59). Type the minute you want the script to execute, and press Enter again. TEF then displays a screen similar to figure 11.2, indicating that TEF will execute the script at the specified time. When the TEF timer reaches the appointed time, TEF starts PROCOMM PLUS and immediately runs the script.

As an alternative, you can also type all the necessary parameters at the DOS command line. For example, to start PROCOMM PLUS at 23:30 (11:30 p.m.) and run the MCI.ASP file, you can type the following command at the DOS prompt and press Enter:

tef mci 23:30

TEF then displays a screen similar to figure 11.2. At 23:30, TEF starts PROCOMM PLUS and executes the script.

TIP | In order for PROCOMM PLUS to start from DOS by any method, either the script must be in the same DOS directory as the PROCOMM PLUS program files, or the DOS environment must contain a variable named PCPLUS that points to the directory containing the PROCOMM PLUS program files. Refer to Appendix A for instructions on using the SET PCPLUS= command in your AUTOEXEC.BAT file in order to add PCPLUS to the DOS environment.

From the Dialing Directory

If you have created a script and want PROCOMM PLUS to run that script when the program connects to a particular computer, type the name of the script in the SCRIPT line of the Dialing Directory's Revise Entry window. PROCOMM PLUS runs the script every time you connect to the remote computer. Refer to Chapter 3, "Building Your Dialing Directory," for more information on using scripts from the Dialing Directory screen.

Most of the example scripts supplied on the PROCOMM PLUS Supplemental Diskette are intended to be used from the Dialing Directory screen.

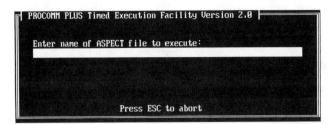

Fig. 11.1.

The TEF screen.

```
Awaiting execution of ASPECT file: MCI.ASP

Current time: 17:09:16.  Execution time: 23:30:00.

Press ESC to abort.  Press F1 to change.  Press F2 to clear the screen.
```

Fig. 11.2.

Timed execution of an ASPECT script.

From Terminal Mode

The third way to execute an ASPECT script is available from the Terminal mode screen. When you press Alt-F5 (Script Files), PROCOMM PLUS displays a narrow window across the middle of the screen. The top border line includes the message SCRIPT SELECTION (Enter for .ASP file list), and the window prompts you to Please enter filename.

You have two alternatives for indicating which script you want to run:

- Type the name of the script file, and press Enter. PROCOMM PLUS assumes that the script has the file name extension .ASP, so you don't need to type it. PROCOMM PLUS executes the script.

- If you are not sure of the script file name's exact spelling, press Enter to see a list of all the script files in the working directory (all files having the .ASP file name extension). PROCOMM PLUS displays up to 15 script file names in a tall thin window (see fig. 11.3). The file name at the top of the list is highlighted (using high-intensity characters, not inverse video). While this second window is displayed, you can use the up arrow, the down arrow, PgDn, and PgUp to scroll the highlight to any ASPECT file name on the working directory. Position the highlight on the name of the script you want to run, and press Enter. PROCOMM PLUS runs the script.

Fig. 11.3.

Selecting a script to run from the Terminal mode screen.

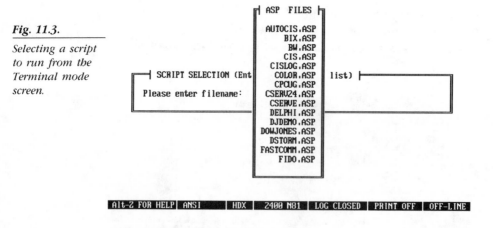

For example, to see a listing of all the screen colors that can be used in ASPECT display commands (refer to "Display and Printer Commands," in this chapter), you can execute the COLOR.ASP script, which is distributed on the Supplemental Diskette. Press Alt-F5 (Script Files), type *color*, and press Enter. The script causes your computer to beep twice and then display 256 different display-attribute combinations.

From within Another Script

You can also run a script from within another script; this technique is sometimes called *chaining* scripts. The ASPECT command for running a script is EXECUTE. You can chain together an unlimited number of ASPECT scripts by using this command. The first script can call a second script, the second script can call a third, and so on. The following ASPECT command plays the script named MYSCRIPT (be sure to enclose the script name in quotation marks):

EXECUTE "myscript"

When one ASPECT script executes another ASPECT script, the first script is no longer active. ASPECT does not have a built-in facility for returning from the second script to the first script. You can, however, easily program around this apparent limitation (refer to the sample scripts given later in this chapter). The end result is that you can break ASPECT scripts into smaller, easier-to-manage modules instead of having to program an entire application in one script. Creating programs in small manageable scripts is sometimes referred to as *modularization*.

Aborting Execution of a Script

From time to time, you may decide that you didn't really want to run that script after all. To abort the script in midstream, press Esc. PROCOMM PLUS displays a small window in the upper left portion of the Terminal mode screen. This window contains the prompt EXIT SCRIPT? (Y/N). Respond Yes and PROCOMM PLUS terminates the script and displays the message COMMAND FILE ABORTED. Respond No and the script resumes.

Editing and Debugging a Script

PROCOMM PLUS provides a built-in editor, PCEDIT, which you can use to create and edit ASPECT scripts. You can also use another editor, if you have a favorite. PCEDIT is a little more tightly integrated into PROCOMM PLUS than any other editor. For example, PCEDIT contains 20 keyboard macros that type ASPECT commands. Because the program's editing features are limited, however, PROCOMM PLUS lets you conveniently attach another editor for use in editing ASPECT scripts.

Refer to Chapter 6, "Using the PROCOMM PLUS Editor," for more information on using PCEDIT and for instructions for attaching another editor to PROCOMM PLUS.

PROCOMM PLUS provides a special command, TRACE ON, to help you find errors in an ASPECT script. When a script is not working in the manner you intended, place a TRACE ON command in a line by itself above the script commands you want to check. When you execute the script, starting just below the TRACE ON command, PROCOMM PLUS displays each ASPECT command as it is executed. You can then compare the actual results of each command with the results you intended.

When you want to check only a portion of a script, place the TRACE ON command above the commands to be checked, and place the TRACE OFF command below the commands to be checked. PROCOMM PLUS then displays only the commands that fall between the two TRACE commands.

Once you determine which command is causing the problem, use your editor to correct the mistake and to delete the TRACE command(s).

An Overview of the ASPECT Commands

PROCOMM PLUS provides more than 90 ASPECT script commands. These commands, used to develop scripts, fall into ten groups (some commands fall into more than one category):

- Communications commands
- Program control commands
- Branching commands
- Variable manipulation commands
- String manipulation commands
- Mathematical commands
- Disk and file commands
- Display and printer commands
- System setup commands
- Other commands

The next sections of this chapter briefly introduce you to these commands and can help you gain a general understanding of the types of ASPECT commands available. Refer to Chapter 11, "ASPECT Script Language Reference," in the PROCOMM PLUS documentation for a complete command reference, which shows you proper syntax and provides explanations of how to use all these commands.

Many ASPECT commands have direct counterparts in PROCOMM PLUS. In other words, a number of the ASPECT commands duplicate the effect of executing certain keyboard commands interactively from the Terminal mode screen. These ASPECT commands, along with the equivalent PRO-COMM PLUS commands are listed in table 11.1.

Table 11.1
ASPECT Communications Commands

ASPECT Command	Equivalent PROCOMM PLUS Command
BREAK	Alt-B (Break Key)
CLEAR	Alt-C (Clear Screen)
*BYE	Alt-X (Exit)/No
CHDIR	Alt-F7 (Change Directory)
DIAL	Alt-D (Dialing Directory)/D (Dial Entry(s))
*DIR	Alt-F (File Directory)
*DLOAD	Alt-D (Dialing Directory)/X (Exchange Directory)
EXECUTE	Alt-F5 (Script Files)
GETFILE	PgDn (Receive Files)
HANGUP	Alt-H (Hang Up)
HELP	Alt-Z (Help)
HOST	Alt-Q (Host Mode)
KERMSERVE	Alt-K (Kermit Server Cmd)
LOG OPEN/CLOSE	Alt-F1 (Log File On/Off)
LOG SUSPEND/RESUME	Alt-F2 (Log File Pause)
MACRO	Alt-0 through Alt-9
*MDIAL	Alt-D (Dialing Directory)/M (Manual Dial)
MLOAD	Alt-M (Keyboard Macros)/L (Load)
PRINTER	Alt-L (Printer On/Off)
*REDIAL	Alt-D (Dialing Directory)/D (Dial Entry(s))/Enter
SENDFILE	PgUp (Send Files)
SET	Alt-S (Setup Facility)
SET	Alt-P (Line/Port Setup)
*SHELL	Alt-F4 (DOS Gateway)
SNAPSHOT	Alt-G (Screen Snapshot)
*TYPE	Alt-V (View a File)

The * in this and subsequent command lists in this chapter indicates that the command is available only in PROCOMM PLUS and not in ProComm.

Communications Commands

PROCOMM PLUS provides 12 ASPECT commands that are involved specifically in the communication of data between your computer and a remote computer. These commands dial a remote computer, transfer data to or from a remote computer, or disconnect your computer from a remote computer. (Refer to Chapter 11, "ASPECT Script Language Reference," in the PROCOMM PLUS documentation for an explanation of the syntax shown in this and subsequent command lists.) These commands are

```
 BREAK
*BYE
 DIAL
 GETFILE
 HANGUP
 KERMSERVE SENDFILE|GETFILE|FINISH|LOGOUT
*MDIAL
*REDIAL
 RGET
 SENDFILE
 TRANSMIT
```

Program Control Commands

The commands in this category let you establish conditions for performance of particular commands in an ASPECT script. You can use program control commands to cause the script to perform a particular group of commands based on the occurrence of a given condition, to stop processing script commands temporarily until a specified time or until the occurrence of a specified event, or to quit processing the script entirely. This group includes the following commands:

```
*BYE
 CASE
*CONNECT
 CWHEN
 EXIT
 IF
 PAUSE
 QUIT
*SUSPEND
 SWITCH
*TERMINAL
 WAITFOR
 WHEN
```

Branching Commands

The ASPECT branching commands enable you to control the sequence in which script commands are executed. Without these commands, scripts execute from top to bottom, just as you read a printed page. You can use these branching commands to cause script execution to branch to another line in the script, to cause script execution to perform commands in another portion of the script and return, or to run another ASPECT script. This group includes the following commands:

EXECUTE
GOSUB
GOTO

Variable Manipulation Commands

These commands enable you to create and use temporary fields, referred to as *variables*. ASPECT variables exist only in memory (RAM) and are used in scripts to hold information temporarily so that the information can be used in some way by the script. With these commands, you can assign a value to a variable, accept input typed on the screen, get the system date or the system time, read information from an input file, and search for text within a variable. The following commands are used to create and manipulate variables in ASPECT scripts:

ASSIGN
*ATGET
*DATE
*FGETC
*FGETS
FIND
*FREAD
GET
*INIT
*KEYGET
*MATGET
*MGET
RGET
*TIME

String Manipulation Commands

A *text string* (or just *string*) is a group of characters in a specific order. The name *Sam Spade* is a string; and so is the telephone number *555-3981*. Communication between two computers by definition involves transfer of text strings in both directions, so PROCOMM PLUS provides a number of ASPECT commands that manipulate these text strings. When you use a text string in an ASPECT command, enclose the text string in double quotation marks. Each string can be no more than 80 characters long.

The following commands are used to perform such operations as assign a value to a string variable, convert a text string to a numeric value, convert a numeric value to a text string, combine text strings, compare text strings, and extract a portion of a string:

ASSIGN
*ATOI
*ITOA
*STRCAT
*STRCMP
*STRCPY
*STRFMT
*SUBSTR

Disk and File Commands

In addition to commands that manipulate text strings, PROCOMM PLUS provides commands that work with DOS files. The following commands enable you to change the current working directory, display a list of files, open and close input and output files, read information from an input file, write to an output file, save a screen snapshot, save the session to a log file, and display a file to a screen:

CHDIR
*DIR
*FCLOSEI
*FCLOSEO
*FGETC
*FGETS
*FOPENI
*FOPENO
*FPUTC

*FPUTS
*FREAD
*FWRITE
 ISFILE
 LOG OPEN
 SNAPSHOT
*TYPE

Mathematical Commands

Occasionally, you want a script to manipulate numeric values. For example, you may want the script to count the number of messages it downloads from an electronic mail system. PROCOMM PLUS provides the following ASPECT commands that perform such operations as add, subtract, multiply, divide, increment, and decrement numeric values:

*ADD
*DEC
*DIV
*INC
*MUL
*SUB

Display and Printer Commands

The commands in this category enable you to take charge of what information PROCOMM PLUS displays on the screen and sends out to the printer. The following commands are in this display and printer command group:

*ATSAY
*BOX
 CLEAR
*CUROFF
*CURON
*FATSAY
 LOCATE
 MESSAGE
 PRINTER
*SCROLL
 TRACE
*TYPE

System Setup Commands

When you use PROCOMM PLUS interactively, you can customize nearly every program parameter. The ASPECT commands in this group enable you to perform the same customization by using an ASPECT script. The commands load a different dialing directory, load a different keyboard macro file, set the current terminal emulation, and change any setting that can interactively be set through the Setup Utility or through the Line/Port Setup screen:

> *DLOAD
> EMULATE
> MLOAD
> SET

Other Commands

The ASPECT script language also provides a number of commands that do not fit neatly into one of the other categories. The commands in this miscellaneous category enable you to cause PROCOMM PLUS to send sound to your computer's speaker, suspend PROCOMM PLUS temporarily and access DOS, execute DOS commands and other DOS programs from within PROCOMM PLUS, activate Host mode, clear the keyboard buffer and the input buffer, and execute a keyboard macro:

> ALARM
> DOS
> HOST
> KFLUSH
> MACRO
> RFLUSH
> RUN
> *SHELL
> *SOUND

Compatibility with ProComm Command Files

In PROCOMM PLUS, you can use with only slight modification script files (known as *command* files) that were created for use with ProComm. The most obvious difference between ProComm command files and PROCOMM PLUS script files is the file name extension. ProComm command files have the file name extension .CMD, and PROCOMM PLUS scripts have the file name extension .ASP. When you want to use a ProComm command file with PROCOMM PLUS, use the DOS REN command to change the file name extension on the ProComm command file from .CMD to .ASP.

Replace any exclamation point (!) intended to represent a carriage return with the code ^M (for Ctrl-M). Replace the vertical bar (|) character, used in ProComm to send the Esc character, with the code for Ctrl-left bracket (^[). Also be sure that all commands are spelled out.

Unless your ProComm command file uses the SET command, you can now run the file as an ASPECT script file without further modification. The SET command, however, has four variations that must be edited or eliminated in order to run in PROCOMM PLUS. The syntax of the SET BACKSPACE and SET CR commands no longer permits use of the word IN. For example, the proper syntax for the old command SET BACKSPACE IN DEST is now SET BACKSPACE DEST, and the proper syntax for the old command SET CR_IN CR is now SET CR CR. The old SET FLOWCTRL ON (or OFF) command is now SET SOFTFLOW ON (or OFF). Finally, the old SET BACKSPACE OUT and SET CR_OUT commands are not supported at all as ASPECT commands in PROCOMM PLUS. Refer to Chapter 11, "ASPECT Script Language Reference," in the PROCOMM PLUS documentation for more information about when to use these SET commands.

Looking at an Example

Although the intention of this chapter is not to teach you how to program, this section briefly examines two simple ASPECT scripts that demonstrate several commonly used programming techniques. Recall from Chapter 7 that DATASTORM also provides a number of useful scripts on the two distributed PROCOMM PLUS disks. Each distributed script serves as an excellent example of ASPECT programming.

Returning again to the Terry's T-Shirts, Inc., scenario introduced in Chapter 9, you are still the national sales manager of this T-shirt manufacturing company. You have recently used PROCOMM PLUS's Host mode to implement a small electronic bulletin board for your sales managers. Now, you want to write a series of scripts that will completely automate the dialing and log-on procedure for each manager.

Your plan is to write an ASPECT script that will display a Dialing menu as soon as PROCOMM PLUS starts. The menu for the Western regional sales manager must contain the following options:

```
(1)   Dial Headquarters' BBS
(2)   Dial Northern Region
(3)   Dial Southern Region
(4)   Dial Eastern Region
(5)   QUIT
```

All Dialing menus will be similar to this one. For instance, the menu for the Northern region will look the same except Dial Northern Region will be replaced by Dial Western Region.

You decide to modularize the scripts as much as possible. You could create one long script with multiple subroutines, but a long script can become unwieldy. In order to make the scripts clearer and easier to work with, you decide to create one script that will display and accept input from the Dialing menu and a completely separate script for each of the first four options on the menu. One script will dial the Headquarters' BBS; another script will dial the Northern Region; and so on. The last option, (5) QUIT, requires only a single ASPECT command so doesn't require a separate script. Each subroutine script runs the menu script again when the sales manager disconnects from the host system.

The menu script you create, named PROFILE.ASP (so that it will run each time PROCOMM PLUS runs), is shown in figure 11.4. This script demonstrates the use of a number of ASPECT commands, including CLEAR, BOX, ATSAY, LOCATE, KEYGET, MESSAGE, and SWITCH. Each line in the script is explained by the comment to its right.

Fig. 11.4.

The Dialing menu script PROFILE.ASP.

```
*****************************************************************************
*  Script to display a menu with 5 options. Each of the first four options  *
*  executes another script. Each of these subroutine scripts, in turn, runs *
*  this script on disconnecting from the remote computer.                    *
*****************************************************************************


;-----------------------------------------------------------------------
;  Draw Menu
;-----------------------------------------------------------------------
CLEAR                                                    ; Clear screen
BOX 3 20 17 60 112                                       ; Draw box
ATSAY 4 21 112  "   TERRY'S PROCOMM PLUS DIALING MENU"   ; Menu title
ATSAY 5 21 112  "----------------------------------------"  ; Box-drawing char.
```

```
ATSAY 7 21 112  "    (1)  DIAL HEADQUARTERS' BBS"    ; 1st menu option
ATSAY 9 21 112  "    (2)  DIAL NORTHERN REGION"      ; 2nd menu option
ATSAY 11 21 112 "    (3)  DIAL SOUTHERN REGION"      ; 3rd menu option
ATSAY 13 21 112 "    (4)  DIAL EASTERN REGION"       ; 4th menu option
ATSAY 15 21 112 "    (5)  QUIT"                      ; 5th menu option
LOCATE 19 33                                         ; Move cursor
MESSAGE "YOUR CHOICE?"                               ; Prompt for input
LOCATE 19 45                                         ; Put cursor after prompt

CHOOSE_OPTION:                          ; Label
   KEYGET SO                            ; Accept input into variable SO
   SWITCH SO                            ; Start SWITCH using user input (SO)
      CASE "1"                          ; What to do when user selects option 1
         EXECUTE "HQ"                   ; Run the HQ.ASP script
      ENDCASE
      CASE "2"                          ; What to do when user selects option 2
         EXECUTE "NORTH"                ; Run the NORTH.ASP script
      ENDCASE
      CASE "3"                          ; What to do when user selects option 3
         EXECUTE "SOUTH"                ; Run the SOUTH.ASP script
      ENDCASE
      CASE "4"                          ; What do do when user selects option 4
         EXECUTE "EAST"                 ; Run the EAST.ASP script
      ENDCASE
      CASE "5"                          ; What to do when user selects option 5
         QUIT                           ; Quit PROCOMM PLUS, return to DOS
      ENDCASE
      DEFAULT                           ; What to do when user presses invalid key
         SOUND 200 25                   ; Beep
         GOTO CHOOSE_OPTION             ; Go to beginning to accept more input
      ENDCASE
   ENDSWITCH
```

The first section of the PROFILE.ASP script, beginning with the **CLEAR** command and ending with the command

```
MESSAGE    "Your Choice?"
```

draws a menu on the screen. This menu is shown in figure 11.5.

Fig. 11.5.

The Dialing menu generated by the script in figure 11.4.

The remainder of the script, beginning with the label CHOOSE_OPTION, waits for the user to select an option from the menu by pressing one of the number keys 1, 2, 3, 4, or 5. Any other key causes PROCOMM PLUS to sound a beep on your computer and continue waiting. If the user presses one of the numbers 1 through 4, PROCOMM PLUS executes the script HQ.ASP, NORTH.ASP, SOUTH.ASP, or EAST.ASP, respectively. When the user is finished using the Dialing menu, the user presses 5, and PROCOMM PLUS returns to DOS.

To select the first option from your Dialing menu, the user presses the number 1. As a result, PROCOMM PLUS executes a script named HQ.ASP. This script, shown in figure 11.6, demonstrates use of the ASPECT commands ASSIGN, ALARM, IF, CLEAR, CURON, CUROFF, LOCATE, MESSAGE, TIME, DATE, FATSAY, EMULATE, SET, EXECUTE, TRANSMIT, and WAITFOR. Each line in the script is explained by the comment to its right.

Fig. 11.6.

A script, HQ.ASP, to dial the headquarters' BBS.

```
****************************************************************************
*   Script to dial the headquarters' BBS. After disconnection or if no    *
*   connection is made, the script executes the PROFILE.ASP script to     *
*   return to the dialing menu.                                           *
****************************************************************************

;--------------------------------------------------------------------------
;  Assign values to variables for phone number, user name, and password
;--------------------------------------------------------------------------
```

```
ASSIGN S0 "ATDT555-1234^M"            ; Dial prefix, number, and suffix
ASSIGN S1 "Sam;Spade^M"               ; Assign user name to variable S1
ASSIGN S2 "Falcon^M"                  ; Assign password to variable S2


;-----------------------------------------------------------------------
;  Set up PROCOMM PLUS and dial the headquarters' BBS
;-----------------------------------------------------------------------
CLEAR                                 ; Clear screen
TIME S8 0                             ; Assign system time to variable S8
DATE S9                               ; Assign system date to variable S9
CUROFF                                ; Turn off cursor (cosmetic)
FATSAY 0 0 112 "DIALING HEADQUARTERS AT %s ON %s" S8 S9    ; Display message
EMULATE ANSI                          ; Emulate an ANSI terminal
SET BAUD 2400                         ; Set transmission speed
SET PARITY NONE                       ; Set parity
SET DATABITS 8                        ; Set databits
SET DUPLEX FULL                       ; Set duplex
SET DISPLAY OFF                       ; So we don't see OK from modem
TRANSMIT S0                           ; Dial


;-----------------------------------------------------------------------
;  Test for busy signal or no answer...
;-----------------------------------------------------------------------
WAITFOR "BUSY" 12                     ; Wait for BUSY from modem
IF WAITFOR                            ; If busy...
    LOCATE 5 1                        ;       locate cursor...
    MESSAGE "BUSY... Returning to menu"  ;    message to user...
    PAUSE 2                           ;       time to read message...
    CURON                             ;       turn on cursor and
    EXECUTE "PROFILE"                 ;        return to menu if line is busy
ELSE
WAITFOR "CONNECT" 30                  ; If not busy, wait for CONNECT
    IF NOT WAITFOR                    ; If no CONNECT from modem...
      TRANSMIT "^M"                   ;       hang up modem...
      LOCATE 5 1                      ;       locate cursor...
      MESSAGE "NO ANSWER... Returning to menu"   ;  message to user...
      PAUSE 2                         ;       time to read message and for
                                      ;          modem to recycle...
      CURON                           ;       turn on cursor...
      EXECUTE "PROFILE"               ;       and back to menu
    ENDIF
ENDIF
```

```
;-------------------------------------------------------------------------
; Log on to BBS. Wait for NO CARRIER and then return to dialing menu
;-------------------------------------------------------------------------
CLEAR                                   ; Get rid of "Dialing Headquarters..."
CURON
SET DISPLAY ON                          ; Now that we're on-line we want
                                        ;    to see incoming characters

TRANSMIT "^M"                           ; Send a carriage return
WAITFOR "First name: "                  ; Wait for "First name:" prompt
TRANSMIT S1                             ; Transmit user name
WAITFOR " correct (Y/N)? "              ; Wait for "correct (Y/N)?" prompt
TRANSMIT "Y^M"                          ; Transmit Y for Yes
WAITFOR "Password: "                    ; Wait for "Password:" prompt
TRANSMIT S2                             ; Transmit password stored in S2
ALARM 2                                 ; Beep-beep twice to indicate log-on
WAITFOR "NO CARRIER" FOREVER            ; Wait for disconnect
EXECUTE "PROFILE"                       ; Return to dialing menu when off-line
```

The HQ.ASP script is intended to dial and log on to the headquarters' bulletin board (refer to the discussion of Terry's T-Shirt, Inc., BBS, in Chapter 9). When the user exits from the bulletin board, the script runs the PROFILE.ASP script to display the Dialing menu again.

The script has four main sections. The first section, which consists of three ASSIGN commands, establishes the values of three variables: S0, S1, and S2. These variables contain the telephone number to be dialed, the user's user name, and the user's password, respectively; and they are for use in sections of the script that dial and log on to the bulletin board.

Beginning at the CLEAR command, the second section of the script sets the terminal emulation, line settings, and duplex, and then dials the remote computer.

The third section of the script begins with the WAITFOR "BUSY" 12 command and ends with the second ENDIF command. This section executes the PROFILE script, returning to the Dialing menu if the program detects a "BUSY" message from the modem or if there is no answer within 30 seconds.

The last portion of the script is similar to the script developed in Chapter 7. The script logs on to the bulletin board and sounds "beep beep" twice to indicate that the user is on-line.

The next to the last line in the script is

```
WAITFOR "NO CARRIER" FOREVER
```

This command causes the script to pause processing indefinitely. As long as the user is on-line to the bulletin board, the script simply passes all keystrokes through to the Terminal mode. The only indication to the user that a script is running is the message HQ.ASP in the left section of the Terminal mode screen status line. As soon as the user disconnects, the user's modem sends the message NO CARRIER. The WAITFOR command then springs to life, and the script continues. The last line of the script executes the PROFILE.ASP script to return to the Dialing menu.

Chapter Summary

This chapter has given you an overview of all the major features of the PROCOMM PLUS script language, ASPECT. The chapter has helped you develop a feel for the overall capabilities of ASPECT and has taught you the basic ways you can create, run, and debug (identify errors in) ASPECT programs. You should now be aware of the wide array of ASPECT commands that are available to you. Use the scripts found in this chapter, as well as the scripts supplied by DATASTORM as examples. It is now up to you to write your own scripts and to put to the best use the powerful tools that make up the ASPECT script language.

This chapter completes your guided magic-carpet ride on PROCOMM PLUS. This book has taken you from the confines of your office into the ever-expanding world of computer communications. You have learned not only how to understand and use the language of communications; more important, you have discovered that with PROCOMM PLUS you can always feel right at home, regardless of what new communications horizons you are exploring.

Installing and Starting PROCOMM PLUS

This appendix explains how to install PROCOMM PLUS and begin using the program. The material covers installation on hard disk systems and on floppy disk systems. This appendix also explains how to upgrade to PROCOMM PLUS from a previously installed version of ProComm.

Understanding System Requirements

Before installing PROCOMM PLUS, you must make sure that your system meets the program's minimum requirements. To run PROCOMM PLUS, you must be using an IBM PC, XT, AT, or PS/2 computer or compatible. Your system must have at least 192K of available memory (RAM) and DOS 2.0 or higher. PROCOMM PLUS can be run from a floppy disk drive or from a hard disk drive. You also must have a modem installed in your computer (an internal modem) or connected to one of your system's serial ports (an external modem). If you have a choice, the modem should recognize the Hayes AT command set (usually referred to as a Hayes-compatible modem), but virtually any modem can be used with PROCOMM PLUS.

Installing PROCOMM PLUS on a Hard Disk System

Two methods are available for installing PROCOMM PLUS on a hard disk system: the Installation Utility and the DOS COPY command. The best method is the PROCOMM PLUS Installation Utility, PCINSTAL.EXE. This method enables you easily to set several PROCOMM PLUS parameters so that the program will run properly on your computer the first time.

Using the Installation Utility

To use the PROCOMM PLUS Installation Utility, place the distributed PROCOMM PLUS Program Diskette into drive A of your computer. At the DOS prompt, type *a:pcinstal*, and press Enter.

The program displays the logo PCINSTALL and a copyright notice. Press any key to display a screen of instructions entitled PROCOMM PLUS INSTALLATION UTILITY at the top left, and PROGRAM FILE INSTALLATION at the top right of the screen. Read these instructions, and then press any key except Esc to continue.

The Installation Utility program first prompts you for the Source drive for PROCOMM PLUS program diskette:. The default response is A. You have already placed the Program Diskette into drive A, so press Enter.

Next, the installation program prompts for the Destination path for PROCOMM PLUS files: and suggests a response of C:\PCPLUS, the standard directory used to hold PROCOMM PLUS program files. Unless you have a good reason to do otherwise, press Enter to accept this destination path. The installation program checks your hard disk for the C:\PCPLUS directory and creates it if it does not already exist. The installation program then begins copying the PROCOMM PLUS program files to the C:\PCPLUS directory on your hard disk.

After the Installation Utility has finished copying the program files to your hard disk, the program displays a second screen of instructions. This screen explains the purpose of the configuration screen that will follow. Press any key except Esc to display the Program Startup Settings screen shown in figure A.1.

```
┌─────────────────────────────────────────────────────────────────────┐
│ PROCOMM PLUS INSTALLATION UTILITY              PROGRAM STARTUP SETTINGS │
│                                                                       │
│      Monitor:                    Monitor:                             │
│        Modem:                                                         │
│     Terminal:                    1- Color                            │
│    Comm Port:                    2- Composite B&W                    │
│    Baud Rate:                    3- Monochrome                       │
│       Parity:                                                        │
│    Data Bits:                                                        │
│    Stop Bits:                                                        │
│       Duplex:                                                        │
│     Protocol:                                                        │
│                                                                       │
│  ┌ MONITOR ────────────────────────────────────────────────────────│
│                                                                       │
│   Specify the type of monitor that you are using.  Select COLOR if you are │
│   using a color monitor with a color graphics card.  PROCOMM PLUS will come │
│   alive in a blaze of exciting colors.  Select COMPOSITE B&W if you have a │
│   black and white composite monitor with a graphics card.  Select MONOCHROME │
│   if you are using a monochrome monitor with a monochrome dislay card. │
└─────────────────────────────────────────────────────────────────────┘
```

Fig. A.1.

The Program Startup Settings screen.

The Program Startup Settings screen is divided into three sections. The first section, at the top of the screen, consists of a one-line title. The middle section of the screen is the largest. This section contains a list of program parameters, which are all blank at first. These parameters include the following:

Monitor
Modem
Terminal
Comm Port
Baud Rate
Parity
Data Bits
Stop Bits
Duplex
Protocol

The bottom portion of the screen displays instructions.

The purpose of the Program Startup Settings screen is for you to fill in this list of parameters with values that match your equipment and your normal operating procedure. You can later modify any or all of these settings by using the PROCOMM PLUS Setup Utility, discussed in Chapter 8 of this book.

When you first display the Program Startup Settings screen, the cursor is blinking to the right of the prompt Monitor on the right side of the screen. Below this prompt, the Installation Utility lists three options: **1** (Color), **2**

(Composite B&W), and 3 (Monochrome). Press a number to select the option that matches the type of monitor and display adapter connected to your system, and press Enter.

TIP Use the 2 (Composite B&W) option when you have any type of two-color monitor connected to a display adapter that generates a multicolor signal. For example, you may have a so-called paper-white monitor connected to a VGA adapter, or you may be using a laptop computer with a monochrome LCD (liquid crystal display) screen. In either of these two cases, use the 2 (Composite B&W) option.

The Program Startup Settings screen initially displays in black and white. If you select the 1 (Color) setting for the Monitor parameter, the screen switches to a blue background with light gray letters.

Once you select a monitor setting, the Installation Utility adds the selection to the list of settings on the left side of the screen. The utility then displays the prompt Modem type on the right side of the screen, followed by a list of modems. Only eight modems are listed in the first group. You can display another seven modems by pressing Enter. The complete list of modems is as follows:

1- USRobotics Courier 2400
2- Hayes Smartmodem 1200/1200B
3- Hayes Smartmodem 2400/2400B
4- Novation Professional 2400
5- Multitech Multimodem 224/224E
6- Universal Data Fastalk 2400
7- Microcom AX/1200 & AX/2400
8- IBM PC 2400 bps Modem
9- Ven-Tel Half Card
10- Ven-Tel Half Card 24
11- IBM PCjr Internal Modem
12- Racal-Vadic VA212
13- Cermetek 1200/1200 SPC
14- Cermetek 2400R/2400 SPC
15- Cermetek Info-Mate 212A

Type the number for your modem and press Enter. The Installation Utility adds the modem type to the list of parameters on the left side of the screen.

TIP | If your modem is not listed, select the one that is closest to your modem. For example, the USRobotics Microlink 2400 uses the same commands as the USRobotics Courier 2400. When in doubt, select the Hayes modem with the same maximum speed. Refer to your modem's manual and to "Setting Modem Options," in Chapter 8, "Tailoring PROCOMM PLUS," to determine whether you need to modify the modem commands inserted by the installation program.

Next, the Installation Utility prompts you to select a Terminal type. This selection determines the default terminal emulation setting that PRO-COMM PLUS will use for each dialing directory entry (terminal emulation is discussed fully in Chapter 10, "Terminal Emulation"). As explained in Chapter 3, "Building Your Dialing Directory," you can select a different terminal emulation for each dialing directory entry. The selection you make now should be the emulation you expect to use the majority of the time. You can, however, use the Setup Utility, discussed in Chapter 8, to change this default setting later if you decide that you made the wrong choice during installation. The available choices are as follows:

1- VT52
2- VT102
3- ANSI
4- HEATH 19
5- IBM 3101
6- ADDS VP
7- ADM 5
8- TVI 910
9- TVI 920
10- TVI 925
11- TVI 950
12- TVI 955
13- WYSE 50
14- WYSE 100
15- 3270/950
16- TTY

Type the number for the terminal emulation you want to use and press Enter.

TIP | If you don't know at this point which terminal emulation to choose, select **3** (ANSI). This emulation is best for use when connecting to PC-based electronic bulletin boards. For information services such as CompuServe, however, the **2** (VT102) emulation is the best choice.

After you select the terminal setting, the Installation Utility prompts you to select a Comm port. Eight options are listed, **1** (COM1) through **8** (COM8). Determine to which serial port your modem is connected or the COM port to which your internal modem is assigned (refer to Appendix B, "Installing a Modem"). Press the number that matches the port you are using, and press Enter.

TIP | This setting is critical. Make sure that you indicate the correct COM port, or PROCOMM PLUS will not communicate with your modem, and your modem will neither dial nor connect to another computer.

Baud rate is next. Choose between the following options:

1- 300
2- 1200
3- 2400
4- 4800
5- 9600
6- 19200

Each of these numbers represents a data transmission rate. Select the rate that matches the maximum capability of your modem. For example, if you have a 1200-bps modem, choose **3** (1200). For error-control and data-compression modems (such as MNP, V.42, and V.42bis-compliant modems), the data transfer rate between your computer and the modem should be higher than the modem's nominal transmission rate. Consult the modem's manual for the proper setting to achieve optimum performance.

The next three parameters are collectively called the line settings: Parity, Data Bits, and Stop Bits. If you plan to connect primarily to PC bulletin boards, choose **1** (NONE) parity, **1** (8 data bits), and **1** (1 stop bit). When you expect to connect mainly to on-line information services or electronic mail services, choose **3** (EVEN) parity, **2** (7 data bits), and **1** (1 stop bit).

Once you have selected the line settings, choose between **1** (FULL), and **2** (HALF) duplex. Most bulletin boards and on-line services require full duplex.

Finally, choose a default file transfer protocol. The following choices are available:

1- NONE
2- MODEM7
3- YMODEM
4- TELINK

5- XMODEM
6- YMDM BAT
7- KERMIT
8- ASCII
9- CIS B
10- WXMODEM
11- YMDM-G
12- YMDM-G BAT
13- IMODEM
14- SEALINK

XMODEM is the most widely supported of the available protocols, but YMODEM is much faster and is supported by most popular bulletin boards.

Like the terminal emulation setting, the baud rate, line settings, duplex setting, and transfer protocol default settings are overridden by a dialing directory entry that has a different settings.

After you select the default file-transfer protocol, the Installation Utility asks Are the settings OK? (Y/N). If you want to correct one of the settings, answer No and start again at the top. When the settings are fine, respond Yes. The Installation Utility asks Create the PCPLUS.PRM and PCPLUS.DIR files? (Y/N). Respond Yes and the Installation Utility creates the two files. PCPLUS.PRM contains the default settings you have selected and all the other PROCOMM PLUS settings that can be customized (using the Setup Utility, discussed in Chapter 8). PCPLUS.DIR is the initial dialing directory (discussed in Chapter 3).

The PROCOMM PLUS Installation Utility displays two final screens of instructions. The first screen describes modifications you should make to the CONFIG.SYS file on your computer. Once you read the first screen, press any key to see the next screen. This second screen explains special modem setup considerations. Refer to "Modifying CONFIG.SYS and AUTO-EXEC.BAT," in this appendix, for more about CONFIG.SYS. Refer also to Appendix B for details on how to set up your modem. After you read the second screen, press any key to exit from the Installation Utility and to return to DOS.

At this point, your hard disk directory C:\PCPLUS contains the files listed in table A.1.

Table A.1
Files Copied to Hard Disk by the Installation Utility

File Name	Description
PCPLUS.EXE	The PROCOMM PLUS program
PCSETUP.EXE	The Setup Utility program
READ.ME	Supplemental information not in manual
PCPLUS.KBD	Keyboard mappings for terminal emulations
DSTORM.ASP	An ASPECT script to dial DATASTORM BBS
PCPLUS.USR	User file for Host mode
PCEDIT.EXE	PCEDIT text editor
PCPLUS.PRM	Parameter file
PCPLUS.DIR	Dialing directory file

The PROCOMM PLUS Supplemental Diskette also includes files that support or can be used with certain special PROCOMM PLUS features. These files are listed in table A.2. along with a description of each file's purpose and a notation of where the file is covered in this book (if the file is covered). You don't need to copy any of these files to your hard disk until you decide to use the function that requires that file.

Table A.2
Files Contained on the Supplemental Diskette

File Name	Description	Discussed in
Utility Programs		
PCKEYMAP.EXE	Keyboard mapping utility	Chapter 10
PCMAIL.EXE	Host mail management utility	Chapter 9
CVTDIR.EXE	Utility to convert ProComm dialing directories to PROCOMM PLUS format	Appendix A
SORTDIR.EXE	Dialing Directory sort utility	Chapter 3
TEF.EXE	Timed Execution Facility	Chapter 11
DT_ATCH.EXE	Program patching utility	Appendix D
Aspect Script Files		
*.ASP	Refer to table 7.3	Chapter 7

Table A.2—*Continued*

File Name	Description	Discussed in
PCPMCI.EXE	Self-extracting compressed file containing 5 scripts for using MCI Mail	

Supplemental Documentation

File Name	Description	Discussed in
README.TOO	Description of disk contents	
PCPLUS.NEW	List of new features and bug fixes	
DT_ATCH.DOC	Description of DT_ATCH.EXE utility	Appendix D
SUPPORT	Announcement of GEnie DATASTORM Roundtable	
PCPLUS.HHP	Sample Host mode help file	Chapter 9

Program Information Files

File Name	Description	Discussed in
PCPLUS.PIF	Windows/Topview PIF file for PCPLUS.EXE	
PCEDIT.PIF	Windows/Topview PIF file for PCEDIT.EXE	
PCSETUP.PIF	Windows/Topview PIF file for PCSETUP.EXE	
PCKEYMAP.PIF	Windows/Topview PIF file for PCKEYMAP.EXE	
PP-PIF.DVP	DESQview DVP file for PCPLUS.EXE	
PE-PIF.DVP	DESQview DVP file for PCEDIT.EXE	
PS-PIF.DVP	DESQview DVP file for PCSETUP.EXE	
PK-PIF.DVP	DESQview DVP file for PCKEYMAP.EXE	

Table A.2—*Continued*

File Name	Description	Discussed in
TASKVIEW.DAT	Program setup information using PROCOMM PLUS with TASKVIEW and OMNIVIEW	

Sample Dialing Directories

File Name	Description	Discussed in
ATLANTA.DIR	Sample Atlanta bulletin boards	Chapter 3
BOSTON.DIR	Sample Boston bulletin boards	Chapter 3
SANFRAN.DIR	Sample San Francisco boards	Chapter 3
WASHDC.DIR	Sample Washington, D.C., bulletin boards	Chapter 3

Using the DOS COPY Command

Instead of using the Installation Utility (PCINSTAL.EXE) to copy PRO-COMM PLUS to your hard disk, you can choose to use the DOS COPY command.

First, create a directory for the PROCOMM PLUS program files. The PRO-COMM PLUS documentation and the examples in this book assume that you create a directory named PCPLUS on your C drive. To create this directory, make the C drive current, type the following at the DOS prompt, and press Enter:

    md \pcplus

Place the PROCOMM PLUS Program Diskette into drive A. Type the following command at the DOS prompt, and press Enter:

    copy a:*.* c:\pcplus

DOS copies all the files from the Program Diskette to the directory C:\PCPLUS. At this point, the directory C:\PCPLUS contains the files:

    PCINSTAL.EXE
    PCPLUS.EXE
    PCSETUP.EXE
    READ.ME
    PCPLUS.KBD

DSTORM.ASP
PCPLUS.USR
PCEDIT.EXE

You can delete the PCINSTAL.EXE file because it is the installation program that you decided not to use. Type the following command and press Enter:

del c:\pcplus\pcinstal.exe

The PROCOMM PLUS Supplemental Diskette also includes files that support or can be used with certain special PROCOMM PLUS features. These files are listed in table A.2 with a description of each file's purpose and a notation of where the file is discussed in this book. You don't need to copy any of these files to your hard disk until you decide to use the particular function that requires the file.

TIP | The DOS method of installing PROCOMM PLUS seems quicker and easier than using the Installation Utility but may leave the program improperly set up for use on your system. The first time you run PROCOMM PLUS, it finds no parameter file and no dialing directory file, so PROCOMM PLUS creates both files using default values. The parameters that would have been set by the Installation Utility are set as follows:

Monitor:	Color
Modem:	Hayes 1200/1200B
Terminal:	ANSI
Comm Port:	COM1
Baud Rate:	1200
Parity:	NONE
Data Bits:	8
Stop Bits:	1
Duplex:	Full
Protocol:	XMODEM

If these settings are not correct, you have to use the Setup Utility and the Alt-P (Line/Port Setup) command, discussed in Chapter 8, to change them.

Installing PROCOMM PLUS on a Floppy Disk System

Unlike many popular PC programs, PROCOMM PLUS runs just as well on a floppy disk system as on a hard disk system. All the files you normally need to run PROCOMM PLUS fit on one disk with room to spare.

Two methods are available for installing PROCOMM PLUS on a floppy disk system: the Installation Utility and the DOS command DISKCOPY. The best method is the PROCOMM PLUS Installation Utility, PCINSTAL.EXE. This method enables you easily to set several PROCOMM PLUS parameters so that the program will run properly on your computer the first time.

Using the Installation Utility

To use the PROCOMM PLUS Installation Utility, place the distributed PROCOMM PLUS Program Diskette into drive A of your computer. At the DOS prompt, type *a:pcinstal*, and press Enter. The program displays the logo PCINSTALL and a copyright notice. Press any key to display a screen of instructions entitled PROCOMM PLUS INSTALLATION UTILITY at the top left and PROGRAM FILE INSTALLATION at the top right of the screen. Read these instructions, and then press any key except Esc to continue.

The Installation Utility program first prompts you for the Source drive for PROCOMM PLUS program diskette. The default response is A. You have already placed the Program Diskette into drive A, so press Enter.

Next, the installation program prompts for the Destination path for PRO-COMM PLUS files and suggests a response of C:\PCPLUS. This directory is the standard directory used to hold PROCOMM PLUS program files on a hard disk. Press the Tab key to delete this directory name. Place a blank formatted disk into your computer's second floppy disk drive, and type the drive's letter followed by a colon, such as *B:* (if your computer has only one drive, type the drive letter followed by a colon—such as *A:*). Press Enter. The installation program then immediately begins copying the PROCOMM PLUS program files to the other disk drive. (When copying to the same disk—A drive to A drive—swap disks only when instructed to do so.)

The PROCOMM PLUS Supplemental Diskette also includes files that support or can be used with certain special PROCOMM PLUS features. These files are listed in table A.2 along with a description of each file's purpose, and a notation of where the file is discussed in this book. You don't need to copy any of these files to your hard disk until you decide to use the particular function that requires the file.

Using the DOS DISKCOPY Command

You don't have to use the Installation Utility to make a copy of the PRO-COMM PLUS Program Diskette. You can choose instead to use the DOS DISKCOPY command.

TIP | Place a write-protect tab over the notch on the PROCOMM PLUS Program Diskette and on the Supplemental Diskette before beginning this procedure. This tab will prevent any accidental corruption of these distributed diskettes during a DISKCOPY operation.

Place the PROCOMM PLUS Program Diskette into drive A. Type the following command at the DOS prompt and press Enter:

    diskcopy a: a:

DOS prompts you to Insert SOURCE disk in drive A: and to Press any key when ready . . . . Because the PROCOMM PLUS Program Diskette is already in drive A, press any key.

DOS begins to copy all the files from the Program Diskette to memory (RAM). When memory is full or when all files have been copied into memory, DOS prompts you to Insert TARGET disk in drive A. Place a blank disk into drive A, and press any key. DOS copies the files from RAM on to the new disk. Depending on the amount of memory in your computer, you may have to repeat the last two steps.

When DISKCOPY has finished making the copy, it displays the message Copy another disk (Y/N)? Respond No to return to the DOS prompt.

At this point, the new working disk contains the files:

    PCINSTAL.EXE
    PCPLUS.EXE
    PCSETUP.EXE
    READ.ME
    PCPLUS.KBD
    DSTORM.ASP
    PCPLUS.USR
    PCEDIT.EXE

You must delete the PCINSTAL.EXE file in order to make room for two files PROCOMM PLUS creates when you run the program the first time (PCPLUS.PRM AND PCPLUS.DIR.) Type the following command and press Enter:

    del a:\pcinstal.exe

The PROCOMM PLUS Supplemental Diskette also includes files that support or can be used with certain special PROCOMM PLUS features. These files are listed in table A.2 along with a description of each file's purpose and a notation of where the file is discussed in this book. If the PROCOMM PLUS working disk you just created is a 360K disk, you won't have room on it for any of the supplemental files. You can copy the files you need to another disk.

Modifying CONFIG.SYS and AUTOEXEC.BAT

DOS reads a file named CONFIG.SYS on the root (main) directory of the boot (start-up) disk each time you start the computer. For PROCOMM PLUS to work properly, the following two lines should appear in this file:

    FILES = 25
    BUFFERS = 20

Be sure that these lines exist in the CONFIG.SYS file. If they don't, use any ASCII text editor (such as PCEDIT) to add these lines. The CONFIG.SYS file is located on the root (main) directory of the disk that contains the DOS system files. If you don't find such a file, create one containing only two lines.

If you change or add CONFIG.SYS, you have to restart the computer for the modification or addition to take effect.

Each time you start the computer, DOS also looks for a file named AUTO-EXEC.BAT on the root directory of the boot disk. If this file exists, DOS executes any DOS commands that it contains. One of the commands normally placed in AUTOEXEC.BAT is the SET command. This command enables you to customize the DOS environment.

When you use the PROCOMM PLUS Dialing Directory screen, the Setup Utility, the Keyboard Mapping screen, keyboard macros, or the translation table, PROCOMM PLUS looks for a number of supporting files on the disk. PROCOMM PLUS looks first for these supporting files in the current working directory. You can, however, cause PROCOMM PLUS to look next to one particular directory by creating a DOS environment variable named PCPLUS. Place the following command in the AUTOEXEC.BAT file:

    SET PCPLUS = C:*dirname*

The parameter *dirname* is the name of the directory that contains the supporting files. Normally, you set PCPLUS equal to the directory that contains the PROCOMM PLUS program files, which is usually C:\PCPLUS. The typical SET command is therefore as follows:

    SET PCPLUS = C:\PCPLUS

Using a text editor, add the appropriate SET line to the AUTOEXEC.BAT file.

Starting PROCOMM PLUS

Now that your program is installed, you are ready to start PROCOMM PLUS. The simplest way to start PROCOMM PLUS from a hard disk is to change to the directory in which the program is installed (C:\PCPLUS if you used the default directory during installation), type *pcplus*, and press Enter.

To start PROCOMM PLUS from a floppy disk system, place your working copy (not the original) of the PROCOMM PLUS program disk into drive A. Make drive A the current drive, type *pcplus*, and press Enter.

PROCOMM PLUS also has three special start-up parameters, referred to as *switches*, that you can add to the start-up command. You can cause PRO-COMM PLUS to refrain from generating sound effects by adding the switch /s to the start-up command. Adding the switch /b causes PRO-COMM PLUS to use screen colors that display best on black-and-white monitors and noncolor LCD screens (unnecessary if you selected **2** (Composite B&W) as the monitor option during installation using the Installation Utility). You can also cause PROCOMM PLUS to run an ASPECT script as soon as the program starts by adding to the start-up command the switch /f followed by the script file name. You can add these switches in any order and in any combination. For example, to cause PROCOMM PLUS to display in black and white and to suppress sound effects, use the start-up command:

    pcplus /b /s

or

    pcplus /s /b

Upgrading ProComm 2.4.3 to PROCOMM PLUS

PROCOMM PLUS lets you upgrade from ProComm 2.4.3 to PROCOMM PLUS and reuse your ProComm keyboard macros, translation table, Host mode log-on message file, and dialing directory.

First, install PROCOMM PLUS as described in the preceding sections of this appendix. Then perform any or all of the following procedures, as desired:

- To reuse your ProComm keyboard macros file, use the DOS REN command to rename the macros file to PCPLUS.KEY. At the DOS prompt, and in the directory that contains the ProComm files, type *ren procomm.key pcplus.key*, and press Enter. Then copy PCPLUS.KEY to the directory that contains your PROCOMM PLUS program files. Once you start PROCOMM PLUS, use the Alt-M (Keyboard Macros) command to display the macros. You may have to edit the macros. Two macro codes have changed. Replace any exclamation point (!) with the code ^M, which means carriage return. Replace any occurrence of the pipe character (|—also called a vertical bar) with the code ^[, which means the Esc key.

- In order to reuse the ProComm translation table in PROCOMM PLUS, rename the PROCOMM.XLT file to PCPLUS.XLT. At the DOS prompt, and in the directory that contains the ProComm files, type *ren procomm.xlt pcplus.xlt*. Then copy PCPLUS.XLT to the directory that contains your PROCOMM PLUS program files.

- You can reuse the Host mode log-on message. Use the REN command to rename the PROCOMM.MSG file to PCPLUS.NWS. At the DOS prompt, and in the directory that contains the ProComm files, type *ren procomm.msg pcplus.nws*. Then copy PCPLUS.NWS to the directory that contains your PROCOMM PLUS program files.

- To reuse the ProComm dialing directory, you must convert the data in the ProComm directory file to the new PROCOMM PLUS format. First, copy the conversion utility, CVTDIR.EXE, from the Supplemental Diskette into the ProComm directory. Then, from the DOS prompt, and in the directory that contains the ProComm files, type *cvtdir*, and press Enter. This utility displays five lines of text including a copyright notice, an explanation of the program's purpose, and the message Press Ctrl-C to abort, any other key to begin.... Press any key except Ctrl-C, and the utility creates a file PCPLUS.DIR, a PROCOMM PLUS dialing directory file, containing all the entries from your ProComm dialing directory file. (If PCPLUS.DIR already exists in the current directory, CVTDIR creates the file PCPLUS2.DIR instead.) Copy this file to the directory that contains your PROCOMM PLUS programs.

You can also convert ProComm command files into PROCOMM PLUS ASPECT script files. Refer to "Compatibility with ProComm Command Files," in Chapter 11, "An Overview of the ASPECT Script Language," for more information.

B

Installing a Modem

One of the basic requirements of computer communications is a modem, the piece of hardware that accomplishes both *mod*ulation and *dem*odulation so that a digital computer can communicate over analog phone lines. Without a modem at each end of the phone line, the two computers cannot communicate. Refer to Chapter 1, "A Communications Primer," for a general discussion of modems.

Modem manufacturers often produce modems in both internal and external versions. Installation of the two types of modems is different.

Connecting an External Modem

An *external modem* is typically a metal or plastic box about 10 inches by 6 inches by 2 inches, with a panel of LED (light-emitting diodes) on the front. An external modem is connected to your PC by a serial cable and powered by an AC adapter. The modem has at least one telephone jack for connecting the modem to the telephone line and often a second telephone jack for connecting a telephone.

In order to connect your external modem to your computer and ready it for communication, you need a couple other parts: a serial port and a serial cable to connect the modem to the serial port.

A relatively new group of external modems are small enough to fit in your pocket and can run on batteries. These mini-modems typically plug directly into a serial port on your computer and are particularly handy for laptop-computer users. This type of external modem requires a serial cable only if the serial port on your computer has a different number of pins than the connector on the modem has.

Choosing a Serial Port

Your computer must have an available serial port (COM port) to which you can connect your external modem. If all the serial ports on your computer are already in use (for example, for a mouse, printer, and plotter), you need to add another serial port to your computer. You can find serial ports on many different types of integrated circuit boards (referred to generically in this discussion as *serial cards*), many of which may contain other types of input/output devices, such as parallel ports and game ports. Make sure that any board you obtain will fit into an available slot in the expansion bus of your computer.

Serial ports are numbered COM1, COM2, COM3, and so on. Nearly all serial cards can be configured as COM1 or COM2, either by flipping a DIP switch or by moving a jumper; but not all serial cards can be configured as COM3 or higher. Before you buy a card, make sure that it will meet your needs.

Another complication is that versions of DOS before 3.3 support only COM1 and COM2, and no version of DOS supports beyond COM4. PRO-COMM PLUS is one of the few communications programs, however, that enables you to use serial ports COM1 through COM8 no matter what version of DOS you have. PROCOMM PLUS goes around the operating system to accomplish this feat. To use a serial port beyond COM2, you need to make sure that the port's base address and IRQ line (Interrupt Request line) match those listed in the Setup Utility Modem Port Assignments screen, shown in figure B.1 (see Chapter 8, "Tailoring PROCOMM PLUS").

Fig. B.1.

The Modem Port Assignments screen.

```
 PROCOMM PLUS SETUP UTILITY                      MODEM PORT ASSIGNMENTS

                          BASE      IRQ
                        ADDRESS    LINE

         A- COM1 ......  0x3F8     IRQ4

         B- COM2 ......  0x2F8     IRQ3

         C- COM3 ......  0x3E8     IRQ4

         D- COM4 ......  0x2E8     IRQ3

         E- COM5 ......  0x3F8     IRQ4

         F- COM6 ......  0x3F8     IRQ4

         G- COM7 ......  0x3F8     IRQ4

         H- COM8 ......  0x3F8     IRQ4

 Alt-Z: Help |    Press the letter of the option to change:   | Esc: Exit
```

TIP When you are using more than one serial port, if at all possible, use either COM1 or COM2 for communications. The most typical configuration is to connect a mouse to COM1 and connect your modem to COM2. Even though PROCOMM PLUS enables you to address serial ports numbered through COM8, you may experience conflicts between your modem and other peripheral devices attached to your computer through other serial ports. For example, COM1 and COM3 typically use the same IRQ line. With a mouse connected to COM1 and a modem connected to COM3, you may have problems running PROCOMM PLUS or other programs.

Choosing a Serial Cable

A *serial cable* is one of the simplest pieces of hardware used in your system, but a serial cable is crucial to successful PC communications. The purpose of a serial cable is to connect a serial device to a serial port—in this case, a modem to the serial port on your PC. A serial cable is, therefore, quite often referred to as a *modem cable.* Not all these cables have the same number of pins on each end. You need to be sure that the cable you buy matches both your computer and your modem.

People sometimes confuse *modem* cables with *null-modem* cables, which serve a different function. To minimize confusion, the discussion that follows refers to the cable that connects a modem to a serial port simply as a *serial cable*. Refer to "Connecting Computers with a Null Modem Cable," later in this appendix, for information about null modem cables.

On the IBM PC, PC/XT, and PS/2 computers, as well as most compatibles, each serial port is a D-shaped connector on the back of the computer and has 25 protruding metal pins. This type of connector is called a *DB-25 M* (male) connector. The connector on the serial cable that attaches the modem to this serial port therefore must have 25 holes to match the male connector's 25 pins. It is called a *DB-25 F* (female) connector. Most external modems have a DB-25 F serial connector, so the typical serial cable must have a DB-25 M connector on one end and a DB-25 F connector on the other end.

The 25 pins on a DB-25 serial port connector correspond to the 25 circuits defined by the Electronic Industries Association Recommended Standard 232, revision C, referred to as the RS-232C standard or just RS-232. When used for asynchronous communications, however, only 9 of the 25 pins are normally used: pins 2, 3, 4, 5, 6, 7, 8, 20, and sometimes 22. The IBM PC AT and many compatibles, therefore, use a 9-pin DB-9 connector instead of the DB-25. (Note: Some 25-pin serial cables also use pin 1, the Frame Ground (FG) circuit; but this circuit is usually not necessary, so 9-pin cables do not use it.) When connecting a modem to an IBM PC AT or AT-compatible, you usually need a serial cable with a DB-9 F on one end (to connect to the computer) and a DB-25 M on the other end (to connect to the modem).

The RS-232 standard defines two types of serial devices: data terminal equipment (DTE) and data communications equipment (DCE). The standard is designed so that DTE devices connect directly to DCE devices. The serial port on your computer is configured as a DTE device, and your modem is configured as DCE device.

Table B.1 lists the names given by the RS-232 standard to the nine circuits commonly used in asynchronous communications, from the point of view of a DTE device. The table indicates the pin to which each circuit is assigned on both a DB-25 and a DB-9 connector. This table also tells how circuits should be configured in a serial cable that has a DB-25 pin connector on one end and a DB-9 pin connector on the other.

Table B.1
RS-232 DTE Circuits in DB-25 and DB-9 Connectors

Circuit Name	Abbreviation	DB-25 Pin	DB-9 Pin
Transmit Data	TD	2	3
Receive Data	RD	3	2
Request To Send	RTS	4	7
Clear To Send	CTS	5	8
Data Set Ready	DSR	6	6
Signal Ground	SG	7	5
Data Carrier Detect	DCD	8	1
Data Terminal Ready	DTR	20	4
Ring Indicator	RI	22	9

The pin assignments for a DCE device are the same as those shown in table B.1, except pins 2 and 3 are reversed. When you connect a DTE device to a DCE device, the Transmit Data (TD) circuit on one end is connected to the Receive Data (RD) circuit on the other end.

TIP | Once you obtain the proper serial cable, be sure to fasten the cable securely to your computer and to the modem, using the screws provided. Many communication problems can be traced to loose cable connections.

Installing an Internal Modem

Each *internal modem* is built on a circuit board that plugs into an empty expansion slot inside your PC. Some internal modems are long enough to fill up a long expansion slot, but others need only a half-size slot.

The exact steps for installing an internal modem depend on the brand and model of PC you have. This section is intended only as a general guide. Read your PC owner's manual and the modem's documentation for the proper procedure.

CAUTION: Before you open or do anything inside your computer TURN OFF ALL POWER AND UNPLUG your computer.

TIP | Avoid discharging static electricity into an internal modem. Before touching the modem, touch a large metal object, such as your computer's chassis. This action will discharge into the metal object any static electricity that has built up on your body.

Before you install the modem, read the section that follows this one to determine whether you need to change the modem's configuration settings. If your modem is configured by DIP switches, you may not be able to reach the switches after the board is installed.

Unlike an external modem, an internal modem is not connected to a serial port. The modem board, however, contains a special chip called the UART (Universal Asynchronous Receiver/Transmitter), which is, in fact, a built-in serial port. The modem, therefore, counts as one of the computer's serial ports. Determine which serial ports are already installed and configure your modem as a different port. For example, if your computer has only one serial port installed, COM1, configure the modem as COM2. But if your computer already has two serial ports, you need to be able to configure the modem as COM3. Nearly all internal modem cards can be configured as COM1 or COM2, either by flipping a DIP switch or by moving a jumper; but not all modem cards can be configured as COM3 or higher. Make sure before you purchase that a particular card will meet your needs. To use a modem card configured as a serial port beyond COM2, you need to make sure that the base address and IRQ line used by the modem card match those listed in the Setup Utility Modem Port Assignments screen, shown in figure B.1 (see Chapter 8, "Tailoring PROCOMM PLUS").

After configuring the modem, turn off the power to your computer and unplug it. Open the computer's case, and locate an empty expansion slot.

Remove the retaining screw of the metal slot cover on the back of the computer; then take off the cover. Slide the card into the slot. Carefully but firmly press the card down until it is securely seated. Replace the retaining screw and cover.

Configuring the Modem

You must configure a modem to take advantage of its features. Some modems, internal or external, are configured using DIP (dual in-line package) switches. Other modems are configured through software commands. Refer to your modem's manual to determine the proper method and configure the modem as specified in the following list. Configured to these settings, your modem will work properly with PROCOMM PLUS.

- The computer controls the Data Terminal Ready (DTR) signal, and the modem follows the actual state of the DTR signal.

- The modem generates verbal result codes.

- Result codes are displayed.

- Keyboard commands are displayed in Command mode.

- Auto Answer is off.

- Modem controls Data Carrier Detect signal (DCD or CD), asserting the DCD signal only when the modem detects a carrier signal from a remote modem, rather than always asserting the DCD signal.

- AT command set is enabled.

- Escape code (+ + +) does not hang up modem.

If your modem is configured using software commands, you may be able to use the **A** (Initialization command) setting, found on the Setup Utility Modem Options screen, to send the proper commands to your modem each time PROCOMM PLUS starts. For example, the initialization command inserted by the PROCOMM PLUS Installation Utility for a Hayes 2400 modem includes the software command to control the DTR signal with the computer and the DCD signal (AT&C1&D2) with the modem. Refer to "Editing the Initialization Command," in Chapter 8, for more information.

TIP | The PROCOMM PLUS Supplemental Diskette contains the ASPECT script FASTCOMM.ASP. Use this script to configure a Fastcomm 2496/Turbo modem. Do not run this script with other types of modems.

Setting Up an Error-Control or Data-Compression Modem

Several manufacturers now produce modems that perform error-control, as opposed to requiring that the communications software perform this chore. When a modem performs error-control, both error detection and the overall speed of transmission are superior to software error checking. Many modems that provide built-in error control also compress the data as it is sent. A compatible modem on the other end decompresses the data.

For error control to work, however, the modems on both ends of the connection have to be using the same error-control protocol. Two different types of error-control protocols have developed a significant following: Hayes and MNP Classes 1 through 4. Fortunately, an industry standard has emerged, that incorporates the two competing error-control schemes: CCITT V.42. Similarly, two popular data-compression algorithms are used by competing manufacturers: Hayes and MNP Class 5; and a new international standard has emerged to replace them both: CCITT V.42bis. Refer to "New Standards," in Chapter 1, for more information about these different error-control and data-compression standards.

Regardless of which type of error-control or data-compression modem you are using, the modem at the other end must be compatible for you to recognize any benefit. Assuming use of compatible modems, you also have to set up the modem and PROCOMM PLUS properly, or your modem will not give you the added transmission speed you are looking for. Keep in mind that the modems will connect at their nominal speed (for example, 2400 bps for a 2400-bps modem); but through data-compression algorithms, data throughput can be much faster. The data must, therefore, flow between your computer and your modem at a rate higher than the connection speed between the two modems. If you set PROCOMM PLUS at the same rate as the connect rate, you defeat the purpose of the modem.

The software commands to set up the modem vary from manufacturer to manufacturer; check your modem manual for the appropriate commands. (Some modems use DIP switches instead of software commands.) In order to achieve maximum throughput, set up PROCOMM PLUS and your modem as follows:

1. Set PROCOMM PLUS's baud rate to the highest rate allowed by the modem. Often this rate will be 9600 for a 2400-bps modem or 19200 for a 9600-bps modem. Change all your dialing directory entries to this higher rate as well. The status line will

display 9600 or 19200, respectively, but the modem will control the actual rate of data transfer from the computer to the modem and back.

2. Use the Setup Utility to set the **H** (Auto baud detect) option on the Modem Options screen to OFF. For example, if you are using a 2400-bps error-control/data-compression modem, the modems will connect at 2400 bps. You have set PROCOMM PLUS's baud rate to 9600 bps, however, and you don't want PROCOMM PLUS automatically to adjust the rate down to 2400.

3. Use the Setup Utility to set the **D** (Hardware flow control (RTS/CTS)) option on the Terminal Options screen to ON. Also, send a software command to your modem to enable RTS/CTS handshaking.

4. Issue the software command to your modem to activate the error-control and data-compression features.

5. Set the modem so that it follows the actual state of the Data Terminal Ready (DTR) signal (Usually the command is AT&D2.) and the modem controls Data Carrier Detect (DCD or CD) signal. (Normally the command is AT&C1.)

6. Save the modem's configuration to nonvolatile memory. (Usually the command is AT&W0.) Even when you turn off the modem, the new settings are stored in this area of memory.

7. Use the Setup Utility to set the **A** (Initialization command) in the Modem Options screen to ATZ^M. This command is the standard AT command to reset nonvolatile memory to the currently stored settings. In other words, when PROCOMM PLUS starts, it sends this command to the modem, and the modem retrieves the setup values you have stored in nonvolatile memory.

8. Use the Setup Utility to change all the connect messages in the Modem Result Messages screen to CONNECT.

TIP DATASTORM distributes a file named MODEMS.ARC on its bulletin board, through PC Vendor A Forum (GO PCVENA) on Compu Serve, and through the DATASTORM Support Roundtable on GEnie. This archived file (several files compressed and combined into one for ease of storage and transmission) contains eight ASPECT scripts that execute the proper setup commands for different brands and models of error-control/data-compression modems. At the time of this writing, the scripts included and the modems supported are as follows:

USR2400E.ASP	USRobotics Courier 2400e
USRHST.ASP	USRobotics Courier HST
MTV32.ASP	Multi-Tech V32
MULTITEC.ASP	Multi-Tech 224E
TELEBIT.ASP	Telebit Trailblazer Plus
HAYESV.ASP	Hayes V-series Smartmodem 9600
ATI.ASP	ATI 2400etc
DUAL.ASP	USRobotics Dual Standard
EVEREX.ARC	Contains EVEREX24.ASP and PROFILE.EVX, two script files for the Evercom 24+

If your modem is listed, you can run the corresponding script to set up your modem. The script also displays instructions for other steps you must take, such as changing DIP switch settings and altering PROCOMM PLUS settings.

Connecting to the Telephone Line

Your modem has at least one phone jack, usually on the back or side. When you have only one jack, connect it using a phone cable with a modular (RJ-11) phone plug to the telephone outlet on the wall. If your modem has two phone jacks, connect the one marked LINE to the jack in the wall. The other jack on your modem, usually marked PHONE, is for connecting a telephone. Sometimes the jacks have no markings. In that case, attach either jack to the wall telephone outlet and plug your phone into the other.

Usually, your modem comes with the necessary cable for connecting your modem to the telephone line, but if not, you can find a cable in electronics, hardware, department, drug, and sometimes even grocery stores. If the telephone jack on the wall is the older square type with four holes, you need an adapter. Adapters of this sort are readily available at many stores.

In many business applications, you may want to attach your modem to a multiple telephone circuit. Multiple handsets are attached to the circuit. Each phone can use any line on the circuit when the user presses one of the buttons on the base of the handset. When someone uses a line on the multiple telephone circuit, a light on every attached handset shows that the line is in use. A person who wants to place a call looks for a button that is not lighted, presses the button, and uses the line. It is possible to use a modem on such a system, but you should do so only if your modem supports it.

If you plan to attach your phone to a line that is available on a multiple telephone circuit, check your modem manual to see whether your modem has a way to support multiple telephone installations (DIP switch 7 on many popular modems). This type of phone connection may be referred to in your modem documentation as RJ-12 or RJ-13. These names are the designations for the two types of modular jacks usually used to connect a single phone line to a multiple telephone system. If your modem does not have such a setting, or if you fail to set it correctly, the lights on the handsets attached to the system will not illuminate for your modem's line, even when the modem is in use and connected to another computer. This fact means, of course, that someone is liable to pick up the line to make a call while your PC is connected, interrupting and possibly disconnecting your transmission.

When you connect your modem to an RJ-12 or RJ-13 jack, make sure that the phone cable contains four wires. You can look at the plug at the end of the cable and count the copper wires. Only two wires are needed for a single phone system (or a multiple phone system with no lighted buttons). But one of the other two wires is used to light the buttons on multiple telephone system handsets when your modem is in use. Of course, before you place a call with your modem on such a system, check to see whether the line you intend to use is free.

Connecting Computers with a Null Modem Cable

An increasingly common situation in offices that use PCs is that someone wants to connect one computer directly to another without going through a modem or telephone line. For example, you may use a laptop computer that has 3 1/2-inch disk drives and a desktop PC with only a 5 1/4-inch disk drive. You want to transfer data from the laptop PC to the desktop PC by using PROCOMM PLUS.

To transfer data between two computers without a modem, you can use a *null modem cable* (sometimes referred to simply as a *null modem*). The easiest way to think of this device is to focus only on pins 2 and 3 of the cable and compare this type of cable to a normal serial cable. Because both PCs' serial ports comply with the DTE configuration, a normal serial cable cannot enable them to communicate. The TDs would be connected and the RDs would be connected. Instead, a null modem cable (among other things) switches pins 2 and 3 on one end of the cable. The result is that TD on one end is now connected to RD on the other, and vice versa. This connection enables communication between the computers attached by the null modem cable without the need for a modem or telephone line.

Null modem cables are usually available in computer stores and by mail order. Sometimes a null modem is sold as an adapter that connects to a normal serial port, converting it to a null modem cable. You can also rewire a normal serial cable to convert it to a null modem cable, as depicted in figures B.2 and B.3. The cable must, however, have a female connector on both ends so that it can be connected to the two serial ports.

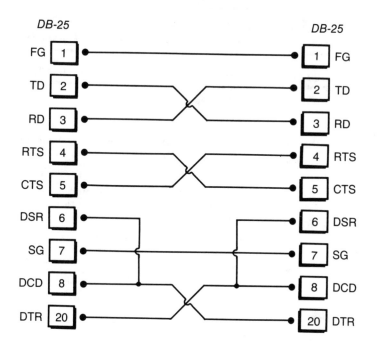

Fig. B.2.

Circuit connections for a null modem cable with DB-25 connectors on both ends.

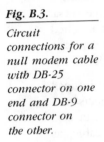

Fig. B.3.

Circuit connections for a null modem cable with DB-25 connector on one end and DB-9 connector on the other.

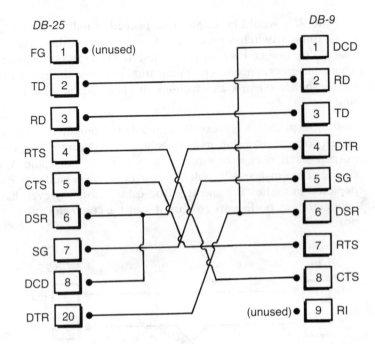

The Host mode is the easiest way to use PROCOMM PLUS to transfer files between computers that are connected by a null modem cable. Activate Host mode on one of the computers and then log on and control the file transfer from the other computer. For this method to work properly, however, you must modify a Host mode setting on the computer that you will place in Host mode. Use the Setup Utility to set the **D** (Connection type) option on the Setup Host Mode Options screen to DIRECT. Refer to "Setting the Connection Type," in Chapter 9, for more information.

Using External Protocols

Even though PROCOMM PLUS supplies 13 built-in file-transfer protocols, it also provides a means for you to invoke up to 3 other protocols from within PROCOMM PLUS. These non-PROCOMM PLUS protocols are referred to as *external* protocols. This appendix explains how to add and use the popular ZMODEM file-transfer protocol. The procedure for adding other external protocols is similar.

ZMODEM is a widely supported and popular file-transfer protocol on PC-based electronic bulletin boards. Like several file-transfer protocols available in PROCOMM PLUS, ZMODEM has the capability to transfer multiple files and to send file name, file size, and data. But ZMODEM has a number of significant advantages over the file-transfer protocols available within PROCOMM PLUS. These advantages include faster file transfers, better error detection using 32-bit CRC checksum, automatic downloading, and the capability to recover from an aborted download.

ZMODEM was developed by Chuck Forsberg of Omen Technology Inc (also the developer of YMODEM). ZMODEM is included in a external protocol driver program named DSZ.COM, which Omen Technology distributes as shareware. The program is available on CompuServe, GEnie, EXEL-PC BBS, and many local PC electronic bulletin boards. You can usually find ZMODEM as an archived file with the name DSZ*mmdd*.ARC or DSZ*mmdd*.ZIP, where *mm* and *dd* represent the month and date of the

DSZ version. For example, DSZ0419.ZIP contains the April 19 version of DSZ. You can also order the program directly from the software publisher:

Omen Technology Inc
P.O. Box 4681
Portland, OR 97208
Voice telephone: (503) 621-3406
FAX: (503) 621-3735

Registration of the program should be sent to the same address.

This appendix is by no means an exhaustive discussion of the program DSZ.COM. The information is intended only to help you quickly get DSZ up and running as a ZMODEM protocol driver for use with PROCOMM PLUS. Read the complete documentation provided by Omen Technology to learn how to take full advantage of this valuable program.

Installing the Protocol Driver

Before you can use an external protocol driver program, you have to install it on your computer. To install DSZ for use with PROCOMM PLUS, copy the main program, DSZ.COM, into the directory that contains your PROCOMM PLUS program files (usually C:\PCPLUS).

Several DOS batch files are distributed with DSZ; these batch files facilitate running ZMODEM from within a communications program like PROCOMM PLUS. The batch files and their purposes are as follows:

Batch File	Purpose
ZMODEMU.BAT	ZMODEM upload
ZMODEMD.BAT	ZMODEM download
ZMODEMAD.BAT	ZMODEM automatic download
ZMODEMDR.BAT	ZMODEM download with crash recovery

Copy these four batch files into the PROCOMM PLUS directory.

After you have copied the necessary files into your PROCOMM PLUS directory, you need to modify your AUTOEXEC.BAT file.

If your modem is connected to a serial port other than COM1, add the following line to your AUTOEXEC.BAT file (refer also to Appendix A):

    SET DSZPORT = n

In this command, *n* is the number of the serial port. For example, if you are using serial port COM2, the command is

SET DSZPORT = 2

To cause DSZ to keep a log of all ZMODEM uploads and downloads, include the following command in the AUTOEXEC.BAT file:

SET DSZLOG = *filespec*

In this line, *filespec* is the complete path and file name of the file you want to contain the DSZ log (for example, C:\PCPLUS\DSZ.LOG).

Add the name of the directory that contains the PROCOMM PLUS program files (and DSZ.COM) to the PATH command in your AUTO-EXEC.BAT file. This command enables DOS to find DSZ.COM even if you use the Alt-F7 (Change Directory) command to switch to a different working directory.

Adding the External Protocol to PROCOMM PLUS

PROCOMM PLUS enables you to add up to three external upload protocols and three external download protocols. You add the protocols through the Setup Utility Protocol Options screen, shown in figure C.1. (Refer to Chapter 8, "Tailoring PROCOMM PLUS," for more on using the Setup Utility).

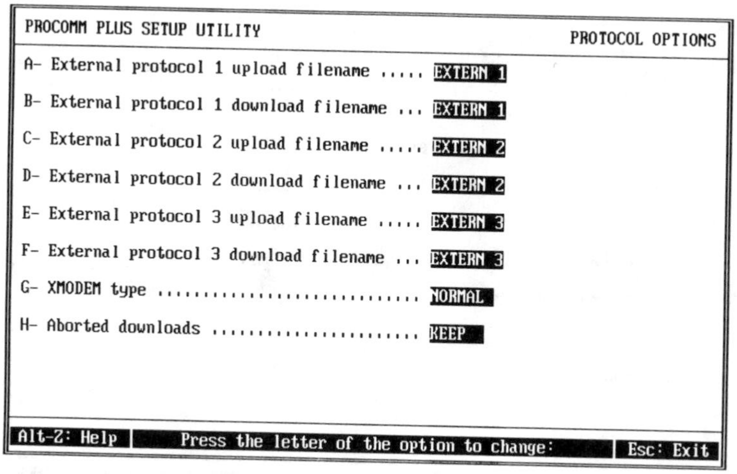

```
PROCOMM PLUS SETUP UTILITY                          PROTOCOL OPTIONS
A- External protocol 1 upload filename ..... EXTERN 1
B- External protocol 1 download filename ... EXTERN 1
C- External protocol 2 upload filename ..... EXTERN 2
D- External protocol 2 download filename ... EXTERN 2
E- External protocol 3 upload filename ..... EXTERN 3
F- External protocol 3 download filename ... EXTERN 3
G- XMODEM type ............................. NORMAL
H- Aborted downloads ....................... KEEP

Alt-Z: Help    Press the letter of the option to change:    Esc: Exit
```

Fig. C.1.

The Setup Utility Protocol Options screen.

Use the first six options listed on the Setup Utility Protocol Options screen to specify external programs that perform file uploading or file downloading. Use options **A**, **C**, and **E** to add upload protocols; and use options **B**, **D**, and **F** to add download protocols. In order to add ZMODEM to PROCOMM PLUS, you add the names of the four distributed batch files to the appropriate options, as explained in the next several paragraphs.

The batch file ZMODEMU.BAT contains only the following two lines:

REM Upload files with ZMODEM
DSZ sz %1 %2 %3 %4 %5 %6 %7 %8 %9

The first line is a *remark* line, which documents the purpose of the batch file. The second line executes the DSZ program with the *sz* parameter, which invokes the ZMODEM upload protocol. Each *%n* parameter (where *n* is an integer 1 to 9) is a DOS replaceable parameter. This batch file enables you to specify up to nine separate file names for uploading, or you can use the DOS wild-card characters (* and ?) when you specify upload file names.

To upload a file by using the ZMODEM protocol from within PROCOMM PLUS, you want PROCOMM PLUS to run the ZMODEMU.BAT batch file from the standard Upload menu. Display the Setup Utility Protocol Options screen, and replace the entry in option **A**, the file name for the first external upload protocol, with

ZMODEMU

Using the same procedure, add the names of the other three batch files to the Protocol Options screen as download protocol file names, options **B**, **D**, and **F**, as shown in figure C.2.

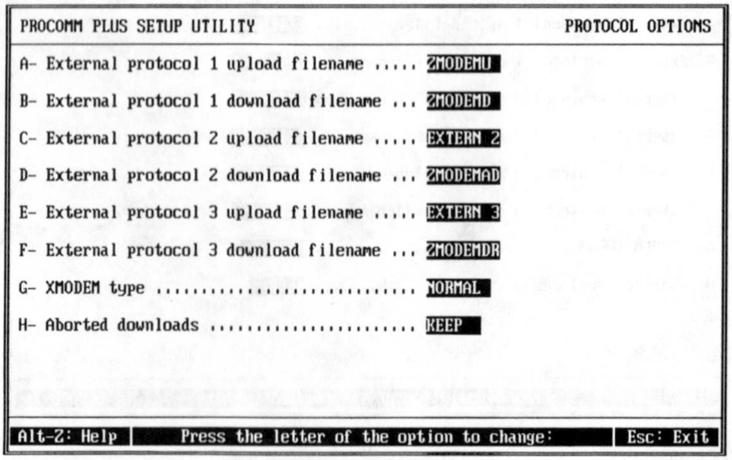

Fig. C.2.

The Protocol Options screen with the ZMODEM files added as the file names of external download protocols.

```
┌─────────────────────────────────────────────────────────────────┐
│ PROCOMM PLUS SETUP UTILITY                        PROTOCOL OPTIONS │
│                                                                   │
│ A- External protocol 1 upload filename ..... ZMODEMU              │
│                                                                   │
│ B- External protocol 1 download filename ... ZMODEMD             │
│                                                                   │
│ C- External protocol 2 upload filename ..... EXTERN 2            │
│                                                                   │
│ D- External protocol 2 download filename ... ZMODEMAD           │
│                                                                   │
│ E- External protocol 3 upload filename ..... EXTERN 3           │
│                                                                   │
│ F- External protocol 3 download filename ... ZMODEMDR          │
│                                                                   │
│ G- XMODEM type ............................. NORMAL             │
│                                                                   │
│ H- Aborted downloads ....................... KEEP              │
│                                                                   │
│                                                                   │
│ Alt-Z: Help │   Press the letter of the option to change:  │ Esc: Exit │
└─────────────────────────────────────────────────────────────────┘
```

Using the External Protocol
To Transfer a File

After you have added a protocol to the Protocol Options screen and
returned to the Terminal mode screen, you can use the added protocol to
transfer a file. To upload a file by using ZMODEMU, first inform the
remote computer that you are about to upload one or more files, using
the ZMODEM file-transfer protocol. Then from the Terminal mode screen,
press PgUp (Send Files). PROCOMM PLUS displays the Upload menu.
ZMODEMU is now option 14, as shown in figure C.3.

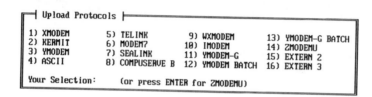

```
┤ Upload Protocols ├
1) XMODEM        5) TELINK       9) WXMODEM       13) YMODEM-G BATCH
2) KERMIT        6) MODEM7      10) IMODEM        14) ZMODEMU
3) YMODEM        7) SEALINK     11) YMODEM-G      15) EXTERN 2
4) ASCII         8) COMPUSERVE B 12) YMODEM BATCH 16) EXTERN 3

Your Selection:      (or press ENTER for ZMODEMU)
```

Fig. C.3.

*The Upload menu
with ZMODEMU
as option 14.*

`Alt-Z FOR HELP` `ANSI` | `FDX` | `2400 N81` | `LOG CLOSED` | `PRINT OFF` | `ON-LINE`

Type *14*, and press Enter. PROCOMM PLUS removes the Upload menu
and displays a small window with the caption ZMODEMU Parameters and the
message Enter parameters. Type from one to nine file names separated by
spaces, or type a file name specification using DOS wild cards, and press
Enter. PROCOMM PLUS temporarily suspends itself and starts the
ZMODEM upload. When the upload is completed, PROCOMM PLUS
returns to the Terminal mode screen.

To download files with ZMODEM, you use one of the other batch files
you installed in PROCOMM PLUS. For straight downloading, you use
ZMODEMD. For automatic downloading, you use ZMODEMAD; and for
downloading with crash recovery, you use ZMODEMDR.

The batch file ZMODEMD.BAT contains the following two lines:

    REM Download file(s) with ZMODEM
    DSZ rz

The first line is documentation. The second line invokes DSZ and starts a ZMODEM download. Notice that unlike ZMODEMU.BAT, this file has no replaceable parameters. File names, file sizes, and so on, are all sent by the remote computer during the download, so you do not need to provide any information to ZMODEM.

To use ZMODEMD, from the Terminal mode screen, press PgDn (Receive Files) to display the Download menu. PROCOMM PLUS displays the Download menu shown in figure C.4, with ZMODEMD, ZMODEMAD, and ZMODEMDR as options 14, 15, and 16, respectively.

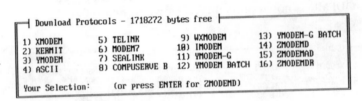

Fig. C.4.

The Download menu with ZMODEMD, ZMODEMAD, and ZMODEMDR as options 14, 15, and 16, respectively.

```
┤ Download Protocols - 1718272 bytes free ├

  1) XMODEM       5) TELINK       9) WXMODEM      13) YMODEM-G BATCH
  2) KERMIT       6) MODEM7      10) IMODEM       14) ZMODEMD
  3) YMODEM       7) SEALINK     11) YMODEM-G     15) ZMODEMAD
  4) ASCII        8) COMPUSERVE B 12) YMODEM BATCH 16) ZMODEMDR

  Your Selection:      (or press ENTER for ZMODEMD)
```

`Alt-Z FOR HELP` `ANSI` `FDX` `2400 N81` `LOG CLOSED` `PRINT OFF` `ON-LINE`

Type *14*, and press Enter to select ZMODEMD. Just press Enter at the Enter parameters prompt without entering any file name. PROCOMM PLUS temporarily suspends itself and runs DSZ. DSZ then begins a ZMODEM download procedure. As soon as the file or files have been received, DSZ exits, and PROCOMM PLUS returns to the Terminal mode screen.

The batch file ZMODEMAD.BAT also contains two lines:

REM Setup for ZMODEM Automatic Downloads
DSZ t

The first line is documentation. The second line invokes DSZ in ZMODEM Automatic Download mode. This mode means that DSZ is acting as a terminal, essentially equivalent to the PROCOMM PLUS Terminal mode screen. As soon as DSZ detects an incoming ZMODEM file transfer from the remote computer, DSZ begins its ZMODEM download procedure without the need for any further instructions from you. This procedure is similar in operation to downloading a file from CompuServe with the **J**

(Enquiry (ENQ)) option on the Setup Utility Terminal Options screen set to CIS B.

Typically, you use ZMODEMAD when you expect to execute several downloads from a host computer during a particular communication session. Before you instruct the host to begin the first ZMODEM download, press PgDn (Receive Files); type *15*, and press Enter to select ZMODEMAD. Press Enter at the blank Enter parameters prompt. PROCOMM PLUS temporarily suspends itself and loads DSZ in Automatic Download mode.

Then execute the proper command to the host computer to initiate the ZMODEM download. DSZ receives the file using the ZMODEM file-transfer protocol without your having to give any additional command. After the first transfer is finished, you can continue with the communication session. Later, when you again instruct the host computer to download a file to your computer, DSZ automatically receives the file. When you are finished with the session and want to return to PROCOMM PLUS, press F1 to exit DSZ. PROCOMM PLUS returns to the Terminal mode screen.

The batch file ZMODEMDR.BAT contains the following two lines:

```
REM Download file(s) with ZMODEM Crash Recovery
DSZ rz -r
```

The first line is documentation. The second line invokes DSZ and starts a ZMODEM download with crash recovery. Like ZMODEMD.BAT, this file has no replaceable parameters. Indeed the *-r* parameter in line 2 is the only difference between ZMODEMDR.BAT and ZMODEMD.BAT. This parameter enables DSZ to pick up a download where a previously aborted download left off.

For example, suppose that you are using ZMODEMD to download a large file, say 200K. After 180K have been downloaded, your teenager picks up the phone to call his girlfriend and prematurely aborts the download. With other file-transfer protocols, you would have to start the download from the beginning (once you are able to regain possession of the phone line). With ZMODEMDR, however, you can continue where you left off.

To continue an aborted download, first instruct the host computer to begin the ZMODEM file transfer again. Then from the Terminal mode screen, press PgDn (Receive Files); type *16*, and press Enter to select ZMODEMDR. Press Enter at the Enter parameters prompt without entering a file name. PROCOMM PLUS then temporarily suspends itself and invokes DSZ. DSZ continues the ZMODEM download at the point where the preceding download was aborted, adding the missing 20K of data to the previously received 180K file.

TIP You can use DSZ from a remote computer by executing the Shell command on PROCOMM PLUS running on the host computer in Host mode and then invoking DSZ from the DOS prompt. To use DSZ this way, you must include the special parameter *CON* in the DSZ command. This parameter causes DSZ to send screen output to the host computer's screen rather than to the normal DOS output device.

PROCOMM PLUS redirects the host computer's DOS output to your screen. If you fail to use this parameter, DSZ attempts to transfer or receive a file and send screen output (messages and the like) through the same device, probably locking up the host computer.

For example, to cause the host computer to download files using ZMODEM, type the following command at the host computer's DOS prompt:

DSZ CON rz

After you press Enter, you will see nothing on your screen to indicate that DSZ is operating on the host system, because the screen output is now being sent to the host computer's screen. To complete the transfer, press PgUp (Send Files), and select ZMODEMU from the Upload menu. At the Enter parameters prompt, type the name(s) of the file(s) you want to send, and press Enter. PROCOMM PLUS then activates DSZ. When the transfer is finished, your screen returns to the Terminal mode, and you again see the DOS prompt from the host computer.

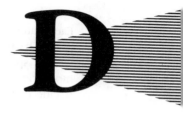

Patching PROCOMM
PLUS

Any experienced computer programmer will tell you that no program is ever really finished. Programmers are continually making small adjustments to their programs in order to improve performance. Many software companies force you to wait until the next formal upgrade to benefit from these small but often significant improvements.

DATASTORM, laudably, anticipates the inevitability of this fine-tuning process. The company provides on the PROCOMM PLUS Supplemental Diskette a special utility program, DT_PATCH, which enables you to update your copy of PROCOMM PLUS with the latest improvements. DATASTORM distributes these updates in files called *patch files*. The process of adding these changes to PROCOMM PLUS is referred to as *patching the program*.

Obtaining Patches

DATASTORM distributes patch files electronically. The most recent patches are always available on the CompuServe PC Vendor A Forum (GO PCVENA) and on the DATASTORM BBS. The patches work only on the specified version of the program. They are distributed in an archived file named DTP*ver*.ARC, where *ver* is the version number. For example, patches for PROCOMM PLUS Version 1.1B are distributed in the file DTP11B.ARC. Use the shareware utility program ARC-E to extract the archived patch files.

Each patch file is an ASCII file, and the file names end with the extension .DTP. The patch file contains a description of the purpose of the patch, as well as special codes that the DT_PATCH program uses to perform the modification. Before using a patch file, always read its contents. You may decide that you don't need to apply a particular patch.

For example, one of the files provided in DTP11B.ARC is the patch file CURSBIOS.DTP. Figure D.1 shows the contents of this file as viewed by using the Alt-V (View a File) command. The purpose of this patch is to eliminate conflicts with some TSR (terminate-and-stay-resident, also known as memory-resident) programs or multitasking shells. A warning, however, informs you that performance of the program in Terminal mode may be detrimentally affected by making this patch. If you have never experienced an apparent conflict between PROCOMM PLUS and a program you use, you have no reason to perform the patch.

Fig. D.1.

*The contents of
the CURSBIOS.DTP
patch file.*

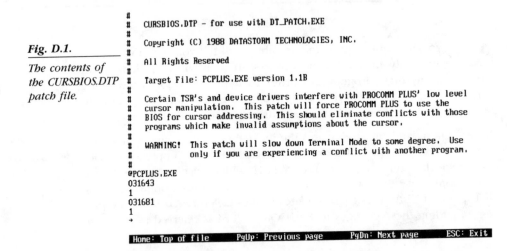

```
#
#   CURSBIOS.DTP - for use with DT_PATCH.EXE
#
#   Copyright (C) 1988 DATASTORM TECHNOLOGIES, INC.
#
#   All Rights Reserved
#
#   Target File: PCPLUS.EXE version 1.1B
#
#   Certain TSR's and device drivers interfere with PROCOMM PLUS' low level
#   cursor manipulation.  This patch will force PROCOMM PLUS to use the
#   BIOS for cursor addressing.  This should eliminate conflicts with those
#   programs which make invalid assumptions about the cursor.
#
#   WARNING!  This patch will slow down Terminal Mode to some degree.  Use
#             only if you are experiencing a conflict with another program.
#
@PCPLUS.EXE
031643
1
031681
1
→
```

Home: Top of file	PgUp: Previous page	PgDn: Next page	ESC: Exit

Some patches fix software problems commonly referred to as *bugs*. For example, the patch file HOSTBAUD.DTP, shown in figure D.2, fixes a bug that may cause a problem in Host mode. If you use PROCOMM PLUS's Host mode often, you probably should apply this patch.

```
#
#   HOSTBAUD.DTP - for use with DT_PATCH.EXE
#
#   Copyright (C) 1988 DATASTORM TECHNOLOGIES, INC.
#
#   All Rights Reserved
#
#   Target Program: PROCOMM PLUS version 1.1B
#
#   There is a bug in version 1.1B that occurs when two callers in a row call
#   in at a baud rate less than the initial baud rate. When Alt-Q is pressed
#   (or the HOST ASPECT command is executed) the initial baud rate is remembered
#   and assumed to be the highest speed the modem supports. Each time a user
#   disconnects, the port is reset to the remembered rate. Under certain
#   circumstances the port and the modem are at different speeds and the modem's
#   connect messages are unreadable so auto baud detection fails. By sending th
#   auto-answer string between each call, the port and the modem are synchronize
#   to the same baud rate and the connect messages work again.
#
#
@PCPLUS.EXE
066079
114
→
```

Fig. D.2.

The contents of the HOSTBAUD.DTP patch file.

```
 Home: Top of file      PgUp: Previous page      PgDn: Next page      ESC: Exit 
```

Running DT_PATCH

Once you decide which patch(es) you want to apply, you use the DT_PATCH.EXE utility program to do the job. DT_PATCH.EXE is one of the files distributed on the PROCOMM PLUS Supplemental Diskette. The file DT_PATCH.DOC is a short file explaining how to use DT_PATCH.EXE.

When you use DT_PATCH, you should never make any change to your original PROCOMM PLUS diskettes. Always apply the patch to a copy. First, copy DT_PATCH.EXE and the patch file into the same directory as PCPLUS.EXE (a *copy* of PCPLUS.EXE, not the original distributed disk). Then type the following at the DOS prompt, and press Enter:

DT_PATCH *filename*

In this command, *filename* is the name of the patch file. Include the full DOS path if the patch file is in another directory. The file name extension .DTP is assumed, so you can leave it off. Add */b* if you are using a black-and-white or LCD screen.

DT_PATCH first displays a logo screen and then the screen shown in figure D.3. This screen instructs you to patch only copies of PROCOMM PLUS, not the original copy. Press Enter to continue the patch procedure. Press Esc to abort.

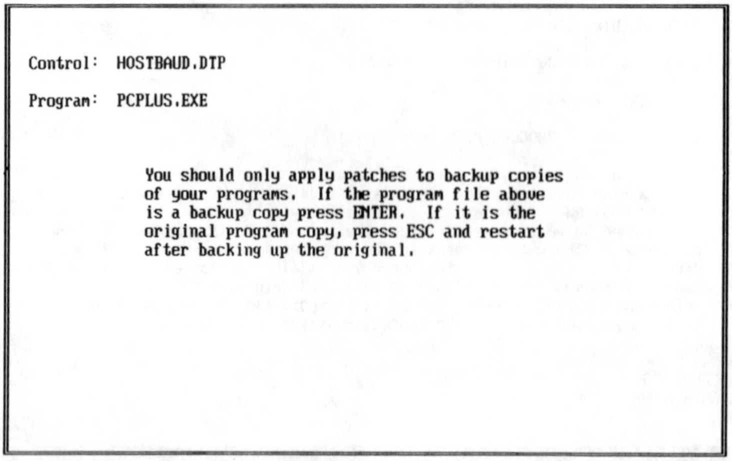

Fig. D.3.

Using DT_PATCH.

```
Control:  HOSTBAUD.DTP

Program:  PCPLUS.EXE



              You should only apply patches to backup copies
              of your programs.  If the program file above
              is a backup copy press ENTER.  If it is the
              original program copy, press ESC and restart
              after backing up the original.
```

While DT_PATCH is making the necessary change to PCPLUS.EXE, it displays the message Patching .... When the patch has been successfully applied, DT_PATCH displays the message Done! and Press any key to continue.... Press any key to return to the DOS prompt.

You repeat this procedure for every patch you want to apply.

CAUTION: Make sure that the patches are intended for the version of PROCOMM PLUS you are using. Read the patch file before applying it. Above all, *never* patch the original copy of PCPLUS.EXE.

TIP One of the files included in DTP11B.ARC is the program NEW-BAUD.EXE. A companion ASCII text file, NEWBAUD.DOC, describes the purpose of this file and tells how to use it. NEW-BAUD enables you to add to PROCOMM PLUS transmission speeds (baud rates) that are not available in the standard distributed version of the program. For example, you can use NEWBAUD to change PROCOMM PLUS's normal 300 bps speed to 600 bps. Read the documentation carefully before using the program. Again, do not modify the original copy of PCPLUS.EXE.

Index

123.BAT batch file, 239

A

Abort (Esc) key, 70, 72-73, 76
ADD (ASPECT) command, 333
ADDS Viewpoint terminal emulation, 307
   keyboard mapping table, 307
ALARM (ASPECT) command, 334
alarms, controlling, 224
Alt-X (Quit) PCEDIT command, 162
analog signals, 16-18
ANSI terminal emulation, 57, 298-299
   escape sequences, 298-299
   keyboard mapping, 311-312
ANSI.SYS
   emulation, 227
   file, 227, 299
Answer mode, 92-93, 206-207, 254, 256
ARC-E program, 381
ARC file extension, 121
archive file, 121
ASCII (American Standard Code for Information Interchange)
   character, 19
   character set, 129, 235-236
   file-transfer protocol, 122, 231-236
      blank lines, 232-233
      carriage return, 234-235

      character pacing, 233-234
      line feed, 234-235
      line pacing, 233-234
      stripping 8th bit, 235-236
      tabs, 232-233
   protocol, 126, 129-131
   text files, 104, 129, 149-151
   transfer options, 231-236
ASCII Transfer Options screen, 130, 231-236
ASP file extension, 151, 175, 187, 320, 322-323, 326, 335
ASPECT script programming language, 4, 167, 177-181, 319-341
   commands, 328-334
      branching, 331
      communications, 330
      disk and file, 332-333
      display and printer, 333
      mathematical, 333
      miscellaneous, 334
      program control, 330
      string manipulation, 332
      system setup, 334
      table, 329
      variable manipulation, 331
   compared to other languages, 321
   creating scripts, 321-322
ASSIGN (ASPECT) command, 331-332

Q

R

T

More Computer Knowledge from Que

For more information, call

1-800-428-5331

All prices subject to change without notice.
Prices and charges are for domestic orders
only. Non-U.S. prices might be higher.

Using 1-2-3 Release 2.2, Special Edition
Developed by Que Corporation

Learn professional spreadsheet techniques from the world's leading of 1-2-3 books! This comprehensive text leads you from work-
sheet basics to advanced 1-2-3 operations. Includes Allways
coverage, a Troubleshooting section, a Command Reference, and
a tear-out 1-2-3 Menu Map. The most complete resource available for Release 2.01 and Release 2.2!

$24.95 USA
Order #1040
0-88022-501-7
850 pp.

Using Harvard Graphics
by Steve Sagman and Jane Graver Sandlar

An excellent introduction to presentation graphics! This well-
written text presents both program basics and presentation fundamentals to create bar, pie, line, and other types of informative
graphs. Includes hundreds of samples!

$24.95 USA
Order #941
0-88022-407-X
550 pp.

Upgrading and Repairing PCs
by Scott Mueller

The ultimate resource for personal computer upgrade, repair,
maintenance, and troubleshooting! This comprehensive text covers all types of IBM computers and compatibles—from the original
PC to the new PS/2 models. Defines your system components
and provides solutions to common PC problems.

$27.95 USA
Order #882
0-88022-395-2
750 pp.

Using DOS
Developed by Que Corporation

The most helpful DOS book available! Que's *Using DOS* teaches
the essential commands and functions of DOS Versions 3 and
4—in an easy-to-understand format that helps users manage and
organize their files effectively. Includes a handy **Command
Reference**.

Order #1035
$22.95 USA
0-88022-497-5, 550 pp.

Que Order Line: **1-800-428-5331**

All prices subject to change without notice. Prices
and charges are for domestic orders only.
Non-U.S. prices might be higher.

Free Catalog!

Mail us this registration form today, and we'll send you a free catalog featuring Que's complete line of best-selling books.

Name of Book _____

Name _____

Title _____

Phone (_____) _____

Company _____

Address _____

City _____

State _____ ZIP _____

Please check the appropriate answers:

1. Where did you buy your Que book?
   ☐ Bookstore (name: _____)
   ☐ Computer store (name: _____)
   ☐ Catalog (name: _____)
   ☐ Direct from Que
   ☐ Other: _____

2. How many computer books do you buy a year?
   ☐ 1 or less
   ☐ 2-5
   ☐ 6-10
   ☐ More than 10

3. How many Que books do you own?
   ☐ 1
   ☐ 2-5
   ☐ 6-10
   ☐ More than 10

4. How long have you been using this software?
   ☐ Less than 6 months
   ☐ 6 months to 1 year
   ☐ 1-3 years
   ☐ More than 3 years

5. What influenced your purchase of this Que book?
   ☐ Personal recommendation
   ☐ Advertisement
   ☐ In-store display
   ☐ Price
   ☐ Que catalog
   ☐ Que mailing
   ☐ Que's reputation
   ☐ Other: _____

6. How would you rate the overall content of the book?
   ☐ Very good
   ☐ Good
   ☐ Satisfactory
   ☐ Poor

7. What do you like *best* about this Que book?
   _____
   _____

8. What do you like *least* about this Que book?
   _____
   _____

9. Did you buy this book with your personal funds?
   ☐ Yes   ☐ No

10. Please feel free to list any other comments you may have about this Que book.
   _____
   _____
   _____

QUE

Order Your Que Books Today!

Name _____

Title _____

Company _____

City _____

State _____ ZIP _____

Phone No. (_____) _____

Method of Payment:

Check ☐   (Please enclose in envelope.)

Charge My: VISA ☐   MasterCard ☐
American Express ☐

Charge # _____

Expiration Date _____

Order No.	Title	Qty.	Price	Total

You can **FAX** your order to **1-317-573-2583**. Or call **1-800-428-5331, ext. ORDR** to order direct.
Please add $2.50 per title for shipping and handling.

Subtotal _____

Shipping & Handling _____

Total _____

BUSINESS REPLY MAIL
First Class Permit No. 9918   Indianapolis, IN

Postage will be paid by addressee

11711 N. College
Carmel, IN 46032

BUSINESS REPLY MAIL
First Class Permit No. 9918   Indianapolis, IN

Postage will be paid by addressee

11711 N. College
Carmel, IN 46032